Advances in Intelligent Systems and Computing

Volume 1027

The series "Advances in Intelligent Systems and Computing" contains publications on theory, applications, and design methods of Intelligent Systems and Intelligent Computing. Virtually all disciplines such as engineering, natural sciences, computer and information science, ICT, economics, business, e-commerce, environment, healthcare, life science are covered. The list of topics spans all the areas of modern intelligent systems and computing such as: computational intelligence, soft computing including neural networks, fuzzy systems, evolutionary computing and the fusion of these paradigms, social intelligence, ambient intelligence, computational neuroscience, artificial life, virtual worlds and society, cognitive science and systems, Perception and Vision, DNA and immune based systems, self-organizing and adaptive systems, e-Learning and teaching, human-centered and human-centric computing, recommender systems, intelligent control, robotics and mechatronics including human-machine teaming, knowledge-based paradigms, learning paradigms, machine ethics, intelligent data analysis, knowledge management, intelligent agents, intelligent decision making and support, intelligent network security, trust management, interactive entertainment, Web intelligence and multimedia.

The publications within "Advances in Intelligent Systems and Computing" are primarily proceedings of important conferences, symposia and congresses. They cover significant recent developments in the field, both of a foundational and applicable character. An important characteristic feature of the series is the short publication time and world-wide distribution. This permits a rapid and broad dissemination of research results.

** Indexing: The books of this series are submitted to ISI Proceedings, EI-Compendex, DBLP, SCOPUS, Google Scholar and Springerlink **

More information about this series at http://www.springer.com/series/11156

Xin-She Yang · Simon Sherratt ·
Nilanjan Dey · Amit Joshi
Editors

Fourth International Congress on Information and Communication Technology

ICICT 2019, London, Volume 2

 Springer

Editors
Xin-She Yang
School of Science and Technology
Middlesex University
London, UK

Nilanjan Dey
Department of Information Technology
Techno India College of Technology
Kolkata, West Bengal, India

Simon Sherratt
University of Reading
Reading, UK

Amit Joshi
Global Knowledge Research Foundation
Ahmedabad, Gujarat, India

ISSN 2194-5357   ISSN 2194-5365 (electronic)
Advances in Intelligent Systems and Computing
ISBN 978-981-32-9342-7   ISBN 978-981-32-9343-4 (eBook)
https://doi.org/10.1007/978-981-32-9343-4

This Springer imprint is published by the registered company Springer Nature Singapore Pte Ltd.
The registered company address is: 152 Beach Road, #21-01/04 Gateway East, Singapore 189721, Singapore

Preface

This AISC volume contains the papers presented at ICICT 2019: Fourth International Congress on Information and Communication Technology in concurrent with ICT Excellence Awards. The conference was held during February 25–26, 2019, London, UK, and collaborated by Global Knowledge Research Foundation, City of Oxford College. The associated partners were Springer, InterYIT, IFIP, and Activate Learning. The conference was held at Brunel University, London. This conference was focused on e-business fields such as e-agriculture, e-education, and e-mining. The objective of this conference was to provide a common platform for researchers, academicians, industry persons, and students to create a conversational environment wherein the topics related to future innovation and obstacles to be resolved for new upcoming projects were discussed, and exchange of views and ideas had taken place. The conference attracted immense experts from various countries, in-depth discussions were taken place, and issues were intended to be solved at international level. New technologies were proposed, experiences were shared, and future solutions for design infrastructure for ICT were also discussed. Research submissions in various advanced technology areas were received and then were reviewed by the committee members, and 92 papers were accepted. The conference was overwhelmed by the presence of various members. Amit Joshi, Organizing Secretary, ICICT 2019, gave the welcome speech on behalf of the conference committee and editors. Our special invited guest—Sean Holmes, Vice Dean International, College of Business, Arts and Social Sciences, Brunel University London, UK—also addressed the conference by a speech. The conference was also addressed by our inaugural guest and speakers—Mike Hinchey, Chair, IEEE UK and Ireland section, and Director of Lero and Professor of Software Engineering at the University of Limerick, Ireland; and Aninda Bose, Sr. Publishing Editor, Springer Nature. Niko Phillips, Group Director, International Activate Learning City of Oxford College, UK, addressed the vote of appreciation on behalf of the conference committee. There were 12 technical sessions in total, and talks on academic and industrial sector were focused on both the days. We are obliged to Global Knowledge Research Foundation for their immense support to make this conference a successful one. A total of 85 papers were presented in technical sessions, and 92 were accepted with strategizing on ICT and intelligent systems. At the closing ceremony, 10 best paper awards by Springer were announced among the best selected and presented paper. Gift vouchers each worth 200 Euro to shop online at www.springer.com were given along with appreciation certificates by Springer and editor of ICICT 2019. On behalf of the editors, we thank all sponsors, press, print, and electronic media for their excellent coverage of this conference.

London, UK   Xin-She Yang
Reading, UK   Simon Sherratt
Kolkata, India   Nilanjan Dey
Gandhinagar, India   Amit Joshi
February 2019

Contents

About the Editors

Xin-She Yang obtained his D.Phil. in Applied Mathematics from the University of Oxford. He then worked at the Cambridge University and the National Physical Laboratory (UK) as a Senior Research Scientist. He is currently a Reader in Modelling and Optimization at the Middlesex University London and an Adjunct Professor at the Reykjavik University (Iceland). He is also an elected Bye-Fellow at the Cambridge University as well as the IEEE CIS Chair for the Task Force on Business Intelligence and Knowledge Management. He is included in the "2016 Thomson Reuters Highly Cited Researchers" list.

Simon Sherratt is a Professor of Consumer Electronics at the University of Reading. His research has primarily been on OFDM, particularly for digital TV and wireless USB. He specializes in DSP system architecture and hardware implementation. Simon is leading the development of the wireless communications and on-body sensors for Reading's part in Sphere. Prof. Sherratt is the current Editor-in-Chief of the IEEE Transactions on Consumer Electronics and an elected member of the IEEE Consumer Electronics Society Board of Governors. Simon is a Fellow of the IET and Fellow of the IEEE.

Nilanjan Dey is an Assistant Professor in the Department of Information Technology, Techno India College of Technology, Kolkata, W.B., India. He holds an honorary position of a Visiting Scientist at Global Biomedical Technologies Inc., CA, USA, and is a Research Scientist at the Laboratory of Applied Mathematical Modeling in Human Physiology, Territorial Organization of Scientific and Engineering Unions, Bulgaria. He is an Associate Researcher of Laboratoire RIADI, University of Manouba, Tunisia. His research focuses on medical imaging, soft computing, data mining, machine learning, rough sets, computer-aided diagnosis, and atherosclerosis. He has published 20 books and 300 international conference and journal papers. He is the Editor-in-Chief of the International Journal of Ambient Computing and Intelligence, International Journal of Rough Sets and Data Analysis, the International Journal of Synthetic Emotions (IJSE), and the International Journal of Natural Computing Research. He is also the Editor of the Advances in

Geospatial Technologies (AGT) book series, Executive Editor of the International Journal of Image Mining (IJIM), and Associate Editor of IEEE Access journal and the International Journal of Service Science, Management, Engineering and Technology. He is a life member of IE, UACEE, and ISOC.

Amit Joshi is Director of Global Knowledge Research Foundation, Ahmedabad, India. He holds an B.Tech. in Information Technology and an M.Tech. in Computer Science and Engineering, who is pursuing research in the areas of cloud computing and cryptography. He has 6 years of academic and industrial experience in the prestigious organizations in Udaipur and Ahmedabad. He is an active member of ACM, CSI, AMIE, IACSIT-Singapore, IDES, ACEEE, NPA, and many other professional societies. He has presented and published more than 30 papers in national and international IEEE and ACM journals/conferences. He has also edited three books on diverse subjects and has organized more than 15 national and international conferences and workshops. For his contribution to the society, he received awards from The Institution of Engineers (India), ULC, in 2014, and from SIG-WNs Computer Society of India in 2012.

PESOHA: Privacy-Preserving Evaluation System for Online Healthcare Applications

Youna Jung

Abstract Online healthcare and wellness applications are one of the alternatives to expensive face-to-face medical services. By using online healthcare applications, patients are able to access clinical interventions at a lower cost and at their convenience. However, the lack of rigorous evaluation system becomes a barrier to further growth. To address the shortcoming, in this paper, we propose the privacy-preserving evaluation system for online healthcare applications, in short PESOHA. The proposed system helps non-IT healthcare professionals easily to create their clinical studies on their online applications and assess the performance of online health applications by using privacy-preserving online monitoring. In this paper, we present the structural architecture of PESOHA and demonstrate its operations for healthcare researchers and patients. The systematic evaluation method of PESOHA will expand the role of online healthcare and wellness applications.

Keywords Online health care · eHealth · Online intervention · Online treatment · Online monitoring · Online application evaluation · Privacy protection

1 Introduction

Over the past decades, many healthcare professionals have identified an urgent need to reduce the cost and increase the accessibility of healthcare practices. Internet-based healthcare applications directly address the need and change many aspects of healthcare and wellness [1, 2]. The advent of online services and the ubiquity of mobile devices have enabled a paradigm shift in how people use healthcare and wellness services from face-to-face interactions to Internet-based online interaction. The online healthcare practices allow participants to access clinical interventions at a lower cost and at their convenience. In addition, most of the applications support anonymous participation in their interventions. As a result, the new form of healthcare

Y. Jung (✉)
Computer and Information Sciences, Virginia Military Institute, 313E Mallory Hall,
Lexington, VA 24450, USA
e-mail: jungy@vmi.edu

© Springer Nature Singapore Pte Ltd. 2020
X.-S. Yang et al. (eds.), *Fourth International Congress on Information
and Communication Technology*, Advances in Intelligent Systems
and Computing 1027, https://doi.org/10.1007/978-981-32-9343-4_1

service has significantly enhanced the Nation's health by improving knowledge on patients' behaviour and health outcomes [3, 4].

Despite the promising future of online healthcare applications, its growth rate is slow to moderate. One of the biggest obstacles is the lack of systematic evaluation platform for online healthcare applications. Among numerous applications, only a few applications have been evaluated [5] and it reveals an urgent need for the rigorous evaluation of these applications for safety and health. However, there is no standard platform or tool exists. Without stringent evaluations, it is impossible for healthcare professionals to check if participants use their applications in a correct way and their applications are effective as designed. If patients keep using online healthcare application with unproven efficiency and reliability, it quickly ruins the applications' reputation and, in turn, reduces the size of the online healthcare market.

To address issues pointed above, we need a software system to engineer the evaluation of online health and wellness applications as a general-purpose solution for the market. The evaluation system must provide the four services below.

1. Systematic authoring of online clinical studies to measure the correctness and effectiveness of online healthcare applications.
2. Automatic delivery of assessments and/or interventions to participants according to a study's workflow.
3. Privacy-preserving online monitoring while participants conduct online assessments and/or interventions on healthcare applications. It is critical that PESOHA considers the privacy and security of FDA-monitored clinical research trials on online healthcare applications.
4. Providing study-related monitoring data for the evaluation of online healthcare applications and clinical studies.

To meet the requirements listed above, in this paper, we propose the privacy-preserving evaluation system for online healthcare applications, in short PESOHA. The proposed system helps us overcome cost, technological and intellectual barriers to rigorous evaluation, thereby accelerating translation from existing applications with unknown efficiency to online applications proven effective in improving health. PESOHA enables non-IT healthcare professionals to create their clinical studies on online applications without knowledge and skills on information technologies, in order to measure the performance of a wide range of online healthcare applications including weight control applications, smoking cessation applications, diabetes control applications, stress management applications, and mental treatment applications. By analysing monitoring data collected by PESOHA, they can optimize their healthcare applications and improve the applications' quality of service. The innovative evaluation method of PESOHA will expand the role of online healthcare applications to improve our Nation's health and wellness.

The rest of this paper is organized as follows. In Sect. 2, we define several concepts and present the structural architecture of PESOHA in Sect. 3. To demonstrate PESOHA's services, in Sect. 4, we present two operational flows for medical researchers and patients with screenshots. In Sect. 5, we introduced related work, and in Sect. 6, we summarize our contributions and suggest future work.

2 Design of Concepts

In this section, we define some building blocks of PESOHA and present its structural design. We explain each component in the system and relationships between components in detail. For a better understanding of PESOHA's services, we present two workflows, one for healthcare professionals and the other one for participants in online clinical studies. PESOHA is built upon several concepts, such as user, study, intervention, assessment, and workflow. It is worthy of defining those concepts before we describe the structural design and operations of PESOHA. The definitions of the concepts are described below.

2.1 User

PESOHA categorizes users based on their roles as shown in Fig. 1. A *system administrator* is in charge of maintaining the PESOHA system. A *researcher* is a healthcare professional who wants to create clinical studies and evaluate online applications by using PESOHA. A researcher supervises *study staff* and *activity staff*. A *study staff* can create and maintain clinical studies while an *activity staff* manages study activities (e.g. assessments or interventions) that belong to a study. *Study participants*, target objects of a particular study, are divided to two different types, *advisor* and *recipient*. An *advisor* is a person who helps patients conduct clinical studies and provides some feedback on patients' performance. In most cases, family members of patients take *advisor* roles. A *recipient* is a patient who participates in a *study* to improve his/her health. Usually, the *recipients* are the customers of a target healthcare application.

A user has eight properties: *UserID*, *Name*, *Address*, *Email*, *Phone*, *Registration date*, *Roles*, and *Permissions*. According to the role that a user is taking, few more properties could be added. For examples, a *Researcher* has an additional property,

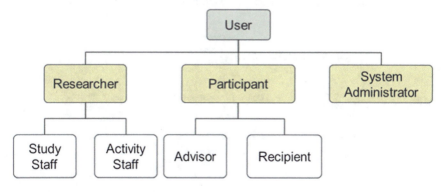

Fig. 1 Role hierarchy in PESOHA

Study, to represent ownership of a clinical study. A *recipient* has two more properties, *Status* and *Enrolment history*. The *Status* property represents whether a patient is participating in a study while the *Enrolment history* is a log of previous participations in past studies.

2.2   Study

A researcher can create a study on online applications to conduct clinical research and evaluate the performance of the applications. To create a study, a researcher must define eight mandatory properties; *StudyID, Name, Period, Status, Study staff, Reuse history, Recipient groups*, and *Workflows*. In addition, he/she can define three optional properties including *Eligibility rules, Assignment rules*, and *Consent questions*. A study must have a unique ID and name. The *Period* is specified with a begin date and an end date of a study. A study will be available only during the period. To represent the *Status* of a study, we define four statuses: *Pending, Activated, Inactive*, and *Deleted* as shown in the lifecycle of study in Fig. 2.

When a study is created, its status is '*Pending*'. At the begin date of the study, its status is transmitted to '*Active*'. Note that a study cannot be modified once activated. If the study finishes its execution (*Normal Deactivation*), or an administrator or a study creator forces to deactivate the study (*Abnormal Deactivation*), the status changes to '*Inactive*'. An inactive study is reactivated if the begin date is changed to the present (*Active*) or a date in the future (*Pending*). Once an active study completes all the activities, the status changes to '*Deleted*'. If a creator or administrator no longer needs an inactive study, he/she can delete the study. A study must have one or more staff including *study creators* who make the study and *study coordinators* who manage an overall workflow of the study. If a study creator reuses an existing study created in the past, the information is saved in the *Reuse history*.

To conduct a study, a study must have at least one recipient group, called *Study Arm*, and each arm must have own *Study Workflow* (see Sect. 2.4 for more details). To filter unqualified participants out and assign qualified one to a particular arm, a

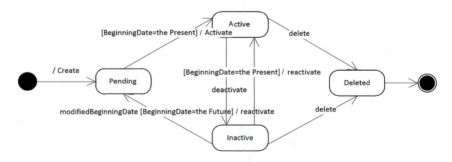

Fig. 2 Lifecycle of online clinical study, intervention, and assessment

study may have its *Eligibility rules* and *Assignment Rules*, respectively. If a clinical study deals with sensitive and private data, a creator must specify questions in the *Consent questions* to get prior consent.

2.3 Intervention and Assessment

A study consists of two types of activities: *Intervention* and *Assessment*. The goal of an intervention is to change the health status of recipients. An intervention has nine properties: *InterventionID, Name, URL, Status, Period, Intervention staff, Reuse history, Intervention Type*, and *Monitoring data types*. Most of the properties are similar to the properties of study, except for a few new properties.

PESOHA handles online interventions and assessments only; thereby, it needs a web address, which is represented as a *URL*. According to its purpose, an intervention's type is defined among four types: *Text tutorial, Essay, Video tutorial*, and *Custom Intervention*. A *text tutorial* is a reading material (e.g. reading useful tips for weight control), while a *video tutorial* is a visual material, such as animations or movie clips (e.g. antismoking campaigns movie clip and online tutorial for teenager driving). Writing an *essay* could be an intervention, e.g. writing diary or notes about health status. A researcher can create a *custom intervention* as a sequence of existing interventions. PESOHA collects different types of data depending on the type of an intervention as shown in Table 1, and the *Monitoring data types* will be specified in the profile of an intervention.

An assessment, another type of study activity, aims at measuring status of a recipient. An assessment has nine properties: *AssessmentID, Name, URL, Status, Period, Assessment staff, Reuse history, Assessment type*, and *Monitoring data types*. Most properties are the same as those of interventions, but assessments have different types: *Survey, Test, External data, External device*, and *Custom assessment*.

A researcher can assign a *survey* to recipients to know his/her thoughts and opinion or *test* (e.g. quiz) to measure improvement in recipients' health status. PESOHA can import recipients' health-related data from an external database (*external data*) or store streaming data from wearable health devices (e.g. KardiaBand for heartbeat tracking and TempTraq for body temperature tracking) (*external device*). Besides

Table 1 Monitoring data types for each intervention type

Data	Completeness		Content	Score
Intervention	Attempt	Progress		
Text tutorial	✓	✓	X	X
Essay	✓	✓	✓	X
Video tutorial	✓	✓	X	X
Custom intervention	✓	✓	✓	✓

Table 2 Monitoring data types for each assessment type

Data	Completeness		Content	Score
Assessment	Attempt	Progress		
Survey	✓	✓	✓	X
Test	✓	✓	✓	✓
External data	✓	✓	✓	X
External device			✓	X
Custom intervention	✓	✓	✓	✓

four basic assessment types, a researcher can create a *Custom assessment*. The type of monitoring data is determined based on an intervention's type as shown in Table 2.

2.4 Workflow

As mentioned above, each study arm has its own workflow. A workflow consists of several workflow steps and must have a start step and an end step. A workflow step is defined with an associated *Activity*, a *Completion condition*, and one or more *Post actions*. An *activity* that a study participant needs to perform is either an intervention or an assessment. To check if a participant successfully finishes an assigned activity, PESOHA uses *Completion condition*. In the *Post action*, we specify one or more next step(s). To move to the next step, the corresponding *Transition condition* of the *Post action* must be satisfied. In most cases, the result of the previous step(s) determines the transition from a step to another.

To specify the order of steps, PESOHA supports four types of routing: *sequential*, *parallel*, *choice*, and *iteration*. In a *sequential* routing, several activities are executed in sequence (e.g. *'First Step A then Step B'*). To perform two or more activities at the same time or in any order, we use a *parallel* routing that commences with an *AND-Split* and concludes with an *AND-Join*. An *AND-Split* allows multiple activities to start their operation simultaneously, and an *AND-Join* converges parallel executing activities into a single common activity. If a researcher needs to synchronize multiple activities at a certain point, he/she must use *AND-Join*. Unlike the parallel routing, a *choice* routing uses *OR-Split* and *OR-Join*. In an *OR-Split*, a single activity among multiple alternatives is selected depending on a corresponding *Transition condition*. An *OR-Join* converges multiple alternative activities to a single activity without synchronization of results. An *iterative* routing involves the repetitive execution of one or a group of step(s) until a transition condition is met. For a better understanding of study workflow, we present two example workflows in Fig. 3.

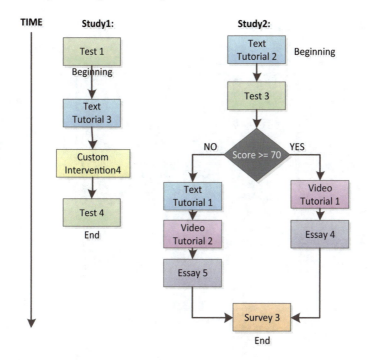

Fig. 3 Example workflows for two groups of study recipients

3   Architecture

PESOHA has three layers: application layer, service layer, and database layer. The application layer is a front end of PESOHA. It receives user requests and inputs and passes to the service layer. In addition, it delivers services provided by the service layer to users. All the data required to perform tasks in the service layer are stored in a relational database in the database layer. The overall architecture of PESOHA is shown in Fig. 4.

To fulfill user requests, the service layer has six manager modules: *User Manager, Study Manager, Workflow Manager, Intervention Manager, Assessment Manager,* and *Monitoring Manager*. The detailed descriptions for each component are below.

- *User Manager*—This module authenticates valid users and handles user information through user creation, modification, and deletion. An administrator can create a user, or a user can register himself/herself in PESOHA (*User Registration*). Once a user account is created, only administrators can modify and delete the user account. To access PESOHA, a user can log in with his/her account credential issued by PESOHA or use external user authentication services [e.g. OpenID services and single sign-on (SSO) services]. Currently, PESOHA supports Google Sign-In and Facebook SSO.

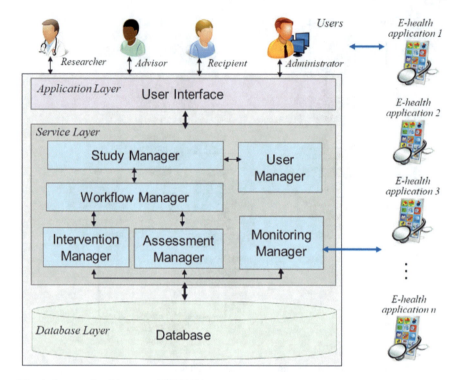

Fig. 4 Structural architecture of PESOHA

- *Intervention Manager* and *Assessment Manager*—These two modules are responsible for creating, modifying, and deleting online interventions and assessments. Researchers and administrators can create a study activity, either an intervention or an assessment, by entering details for an activity's properties. If PESOHA has a similar activity already, they can reuse a past activity. Once an activity is created, these modules must maintain the status of the activity according to the lifecycle shown in Fig. 2. Note that a creator can modify and delete his/her own interventions and assessments while administrators can manage all the activities in PESOHA.
- *Study Manager*—This module creates, modifies, and deletes studies by interacting with other managers. To create a study, the *Study Manager* first needs to receive information about all the participants in a new study from the *User Manager*. Once a study becomes *active*, it fetches one or more corresponding workflow(s) from the *Workflow Manager*. To carry on a study, it delivers interventions and assessments to participants according to a workflow assigned to a participant's study arm.
- *Workflow Manager*—This module maintains all the study workflows. A study creator and administrator can create, modify, and delete workflows. As explained earlier, each study arm must have its own workflow and a workflow step has an associated activity. To get information about interventions and assessments that are

linked to a workflow, the *Workflow Manager* communicates with the *Intervention Manager* and the *Assessment Manager*.

- *Monitoring Manager*—If the *Study Manager* delivers a link to an activity, participants access an online intervention or assessment at their convenience. Once a participant starts to use, some usage and/or user data are collected by online monitoring code embedded in the intervention/assessment and sent to the *Monitoring Manager* in PESOHA. The types of data to be monitored are determined based on the type of activity (see Tables 1 and 2). As mentioned earlier, PESOHA must protect the privacy of patients who are using online healthcare and wellness applications. Towards this, PESOHA uses the privacy-preserving online monitoring (PPoM) service, an online monitoring service that gathers authorized user/usage data that users allow to monitor only [6, 7]. The PPoM service allows participants to specify their privacy preferences on data monitoring. It means that patients can determine which data can be monitored. Then, the *PPoM Service* selectively collects data based on patients' preference, not preferences of healthcare service providers.

4 Demonstration

PESOHA provides different services to researchers and participants as shown in Figs. 5 and 6, respectively. For researchers, it provides intuitive user interfaces to create and run a clinical study. In addition, it provides monitoring data related to the study for further analysis of performance and usability of online healthcare applications. Once a researcher creates a study, PESOHA sends an invitation to all the potential participants. If a potential participant accepts the invitation, PESOHA starts an operation for them. Note that online health applications must meet the following requirements to use PESOHA.

- Must support HTML 4 or higher and Flash to call JavaScript functions.
- Must have the membership-based authorization.
- Must define all the HTML objects to be monitored with unique object IDs.
- Must embed the PPoM monitoring code [6] into each webpage to be monitored.

To participate in studies operated by PESOHA, a participant's browser must have the following capability.

- Must enable JavaScript.
- For mobile browsers, must support JQuery.
- Must allow HTTP POST method and make sure that Firewall does not block the HTTP POST messages.

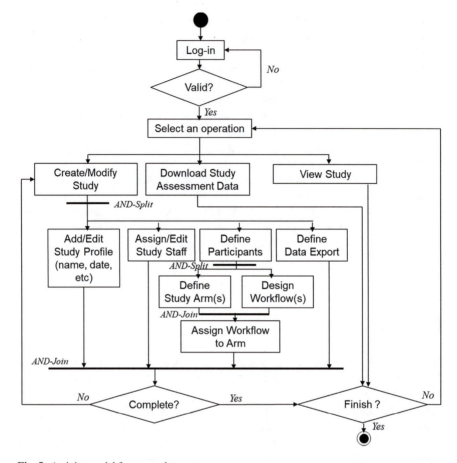

Fig. 5 Activity model for researchers

4.1 Operation of Researchers

A researcher who is registered in PESOHA can create a clinical study to evaluate online health applications by analysing patients' usage and user data. Towards this, a researcher logs in PESOHA and clicks the 'add' button on the 'STUDIES' tab, as shown in Fig. 7. To create a new study, he/she needs to enter all the required data, including *Study Name*, *Status*, *Begin Date*, *End Date*, *IRB approval status*, *Study Staff* (e.g. *creators* and *coordinators*), *Eligibility Rule*, *Assignment Rules*, *Study Arms*, and *Participants*. After creating a *Participants Pool*, a researcher can define a *Study Arm* with a subset of study participants and a corresponding workflow. To create a workflow, a researcher first creates all the workflow steps and then specifies routing types for each step. For a *sequential* routing, we need to specify one next activity as *Post Action*. For a *parallel* routing, you must identify an *AND-Join* and an *AND-Split* step. If a workflow step is a converged activity that synchronizes all the

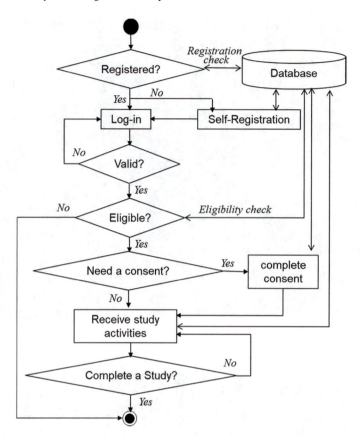

Fig. 6 Activity model for participants

results from multiple previous steps, you must define the step as an *AND-Join* step by marking on the 'AND-Join' checkbox shown in Fig. 8. To represent an *AND-Split* step, you must specify two or more next activities by adding several *Post Actions*. To represent a *choice* routing, you need to add two or more *Post Action*s and each *Post Action* must have a different transition condition. To identify an *OR-Join* step, one or more activities must have the same single *Post Action* without marking the 'AND-Join' checkbox. For an *Iteration* routing, the *Post Action* of an activity must be the activity itself.

When assigning participants to a study arm, a researcher can choose one method among three: (1) *Manual assignment*, (2) *Rule-based assignment*, or (3) *Randomized assignment*. The *Manual* assignment is performed by study staff, while the *Rule-based* and the *Randomized* assignments are performed by PESOHA. At the *Begin Date*, PESOHA runs a study and starts to deliver study activities to participants according to an assigned workflow. If a study is complete, study staff can download all the monitoring data collected for the study. To do so, staff first retrieves the study

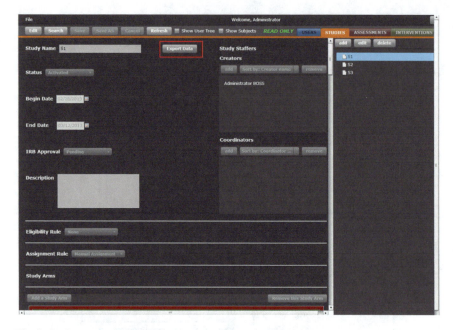

Fig. 7 Study creation in PESOHA

Fig. 8 Creation of a workflow activity in PESOHA

by clicking the study name on the study navigation panel on the right side (*View Study* in the Fig. 5) and then clicks the 'Export Data' next to the 'Study Name'.

4.2   Operation of Patients

A patient who wants to participate in a study must register in PESOHA first. Once a patient completes registration, PESOHA stores the patient's profile in the database, checks his/her login credential, and verifies the patient's eligibility for a study based on the *Eligibility rules* that a study creator specified. If a study has an activity that handles sensitive data, PESOHA requires the patient to sign in a consent form. If a patient gives consent, PESOHA assigns the patient into the study as a *recipient* and starts to send emails containing links to online interventions or assessments. Once a patient conducts all the activities in a study, PESOHA informs him/her of the end of the study and changes the *Status* of the patient in the database.

5 Related Work

With the advent of seamless network connectivity and Internet-based rich interaction methods (e.g. messaging, posting, chatting, file sharing, and gaming), many online healthcare and wellbeing applications have been developed. Currently, online healthcare applications have been used for a wide range of purposes including medical research [8], remote education [9], online counseling [10], participant recruitment [11], health promotion [12], and Internet-based treatment [13].

To verify the effectiveness of such applications and improve the quality of services, it is critical to evaluate the online healthcare applications. Sherrington et al. [14] reviewed Internet-delivered medical interventions for obese patients. They found twelve medical studies and evaluated nine databases by manually identifying relevant keywords, such as *Internet, web, online, eHealth, nutrition, diet, weight, weight loss, overweight, obesity,* and *clinical trial.* After gathering the required data, they analysed weight loss at third, sixth, and twelfth month by calculating the retention rates for each intervention using BMI values. This work evaluated online health applications, but for a specific type of applications (obesity control applications) only. In addition, they manually defined the search keywords and analysis criteria such as retention rates.

Kushniruk et al. [15] evaluated the usability of online health applications by analysing usage data (e.g. page visits and click events). However, this work focused on measuring usability and do not consider the effectiveness of the applications. In 2017, Rogers et al. [16] evaluated the availability of Internet-delivered health interventions using the Preferred Reporting Items for Systematic Reviews and Meta-Analyses (PRISMA) guidelines [17]. They identified currently available health interventions on mental health, weight control, disease prevention and management, and childhood health management, but do not consider the effectiveness of each intervention (e.g. how much does an intervention improve the health of target patients?).

As shown above, most of the evaluations have been limited to specific applications or collected data by collaboration with IT experts. If health professionals use PESOHA, they can easily create a clinical study to evaluate the correctness and the effectiveness of an online assessment or intervention without any help from IT professionals. In addition, they can receive all the data, not only usage data but also user data including health-related data without concern of privacy loss.

6 Conclusion

There is an urgent need for an evaluation system for prevalent online healthcare and wellness applications to verify the performance of the applications and, in turn, improve the quality of online healthcare services. To address the need, we propose a privacy-preserving evaluation system for online healthcare applications, in short PESOHA, which enables healthcare professionals to do the following tasks:

- Intuitive creation of clinical studies consisting of online interventions and assessments to evaluate the correctness and effectiveness of online healthcare applications.
- Automatic delivery of created studies to patients based on the patients' status.
- Privacy-preserving online monitoring on online interventions and assessments to collect patients' user data, including health-related data, as well as usage data.
- Completion check for all recipients and downloading monitoring data for clinical studies.

To verify the usefulness of PESOHA, we plan to conduct a field test with many existing online healthcare and wellness applications. Once we obtain acceptable results from the test, we plan to industrialize PESOHA as a general-purpose evaluation platform for e-health applications. To extend its service to mobile applications, we first need to develop a mobile monitoring service that concerns patients' privacy preference and then upgrade PESOHA to deal with not only online applications but also mobile applications.

References

1. Bennett, G.G., Glasgow, R.E.: The delivery of public health interventions via the internet: actualizing their potential. Annu. Rev. Public Health 30(1), 273–292 (2009)
2. Eysenbach, G.: What is e-health? J. Med. Internet. Res. 3(2)
3. Stinson, J., et al.: A systematic review of internet-based self-management interventions for youth with health conditions. J. Pediatr. Psychol. 34(5), 495–510 (2009)
4. Beal, T., et al.: Long-term impact of four different strategies for delivering an online curriculum about herbs and other dietary supplements. BMC Med. Educ. 6, 39 (2006)
5. Means, B.: Evaluation of Evidence-Based Practices in Online Learning: A Meta-Analysis and Review of Online Learning Studies. In: E. U.S. Department of Education Office of Planning, and Policy Development Policy and Program Studies Service, pp. 94. US Department of Education, Washington D.C. (2010)
6. Jung, Y.: Toward usable and trustworthy online monitoring on e-health applications. Int. J. Adv. Life Sci. 8(1 and 2), 122–132 (2016)
7. Kim, M., Jung, Y.: A development of privacy-preserving monitoring system for e-health applications. In: Proceeding of the 5th International Conference on Global Health Challenges, pp. 64–70 (2016)
8. Daley, E.M., McDermott, R.J., McCormack, K.R.B., Kittleson, M.J.: Conducting web-based survey research: a lesson in internet designs. Am. J. Health. Behav. 27(2), 116–124 (2003)
9. Bernhardt, J.M., Hubley, J.: Health education and the internet: the beginning of a revolution. Health Educ. Res. 16(6), 643–645 (2001)

10. Barak, A., Bl, Klein, Proudfoot, J.: Defining internet-supported therapeutic intervention. Ann. Behav. Med. **38**(1), 4–17 (2009)
11. Duncan, D.F., White, J.B., Nicholson, T.: Using internet-based surveys to reach hidden populations: case of nonabusive illicit drug users. Am. J. Health. Behav. **27**(3), 208–218 (2003)
12. Evers, K.E.: eHealth promotion: the use of the internet for health promotion. Am. J. Health. Behav. **20**(4), 1–7 (2006)
13. Ybarra, M.L., Eaton, W.W.: Internet-based mental health interventions. Mental Health Serv. Res. **7**(2), 75–87 (2005)
14. Sherrington, A., Newham, J.J., Bell, R., Adamson, A., McColl, E., Araujo-Soares, V.: Systematic review and meta-analysis of internet-delivered interventions providing personalised feedback for weight loss in overweight and obese adults. Obes. Rev. **17**(6), 541–551 (2016)
15. Kushniruk, A.W., borycki, E.M., Kuwata, S., Ho, F.: Emerging approaches to evaluating the usability of health information systems. Stud. Health. Technol. Inform. **169**, 915–919 (2010)
16. Rogers, M. A., Lemmen, K., Kramer R., Mann J., Chopra V.: Internet-delivered health interventions that work: systematic review of meta-analyses and evaluation of website availability. J. Med. Internet Res. **19**(3) (2017)
17. Moher, D., Liberati, A., Tetzlaff, J., Altman, D.G.: The PRISMA group preferred reporting items for systematic reviews and meta-analyses: the prisma statement. PLoS Med **6**(7), e1000097 (2009)

Computer Vision and Hybrid Reality for Construction Safety Risks: A Pilot Study

Rita Yi Man Li and Tat Ho Leung

Abstract Construction sites are among the most hazardous venues. While most of the previous research has shed light on the human aspect, we propose to utilise the fast R-CNN object detection method to detect the construction hazard on sites and employ mixed reality to enable the artificial intelligence to detect the hazard. Fast region-based convolutional neural network object detection acquires expert knowledge to identify objects in the image. Unlike image classification, the complexity of object detection always implies an increase in complexity which demands solutions with regard to speed, accuracy and simplicity.

Keywords Construction hazard · Computer vision

1 Construction Hazard

Construction sites have been considered as some of the most hazardous venues. Many previous researches have studied the causes of construction accidents and various safety measures. One company in Hong Kong aspires to improve safety via the utilisation of virtual reality [10, 11]. While most of these studies have highlighted the human aspect, we propose to utilise the fast R-CNN object detection method to detect the construction hazard on sites and employ mixed reality to enable the artificial intelligence to detect the hazard.

R. Y. M. Li (✉)
Hong Kong Shue Yan University, Hong Kong, China
e-mail: ymli@hksyu.edu

T. H. Leung
University of Manchester, Manchester, UK

© Springer Nature Singapore Pte Ltd. 2020
X.-S. Yang et al. (eds.), *Fourth International Congress on Information and Communication Technology*, Advances in Intelligent Systems and Computing 1027, https://doi.org/10.1007/978-981-32-9343-4_2

2 Object Detection—Related Works

Object detection has been prominently adopted within image recognition processes and computer vision tasks such as pose estimation [6, 8], vehicle detection [14, 25], surveillance [7, 13], face detection, face recognition [12] and pedestrian detection [23, 26]. In object detection, an algorithm is required to solve two basic problems which are image classification and localisation.

This paper attempts to construct an object detection model to identify hazardous objects which may exist in construction sites. The photograph database we have built recognises and locates similar objects such as holes and wires which can then be displayed in an MR environment projected on HoloLens. Once the database sets up, it can establish the prediction API for training iteration (Figs. 1, 2, 3, 4 and 5).

Fig. 1 Sample object input of the 'hole' figures

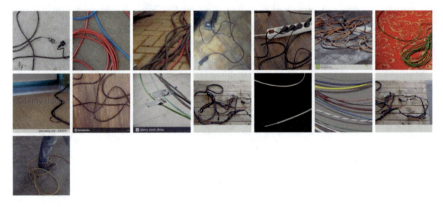

Fig. 2 Sample object input of electrical wire figures

Predictions

Tag	Probability
hole	90.1%

Fig. 3 Test for hole 1

Predictions

Tag	Probability
hole	94.1%

Fig. 4 Test for hole 2

Predictions

Tag	Probability
electrical wire	99.8%

Fig. 5 Test for wire

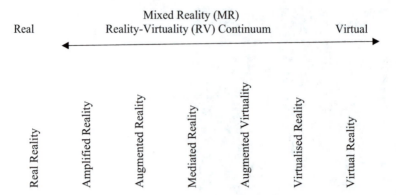

Fig. 6 Order of reality concepts ranging from reality to virtuality [15, 20]

Mixed Reality

In recent years, technological breakthroughs in mixed reality (MR) have allowed us to apply such holograms in various ways. Using the novel techniques in MR, real-life situations and computer-generated visual information are combined to create a hybrid expression which enhances the technological developments and practices in the architecture, engineering and construction (AEC) industries [1, 19].

Basically, MR consists of a variety of realms from real reality to virtual reality as precisely shown in Fig. 6. According to Milgram and Colquhoun [15], the 'real' and 'virtual' environments do not imply merely conceptual alternatives; rather, they are poles lying at opposite ends of a reality–virtuality (RV) continuum. Essentially, within the RV continuum, this is where the MR technologies take place [20]. We can use the mixed reality to see the real hazard but with a virtual label, say, for example, 'hole'.

Although researches in MR have been carried out continually for several decades, the application of MR technologies is becoming mature in practical use [19]. This includes various degrees of applications such as the collaborative web space [2], scientific visualisation [18], the augmented reality (AR) system [2, 3, 5, 21, 22], displays on unmanned air vehicles [16] and 3D video-conferencing systems [17]. More recently, incredibly specific topics in MR-based technological developments have been initiated by global IT leaders such as Microsoft.

In the construction sector, the AR system has widely been adopted in practices [4, 9, 24]. AR-based devices such as head-mounted displays (HMDs) and portable tracking systems are used in different situations. Meanwhile, this paper aligns its concern specifically on the object detection embedded in MR environments, which can provide extraordinary enhancements to construct safety training and management. After successfully detecting a hazardous object, we propose to utilise the mixed reality hologram for construction workers to detect the existence of hazard or safety training, which can then allow them to have an objective criteria on whether or not that is a risk, and enhance safety on sites.

3 Conclusion

It can be advantageous to apply MR technologies into construction practices as such technologies entail a wide range of benefits ranging from efficiency enhancement to safety improvement. This paper highlights the critical architecture in object detection in the fast R-CNN method; hence, with the assistance of Azure, it can adapt the pre-structured scripts to save time in code development. At the same time, it performs effectively with a sufficient database installed and accurate training tests. The custom vision tests 1–3 demonstrate that it is capable of training and recognising the hazardous objects and their locations which may exist in construction sites. If we go one step forward, we can deploy it to HoloLens and consequently identify the hazard on sites.

Acknowledgements Ocular behaviour, construction hazard awareness and an AI chatbot UGC/FDS15/E01/18.
Willingness to share construction safety knowledge via Web 2.0, mobile apps and IoT, RGC grant, UGC/FDS15/E01/17.

References

1. Anders, P.: Cynergies: technologies that hybridize physical and cyberspaces. Paper presented at the 2003 ACADIA. Indianapolis, Indiana (2003)
2. Billinghurst, M., Kato, H.: Collaborative mixed reality. Paper presented at the 1st ISMAR. Yokohama, Japan (1999)
3. Broll, W., Lindt, I., Ohlenburg, J., Wittkamper, M., Yuan, C., Novotny, T., Mottram, C., Gen Schieck, A.F., Strothman, A.: ARTHUR: a collaborative augmented environment for architectural design and urban planning. J. Virtual Reality Broadcast. **1**(1), 1–10 (2004)
4. Dunston, P.S., Shin, D.H.: Key areas and issues for augmented reality application on construction sites. In: Wang, X., Schnabel, M.A. (eds.) Mixed Reality in Architecture, Design and Construction. Springer, Dordercht, The Netherlands (2009)
5. Dunston, P.S., Wang, X.: Mixed reality-based visualization interfaces for architecture, engineering, and construction industry. Ind. J. Constr. Eng. Manag. **131**(12), 1301–1309 (2005)
6. Fang, Q., Li, H., Luo, X., Ding, L., Luo, H., Li, C.: Computer vision aided inspection on falling prevention measures for steeplejacks in an aerial environment. Autom. Constr. **93**, 148–164 (2018)
7. Fang, Q., Li, H., Luo, X., Ding, L., Luo, H., Rose, T.M., An, W.: Detecting non-hardhat-use by a deep learning method from far-field surveillance videos. Autom. Constr. **85**, 1–9 (2018)
8. Fang, W., Ding, L., Luo, H., Love, P.E.D.: Falls from heights: a computer vision-based approach for safety harness detection. Autom. Constr. **91**, 53–61 (2018)
9. Hammad, A.: Distributed augmented reality for visualising collaborative construction tasks. In: Wang, X., Schnabel, M.A. (eds.) Mixed Reality in Architecture, Design and Construction. Springer, Dordrecht, The Netherlands (2009)
10. Li, R.Y.M.: An Economic Analysis on Automated Construction Safety: Internet of Things, Artificial Intelligence and 3D Printing. Springer, Singapore (2018)
11. Li, R.Y.M., Poon, S.W.: Construction Safety. Springer, Berlin (2013)
12. Li, J., Zhang, D., Zhang, J., Zhang, J., Li, T., Xia, Y., Yan, Q., Xun, L.: Facial expression recognition with Faster R-CNN. Proc. Comput. Sci. **107**, 135–140 (2017)

13. Li, X., Ye, M., Liu, Y., Zhang, F., Liu, D., Tang, S.: Accurate object detection using memory-based models in surveillance scenes. Pattern Recognit. **67**, 73–84 (2017)
14. Li, S., Lin, J., Li, G., Bai, T., Wang, H., Pang, Y.: Vehicle type detection based on deep learning in traffic scene. Proc. Comput. Sci. **131**, 564–572 (2018)
15. Milgram, P., Colquhoun, H.: A taxonomy of real and virtual world display integration. In: Ohta, Y., Tamura, H. (eds.) Mixed Reality: Merging Real and Virtual Worlds. Springer, Berlin, Heidelberg, Germany (2001)
16. Rackliffe, N.: An augmented virtuality display for improving UAV usability. MITRE technical papers. MITRE (2005)
17. Regenbrecht, H., Lum, T., Kohler, P., Ott, C., Wagner, M., Wilke, W., Mueller, E.: Using augmented virtuality for remote collaboration. Presence teleop. Virt. Environ. **13**(3), 338–354 (2004)
18. Schmalstieg, D., Fuhrmann, A., Hesina, G., Szalavari, Z., Encarnacao, M.L., Gervautz, M., Purgathofer, W.: The studierstube augmented reality project. Presence Teleop. Virt. Environ. **11**(1), 33–54 (2002)
19. Schnabel, M.A.: Framing mixed realities. In: Wang, X., Schnabel, M.A. (eds.) Mixed Reality in Architecture, Design and Construction. Springer, Dordrecht, The Netherlands (2009)
20. Schnabel, M. A., Wang, X., Seichter, H., Kvan, T.: From virtuality to reality and back. Paper presented at the 12th IASDR. Hong Kong (2007)
21. Seichter, H., Schnabel, M.A.: Digital and tangible sensation: an augmented reality urban design studio. Paper presented at the 10th CAADRIA. New Delhi, India (2005)
22. Seichter, H.: Sketchand+: a collaborative augmented reality sketching application. Paper presented at the 8th CAADRIA, Bangkok, Thailand (2003)
23. Silver, D., Hubert, T., Schrittwieser, J., Antonoglou, I., Lai, M., Guez, A., Lanctot, M., Sifre, L., Kumaran, D., Graepel, T., Lillicrap, T., Simonyan, K., Hassabis, D.: A general reinforcement learning algorithm that masters chess, shogi, and go through self-play. Science **362**(6419), 1140–1144 (2018)
24. Tonnis, M., Klinker, G.: Augmented 3D arrows reach their limits in automotive environments. In: Wang, X., Schnabel, M.A. (eds.) Mixed Reality in Architecture, Design and Construction. Springer, Dordrecht, The Netherlands (2009)
25. Xu, Y., Yu, G., Wang, Y., Wu, X., Ma, Y.: Car detection from low-altitude UAV imagery with the Faster R-CNN. J. Adv. Transp. (2017)
26. Zhao, Z.-Q., Bian, H., Hu, D., Cheng, W., Glotin, H.: Pedestrian detection based on fast R-CNN and batch normalization. Paper presented at the ICIC 2017. Liverpool, UK (2017)

Data Reduction Using NMF for Outlier Detection Method in Wireless Sensor Networks

Oussama Ghorbel, Hamoud Alshammari, Mohammed Aseeri, Radhia Khdhir and Mohamed Abid

Abstract Nowadays, in wireless sensor networks (WSNs) field, the advances in electronics, wireless communications, and data processing have become an important reality. Wireless sensor networks are employed to eliminate problems occurred in health care, monitoring, agriculture, etc. Data reduction is considered as the best method that reduces dimensionality of the database. So, it helps outlier detection technique to classify data during training. In our work, we have constructed a newest data reduction process using non-negative matrix factorization (NMF). This method finds out the nature of data it is regular or outlier. Also, it can give a highest performance. Compared to various methods like Fisher Discriminant Analysis (FDA) and Principal Component Analysis (PCA), our method based NMF are considered as the most efficient and accurate for detecting outlier in WSNs. As real datasets, we use LUCE, Intel Berkeley and Grand St-Bernard. So based on the obtained results, our method is considered as a perfect one used in wireless sensor networks.

Keywords Outlier detection method · Wireless sensor networks (WSNs) · Non-negative matrix factorization (NMF) · Data classification

O. Ghorbel (✉) · H. Alshammari
Jouf University, Sakakah, Kingdom of Saudi Arabia
e-mail: oaghorbel@ju.edu.sa

H. Alshammari
e-mail: hhalshammari@ju.edu.sa

O. Ghorbel · R. Khdhir · M. Abid
National Engineers School of Sfax, Sfax University, Sfax, Tunisia
e-mail: rkhdhira@gmail.com

M. Abid
e-mail: mohamed.abid@enis.rnu.tn

M. Aseeri
King Abdulaziz City for Science and Technology (KACST), Riyadh, Kingdom of Saudi Arabia
e-mail: masseri@kacst.edu.sa

O. Ghorbel · M. Abid
Digital Research Center of Sfax, B.P. 275, 3021 Sakiet Ezzit, Sfax, Tunisia

© Springer Nature Singapore Pte Ltd. 2020   23
X.-S. Yang et al. (eds.), *Fourth International Congress on Information and Communication Technology*, Advances in Intelligent Systems and Computing 1027, https://doi.org/10.1007/978-981-32-9343-4_3

1   Introduction

Wireless sensor networks are considered as many sensors interconnected between each other that generate data [1]. These latter are continually which is doubtful and inaccurate. For outlier detection applications, to classify data, a powerful processing becomes necessary [2]. Including precision agriculture, health care, traffic control, monitoring, etc., various applications in WSN need an exact data to give confident details for final user [3]. A new field has founded by non-negative matrix factorization integrated in most application like outlier detection, data denoising, feature extraction, etc [4].

To reduce a processing load of data and increase the robustness of the classifier, dimensionality reduction techniques are important to be used. So, these reduction approaches are developed like Fisher Discriminant Analysis (FDA) and Principal Component Analysis (PCA). They have been used to clarify the variability between the variables. Generally, FDA gives an optimal lower-dimensional representation in terms of discrimination among classes, but PCA provides a high-dimensional process data via reconstruction in a reduced dimension. However, for classification problems, the PCA performs better than the FDA.

In our work, dimensionality reduction-based NMF is considered as the main contribution for outlier detection in wireless sensor networks.

As an organization in this article, Sect. 2 describes the related works. The different category of outlier detection in wireless sensor networks is depicted in Sect. 3. The different data reduction techniques and mathematical fundamentals of NMF are described in Sect. 4. Section 5 displays the obtained experimental results, and Sect. 6 concludes the outcomes of the paper.

2   Related Works

Wireless sensor networks are sensing equipment connected to each other through links-based wireless. They collect large numbers of high-fidelity data from different locations [5]. They process and transmit data to gateway nodes. Outlier detection technique is inspected as a pattern that does not matching the expected trend in analyzed data. In WSNs, efficient information about state of the network is acquired [6]. It can provide correct detection outlier data like failed nodes, unexpected environmental events, and residual WSN lifetime [7].

In various domains, to monitor the cultivation process, we need fundamental task for development of methods. Monitoring values in various applications is considered as the principal key to analyze events [8]. For data representation, a fundamental approach untitled non-negative matrix factorization has recently used in WSNs. It attracts the attentions of the researchers. As dimensionality reductions, it gives an extraordinary performance [9]. Real-world datasets contain a big extent of data. Storage space and computational time incurred by batch algorithms are prohibitive. At

Fig. 1 Decomposition
class-based NMF

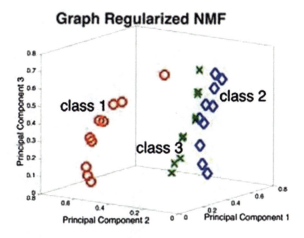

the arrival of new data, this method has to execute computation from zero for data
classification. Then, when values are affected with outliers, they cannot learn reasonable basis vectors. So, they cannot recover the underlying subspace dependably.
To separate true outliers from others caused by measurement errors are inspected as
difficult. This action is considered as a challenge in outlier detection. We noted that
features tend to be correlated in high-dimensional space. So, if data is considered
as outlier that fact it is noted a visible features. Then, the data point will be shown
as an accidental outlier in other subspaces [10]. However, this space is considered
as exponential. It is difficult to treat it for various practical problems and the use of
NMF method is noted as a best solution.

Depending of outlier's types, this latter can used in different ways. These types
can be classified in different class as shown in Fig. 1 as an example based on NMF.
Every type can be represented by a specific class, but in our case the decomposition
are in two classes which are normal or outlier.

3 Outlier Detection Technique

To find wrong and inconsistent values, errors, noise or duplicate values, outlier detection method is used. These abnormal values reduce the performance of the system
and affect the quality of data. In WSN, three sources types of outliers are considered:
errors, events, and malicious attacks. Outlier technique is used recently in WSNs.
It is an impressive method applied in various domains and applications like target
tracking, environmental monitoring, surveillance monitors, medical monitoring, etc.,
[11] as shown in Fig. 2.

In wireless sensor networks, the use of outlier technique is requisite. This latter
gives the highest detection rate [12]. It keeps low false alarm rate (reliable data
incorrectly presented as outliers). Usually, ROC curve (know as receiver operating

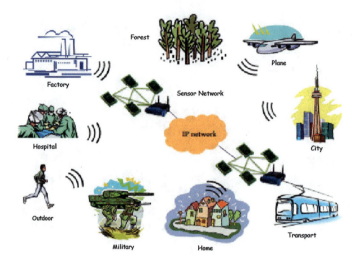

Fig. 2 Various applications field in wireless sensor networks

characteristic) is used to show the trade-off between the false alarm rate and detection rate. So, most of the problems in detection of outliers based on the field of WSNs can be summarized as follows: Modeling error and normal data effectively, Identifying outlier source, High communication cost, and Dynamic network topology.

4 Data Reduction Techniques

4.1 Non-negative Matrix Factorization (NMF)

We start by providing mathematical foundation of NMF. This latter supports the background of outlier detection scheme adopted in this work. The magnitude spectra for collected information approximated by NMF detailed by column (D) which is a linear combination of various spectral (K) basis vectors for matrix $D \in \mathbb{R}_+^{F \times N}$. It presents the product of matrices W and H (know as two non-negative low-rank matrices), such that:

$$D \approx WH$$
$$\text{where } W \in \mathbb{R}_+^{F \times K}, \ H \in \mathbb{R}_+^{K \times N}$$
$$\text{and } K \leq \min(F, N) \tag{1}$$

where W columns is a corresponding columns.

For dimensionality reduction, NMF method is chosen as the best one. The data vectors pass from (F) that represent the dimensional space to (K) that represent the dimensional space as $K < F$ [13].

For (Eq. 1), it represents an iterative minimization of cost function like Euclidean distance [14]. Based on equation demonstrated below in (Eq. 2), this formula gives good results in various domains and especially in feature extraction [15].

$$D_{kl}(D||WH) = \sum_{ij}\left(D_{ij}\log\frac{D_{ij}}{(WH)_{ij}} - (D - WH)_{ij}\right) \qquad (2)$$

In finding an optimum value, we use the formula presented below that is a divergence between (D) and (WH).

$$W \leftarrow W * \frac{\frac{D}{WH}H^T}{1H^T} \quad H \leftarrow H * \frac{W^T\frac{D}{WH}}{W^T 1} \qquad (3)$$

Note that 1 presents the all-ones matrix $(F \times N)$.

4.2 PCA: Principal Component Analysis

Every day, an important correlation between gathered data is considered as temporal. So, we propose to decline this latter to fit the sensors with limited resources such as the memory.

As statistical tools, PCA is used to analyze data without losing information. In high-dimensional data, it is considered as the best technique to find outliers [16]. Principal component analysis creates a new coordinate using linear transformation. Then, eigenvalues and eigenvectors of matrix are determinate. So, to retain the lower order of principal components, PCA keeps the characteristics of the dataset while reducing dimensionality. This datasets contain important data aspects. Unlike other linear transformations, PCA depends on the dataset vectors. This latter considers maximizing the following objective:

$$\max_w \text{Var}(W^Tx) = \text{Cov}(W^Tx, W^Tx) \approx W^TC_w$$
$$\text{where} \quad C = \frac{1}{N}\sum_{k=1}^{N} X_kX_k^T$$

4.3   FDA: Fisher Discriminant Analysis

Considered as one of the oldest techniques, Fisher Discriminant Analysis (FDA) is used as important technique for dimensionality reduction. This latter date back to statistical pioneer named Ronald Fisher [17]. This technique is a pre-processing step for machine learning applications and classification. FDA main idea is to maximize the dispersion provided between classes at the same time. But, reduce the dispersion in class in order to reduce computational costs as presented in the equation below:

$$J(W) = \frac{W^T S_B W}{W^T S_W W} \tag{4}$$

where (S_W) is within classes scatter matrix and (S_B) is between classes scatter matrix.

5   Experimental Results

As a validation of the proposed models, we use various samples collected from three WSN deployments presented below:

- IBRL: dataset collected at University of Berkeley. It is composed of 54 sensors from Intel Berkeley Research Laboratory [18].
- GStB: dataset collected using WSNs deployment composed of 23 sensors from Grand St-Bernard. This latter located between Italy and Switzerland [19].
- LUCE: dataset used by (EPFL) that contain 110 sensor nodes at Lausanne Urban Canopy Experiment project [20].

The goal of NMF technique is determined in this section. In experimental case, normal data gathered by WSNs deployment was used as presented previously. Using Matlab, our method is simulated and a related neighborhood is considered. NMF performance method was compared to others various kernel methods. First, the experiment patterns are obtained using NMF method. Second, a suitable projection of dimensionality (p) is obtained by chooses a training set. Third, the distance of separated data from patterns is calculated. At the end, outliers are detected. This latter has an exceeded distance compared to the constructed threshold value taking into account the projection dimensionality (p). Based on real data, we present our obtained results below.

From comparison obtained by Table 1, we observe that the results obtained by our methods based NMF are more competitive in term of outliers detection compared to PCA and FDA. The obtained value 0.9912 is obtained by IBRL dataset. For GStB dataset is 0.9894 and for LUCE the value is 0.9906. Based on previous experiments, we observe that our method gives an interesting result to detect outliers. Compared to PCA and FDA, the proposed algorithm presents an important advantage in dimensionality reduction. Consequently, we conclude that the proposed technique is very sensitive to DR and FPR as proved by Table 2 below.

Table 1 NMF method for datasets compared to PCA and FDA

	IBRL	GStB	LUCE
NMF	0.9912	0.9894	0.9906
PCA	0.9805	0.9710	0.9798
FDA	0.9697	0.9689	0.9715

Table 2 DR and FPR-based NMF for IBRL dataset

Nodes						
	N1	N2	N3	N4	N5	Average
Detection rate (%)	99	100	99	98	99	99
False positive rate (%)	1	0	1	2	1	1

6 Conclusion

Outlier is considered as inconsistent or duplicate data detected to minimize the recovery time and functionality of the system. A novel outlier technique was used known as non-negative matrix factorization. After applying the NMF method, the system models analyse the obtained data as a normal or as outlier. In the experiment, three datasets are used as presented previously. Also, we use two other methods named PCA and FDA to detect outlier, but our proposed NMF method is noted as the bestest to extract outliers perfectly. At the end, the results of our work based NMF method are competitive compared to other methods to detect abnormal data in wireless sensor networks field.

References

1. Zhang, Y., Meratnia, N., Havinga, P.: Outlier detection techniques for wireless sensor networks: a survey. IEEE Commun. Surv. Tutorials **12**, 159–170 (2010)
2. Cai, D., He, X., Han, J., Huang, T.S.: Graph regularized nonnegative matrix factorization for data representation. IEEE Trans. Pattern Anal. Mach. Intell. **33**(8), 1548–1560 (2011)
3. Deng, X., Tian, X., Chen, S., Harris, C.J.: Fault discriminant enhanced kernel principal component analysis incorporating prior fault information for monitoring nonlinear processes. Chemom. Intell. Lab. Syst. **162**, 21–34 (2017)
4. Rassam, M.A., Zainal, A., Maarof, M.A.: An Adaptive and Efficient Dimension Reduction Model for Multivariate Wireless Sensor Networks Applications. Applied Soft Computing (2013)
5. Nakanishi, T.: A generative wireless sensor network framework for agricultural use. Makassar Intl. Conf. Electr. Eng. Inform. (MICEEI), 205–211 (2014)
6. Yang, S., Yi, Z., Ye, M., He, X.: Convergence analysis of graph regularized non-negative matrix factorization. IEEE Trans. Knowl. Data Eng. **26**(9), 2151–2165 (2014)
7. Kapitanova, K., Sonand, S.H., Kang, K.D.: Event detection in wireless sensor networks. Second international conference, ADHOCNETS 2010, Victoria, Canada 2010
8. Zhang, Y., Hammb, N.A.S., Meratnia, N., Steinb, A., Voorta, M., Havinga, P.J.M.: Statistics-based outlier detection for wireless sensor networks **26**(8) (2012)

9. Yang, G., et al.: Research on anomaly detection based on MNF in hyper spectral imagery. Appl. Mech. Mater. **644–650**, 1085–1088 (2014)
10. Wang, D., Nie, F., Huang, H.: Fast robust non-negative matrix factorization for large-scale human action data clustering. In: Proceedings of the Twenty-Sixth International Joint Conference on Artificial Intelligence IJCAI-2017
11. Liu, j., Wang, C., Gao, J., Han, J.: Multi-view clustering via joint nonnegative matrix factorization. In: Proceedings of SDM, SIAM, vol. 13, pp. 252–260. SIAM (2013)
12. Hsu, D., Kakade, S.M., Zhang, T.: Robust matrix decomposition with sparse corruptions. IEEE Trans. Info. Theory (2011)
13. Guan, N., Tao, D., Luo, Z., Yuan, B.: Online nonnegative matrix factorization with robust stochastic approximation. IEEE Trans. Neural Netw. Learn Syst. **23**(7), 1087–1099 (2012)
14. Fevotte, C., Dobigeon, N.: Nonlinear hyper spectral unmixing with robust nonnegative matrix factorization. ArXiv preprint (2014)
15. Ludeña-Choez, J., Gallardo-Antolín, A.: Speech denoising. Using non-negative matrix factorization with Kullback-Leibler. Divergence and sparseness constraints. In: Torre Toledano, D. et al. (eds.) Advances in Speech and Language Technologies for Iberian. Languages: Iber SPEECH (2012)
16. Chitradevi, N., Palanisamy, V., Baskaran, K., Nisha, U.B.: Outlier aware data aggregation in distributed wireless sensor network using robust principal component analysis. In: International Conference on Computing Communication and Networking Technologies, pp. 1–9 (2010)
17. Nor, N.M., Hussain, M.A., Che Hassan, C.R.: Process monitoring and fault detection in nonlinear chemical process based on multi-scale kernel fisher discriminant analysis. Comput. Aided Chem. Eng. **37**, 1823–1828 (2015)
18. Intel Berkeley Research lab: http://db.csail.mit.edu/labdata/labdata.html. 28 Feb and 5 Apr 2004
19. GStB, Grand-St-Bernard dataset. http://lcav.epfl.ch/cms/lang/en/pid/86035 (2007)
20. Sensor Scope. http://sensorscope.epfl.ch/index.php/MainPage

Measuring Customer Satisfaction on Software-Based Products and Services: A Requirements Engineering Perspective

Ezekiel Uzor Okike and **Seamogano Mosanako**

Abstract Overall customer satisfaction is the most critical quality measure for all bespoke and commercial-off-the-shelf (COTS) software systems deployed in user environments. This study measured the satisfaction of users of a student administration software which was acquired and reconfigured by the IT department at an African university. As a student administration software, the product has all the necessary functionalities. However, most users of the software struggle to effectively and confidently use the software on their own without assistance from the IT department of the university. Although the software is required for capturing and processing students' academic records by all academic staff, it appears there is a general feeling of boredom by many staff toward the software usage at this university. Therefore, this study aims to investigate the level of satisfaction of users of the software and if there is any correlation between satisfaction of use and user participation in the requirements engineering or reconfiguration process of the software. The study was conducted purely as a private academic initiative undertaken in order to investigate the perceived complaints among the users of the software and to propose professional solution to the problem. The study observed all the necessary steps required for conducting a scientific survey study of this nature including ethical considerations of confidentiality of information obtained from participants, optional user participation in the study (as indicated on the survey instrument), and proper acknowledgment of all literature sources referenced in the present study. The study employed a quantitative approach with a sample of 40 participants who were randomly selected from four different faculty units of the university. The Qualification Weighted Customer Opinion with Safeguard (QWCOS) model was used as initial external measurement (EM) of customer satisfaction. A Flexible Qualification Weighted Customer Opinion with Safeguard (FQWCOS) model was also applied to measure EM. Both results of EM were compared. In all, results from the study suggest that 85% of the sample were not very satisfied with the software, 62.5% will not recommend the software to

E. U. Okike (✉) · S. Mosanako
University of Botswana, Gaborone, Botswana
e-mail: euokike@gmail.com

S. Mosanako
e-mail: mosanakos@ub.ac.bw

© Springer Nature Singapore Pte Ltd. 2020
X.-S. Yang et al. (eds.), *Fourth International Congress on Information and Communication Technology*, Advances in Intelligent Systems and Computing 1027, https://doi.org/10.1007/978-981-32-9343-4_4

others, and over 80% of the users did not participate during the requirements reconfiguration of the software before its deployment. There is a significant relationship
between participating in requirements configuration and user satisfaction. Furthermore, both QWCOS and FQWCOS yield the same EM values. To improve customer
satisfaction on software products and services, a 2D4E framework based on requirements and reconfiguration engineering (RRE) and total quality management (TQM)
is proposed in this study.

Keywords Requirements analysis · Software quality · Customer satisfaction ·
Quality-in-use assessment · COTS software

1 Introduction

The IEEE Glossary of Software Engineering Terminology [10] defines a requirement
as

1. A condition or capability needed by a user to solve a problem or achieve an
   objective.
2. A condition or capability that must be met or possessed by a system or system component to satisfy a contract, standard, specification, or other formally
   imposed document.
3. A documented representation of a condition or capability in (1) or (2).

From this definition, requirements engineering (RE) involves applying an engineering approach in gathering or eliciting requirements, modeling requirements,
analyzing requirements, communicating the requirements to both stakeholders and
developers, agreeing on the requirements, and finally, evolving requirements for the
purpose of building a new information system (IS). The application of engineering
approach in software construction implies that each activity engaged in during the
phases of the construction process from the beginning to the delivery and the subsequent maintenance of the product "is understood and controlled, so that there are
few surprises as the software is specified, designed, built, and maintained" [6]. For
the information systems development life cycle (ISDLC) analysts and developers
have reasonable understanding of "the what" to create and "the how" to create based
on the customers' needs (requirements). It is the primary responsibility of systems
analysts to discover, document, and communicate requirements to both the developers and system stakeholders so as to make it clear what the customers' needs and
agreed requirement are. Thus, RE defines a reference model of the information system being built from the agreed customers' requirements' point of view. It has also
been rightly proved that the success of an information system project (ISP) is directly
proportional to the quality of requirements elicited and used in the project [12, 16].
Therefore, requirements engineering (RE) is the cornerstone of information system
projects. The importance of RE in information systems development (ISD) has been
adduced to the following reasons [2]:

1. The pace of product development—Most often, product users frequently ask for new versions of the product due to the dynamics of business environment and requirement (DBER).
2. Turnover and technology changes propel changes in technological and human professionals experience, thus making job change more common than ever before.
3. The impact of outsourcing and offshoring on the product life cycle (PLC)—This enables organizations to be able to create specifications for implementation using their own staff.

The key components of requirements engineering include:

1. Requirements management—This involves putting the process for managing requirements in place through a requirements management plan (RMP) which defines the process, the procedure, and the standard for doing RE. Following the RMP, the process of defining and maintaining the requirements for a particular information systems construction is established. Industrial tools [8] exist for tracking requirements process.
2. Requirements process—This includes the steps in the requirements specification, namely elicitation, analysis, modeling, documentation, communication, and validation.
3. Requirements elicitation—This involves interacting with all stakeholders to discover the requirements of the intended information system from the users and stakeholders' perspectives.
4. Requirements analysis—Elicited requirements are further analyzed to ensure they are correct and are clearly understood by all stakeholders.
5. Requirements documentation—The specification of requirements is documented. This documentation is the basic reference point of the agreed requirements of the intended information system. It gives understanding to both stakeholders and designers, stating what a system is supposed to do as opposed to how it should be done.
6. Requirements validation.
7. Requirements traceability—Traceability implies that the source of every requirement as well as its realization can be ascertained.
8. Requirements sign-off—A sign-off is a formal process where the key stakeholders and the analysts formally agree to the requirements.

1.1 Statement of the Problem

Software products' quality is basically linked to the use of appropriate requirements engineering procedures. If there is a gap in requirements engineering during software development or during the reconfiguration of commercial-off-the-shelf (COTS) software product, it certainly affects the customers' overall satisfaction. Also, if there is

Fig. 1 Knowledge gap in software design

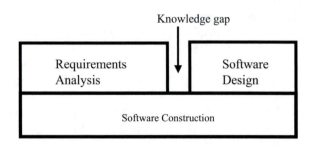

a gap between the developers and the stakeholders, it will lead to developing systems with wrong functions and properties which are among the top ten risk items to guard against in software development [3]. Moreover, overall customer dissatisfaction on a product or service is often the result of one or all of the following:

1. The fact that many software projects appear to proceed with no meaningful requirements engineering [16].
2. Building software on ambiguous, inconsistent, incomplete, and confused requirements specifications.
3. Reconfiguring COTS without appropriate requirements engineering (RE).
4. The existence of a knowledge gap between requirements analysis and software design [5].

This last concern could be the cause of the other three concerns.

Knowledge Gap in Software Development

In the course of developing an information system or reconfiguring a commercial-off-the-shelf (COTS) software product, reference is usually made to the unified software development life cycle (SDLC) [11]. The basic SDLC stages include Planning, Analysis, Design, Implemetation and Maitenance. However, a knowledge gap might exist between requirements analysis (RA) and software design (SD), if necessary requirements engineering is neglected during Systems Analysis (SA) [5]. Figure 1 illustrates this gap.

Requirements analysis or engineering involve domain experts in the analysis of an application domain and the formulation of the requirements for the software solution. On the other hand, software design involves software architects and developers who design, develop, and test the software application that satisfies user requirements. The "knowledge gap happens when the architects and developers are divorced from the original business domain analysis and only see the result of this analysis" [ibid.]. To effectively model a domain, analysts engage stakeholders in meetings and dialogue with domain experts so that appropriate user requirements are elicited, documented, analyzed, communicated, and agreed upon as the basis for building the information or reconfiguring a generic product to suit an environment. However, much of the communicated information may not be all utilized into the resulting domain model. In addition, the developer team may find some ambiguities and inconsistencies in the domain model as the software is developed. At this point, the ideal action is for the

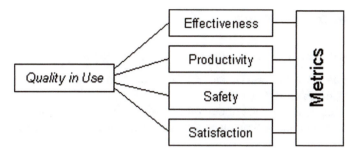

Fig. 2 ISO 9126 quality-in-use metric

analysts to reconsult the domain experts for clarifications of misconceptions instead of developers making assumptions of their own to resolve any misconception in the domain model. With the later approach, the software that results from the information system may not have satisfied the users' need as appropriate. Therefore, customer satisfaction as an external quality software attribute (Fig. 2) needs to be constantly evaluated in software products' user environment.

The focus of this study is the measurement of quality in use with reference to customer satisfaction in a user environment. The study was conducted using a student administration software as case study.

1.2 Aims and Objectives

The aims of this study are:

1. To investigate the level of satisfaction of users of the student administration software.
2. To investigate if there is any correlation between satisfaction of use and user participation in requirements engineering or reconfiguration process of the software.
3. To use the Qualification Weighted Customer Opinion with Safeguard (QWCOS) [19] and the Flexible Qualification Weighted Customer Opinion with Safeguard (FQWCOS) models [15] as external measures of customer satisfaction for the student administration software based on the weighted opinions of the software users, and subsequently measure the overall satisfaction index among sampled users of the software.
4. To propose a 2D4E framework for improving software products and service quality with focus on customer satisfaction.

1.3 Research Questions

The following research questions were investigated in this study:

1. What is the level of satisfaction of users of the student administration software at the African university?
2. How likely is it that users of the software will recommend it to other users?
3. Are user requirements through involvements positively related to user satisfaction with experience from the software?

1.4 Hypotheses

1. H0 Users of the software are very satisfied with the product as a student administration software.
   H1 Users are not very satisfied with the software.
2. H0 Users of the software will strongly recommend it to others.
   H1 Users of the software will not strongly recommend it to others.
3. H0 User requirements through involvements are positively related to user satisfaction.
   H1 User requirements through involvements are not positively related to user satisfaction.

The rest of this paper counts of four sections. Section 2 presents a review of related work. Section 3 presents the methodology of this study. Section 4 presents the empirical study, while Section 5 presents the summary and conclusion.

2 Literature Review

The history of software engineering (SE) suggests that the 1990s and beyond is the era of profound improvements in software technology and systems. In this era, information systems (ISs) provide enough state-of-the-art components, functionalities, user-friendly interfaces, portable mobile versions or applications, ubiquitous business applications, and a preponderance of quality expectations and measurement standards in software products and services. This era is quite unique portraying a radical shift from previous eras which focused on functions (1960s), schedules (1970s), and costs (1980s) [13].

Therefore, in the current era the importance of quality in software products and services is recognized, as measurement plays a critical role in effective and efficient software development as well as provides the scientific basis for software engineering that makes it a truly engineering discipline [9]. Furthermore, quality must be defined and measured for improvement to be achieved in terms of the product and the service

Fig. 3 Internal software
quality attributes. *Source*
[15]

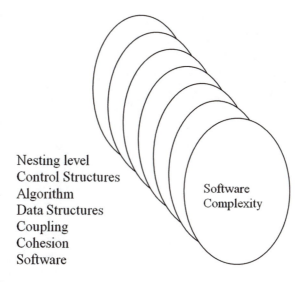

Nesting level
Control Structures
Algorithm
Data Structures
Coupling
Cohesion
Software

Software
Complexity

provided by software. In this regard, product quality deals with the resulting software artifact. This is usually measured from code-level perspectives in what is generally referred to as software complexity in terms of the attributes of cohesion, coupling, nesting level, data structures, algorithms, and control structures as shown in Fig. 3 [7].

Cohesion and coupling are the two most important quality attributes of software complexity [4].

Product quality can also be measured from the perspective of the external assessment of a software product by its users in a user environment taking into cognizance the external software quality attributes as shown in Fig. 4.

A third dimension of measuring software quality concerns the assessment of quality-in-use attributes as shown in Fig. 2.

This study is focused on measuring quality in conformance to customers' expectations and requirements from overall customer satisfaction on commercial-off-the-shelf (COTS) software products and services using a student administration software as case study. The study proposes a 2D3E framework for requirements and reconfiguration engineering (RRE) of COTS based on total quality management (TQM) philosophy which could improve quality and hence the satisfaction of users of software products and services deployed in user environments.

2.1 Metrics and Measurement in Software Quality Engineering

The terms measurement and metric are often used interchangeably. Spacy [18] notes that a metric is a system or standard for measuring something, while a measurement

Fig. 4 External software
quality attributes

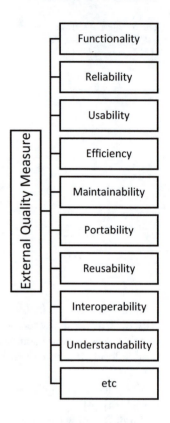

is a numerical observation. Each of the term is significantly important as both serve as building blocks for the assessment of performance and achievement in many disciplines in what is often referred to as key performance indicator (KPI). The KPI is a measurable value that demonstrates how effectively an organization is achieving its business objectives [20]. Thus, the KPI is a performance metric often applied over a specific time period and compared with past performance metrics.

Software metrics and measurements fulfill very important functions in an organization. Some of these include [14]:

1. Assisting in the control of software process and providing appropriate feedback.
2. Assisting in the overall evaluation of the objectivity of a software process.
3. Useful in setting software improvement goals.
4. Useful in cost and schedule estimates.
5. Useful in improving overall software quality.
6. Useful in setting design standards.
7. Useful in forecasting productivity trends.

2.2 Measuring Software Quality Based on Overall Customer Satisfaction and Requirements

In software engineering, the concept of quality defines quality in conformance to customers' expectations and requirements. Therefore, overall customer satisfaction is the most crucial quality-in-use metric. Customer (user) satisfaction in the context of this paper means the act of satisfying the needs of the user of a software product or service. Hence, satisfying a user implies fulfilling the desires or needs of the user with respect to the product or service. In general, the business domain views customer satisfaction as the core of human experience, indicating how individuals like a company's products. Therefore, "a high level of customer satisfaction is a strong predictor of customer retention, customer loyalty, and product or (service) repurchase" [17]. Moreover, providing customer satisfaction could be a quality criteria or requirement for software products or services. In this case, customer satisfaction means meeting all the customer requirements. According to Asher [1], customer satisfaction for a product or service may be measured in terms of:

1. The delivery of all promise about the product or service to the user.
2. The user getting good service from the product or service.
3. The product or service capability to meet all the user needs.
4. The providers' ability to deliver the product or service on schedule.
5. Solving the user's problem rather than that of the product or service.
6. The providers' ability to meet all current user needs and providing to meet potential user needs.
7. Providers' ability to send notifications to the user of any changes to the product or service.
8. The ability of provider team members to understand the needs of the users.
9. The user's acceptance and dependence on the quality of the product or service.
10. Providers' interests in how users use the product or service.

3 Measuring Customer Satisfaction on a Student Administration Software

3.1 The Method

The study employed a quantitative approach. The university was divided into four faculty-based strata using stratified sampling techniques, namely education, humanities, science, and technology. From each stratum, ten participants (academic staff members who compulsorily use the software and who were willing to participate in the study) were randomly selected. From this sample, questionnaires to capture the satisfaction of the software users were administered. All academic staff of this

Table 1 Statistical components of QWCOS

I	Freq (F) (n_i)	Mean (μ)	RF (n_i/N)	StD (S_i)	O_i (z)
1.	n_1	μ_{i1}	n_1/N	S_1	Z_1
2.	n_2	μ_{i2}	n_2/N	S_2	Z_2
3.	n_3	μ_{i3}	n_3/N	S_3	Z_3
...	...	...	...	...	...
40.	N	μ_{iN}	N_n/N	S	Z_n

$N = 40$

university have used the software over time except in cases where a staff has just been employed at the time of the study. A Qualification Weighted Customer Opinion with Safeguard (QWCOS) model as shown in Eq. (1) was used as initial external measurement (EM) of customer satisfaction. QWCOS techniques "estimate the result of the external measurement (EM), which is based on O_i (the normalized score of customer opinion), the E_i (qualifications of customer i), and the use of a number of control questions (safeguard questions S), where S_T is the total number of control questions, from which the customer i (from a total number of n customers) has responded correctly as at S_i" [19]:

$$\text{EM} = \text{QWCOS} = \sum_{i=1}^{n}\left(O_i.E_i.\frac{S_i}{S_t}\right) / \sum_{i=1}^{n}\left(E_i.\frac{S_i}{S_t}\right) / \sum_{i=1}^{n}\left(E_i.\frac{S_i}{S_t}\right) \quad (1)$$

In this study, we provide useful information about the statistical implications of QWCOS (Table 1). A variant measure of customer satisfaction with Flexible Qualification Weighted Customer Opinion with Safeguard (FQWCOS) was proposed in [15]. The measure was equally used to measure EM in order to compare results with QWCOS. FQWCOS is defined in Eq. (5).

From Eq. (1):

i. O_i (the normalized score of customer opinion) may be obtained as

$$Z = \frac{\mu_i - \mu}{S_i} \quad (2)$$

ii. $\mu_i = \dfrac{\Sigma \mu_i}{N}$ 　　　　　　　　　　　　　　　　　　　　　　(3)

iii. $\text{RF} = \dfrac{S_i}{S_T}$ 　　　　　　　　　　　　　　　　　　　　　　(4)

Define external measure (M_E) for FQWCOS as a quadruple $(\Sigma, O_i, \alpha, \beta)$ [15]. Hence,

$$M_{\mathrm{E}} = \sum_{i=1}^{n} \left(O_i.\alpha_j.\beta\right)/ \sum_{i=1}^{n} \left(\alpha_j.\beta\right) \tag{5}$$

where

O_i	Normalized score of users i opinion using mean standard error (S.E.) or standard deviation (StD)
α	Flexible qualifier of user
β	Relative frequency

Hence, ME O_i (best measured using S.E.; see Table 3).

4   Results and Discussion

From Table 1, 40 randomly chosen users of the software were asked to rate their level of satisfaction on a 5-point scale from the lowest 1 to the highest 5 (research questions 1 and 2). Six respondents out of 40 respondents (representing 15%) were satisfied with the software, while 34 respondents (85%) were not. Seven respondents (17.5%) truly enjoyed using the software, while 33 respondents (82.5%) did not enjoy using the software. Interestingly, 15 respondents (37.5%) would recommend the software to others, while 25 respondents (62.5%) would not recommend it to other institutions. At a later stage in this study, the same set of users were asked to rate if they were pretty satisfied with the software on the same 5-point scale. A total of 19 respondents (47.5%) were pretty satisfied with the software, while 21 respondents (52.5) were not. Moreover, there is significant relationship between user satisfaction and participating at requirements reconfiguration process of the software (research question 3). With respect to requirements captured by designers who reconfigured the software for use at the university (research question 3), users were asked if they made any inputs into the software requirements during the reconfiguration process? Only 1 person did, while 20 respondents did not (50%), and 19 respondents (47.5%) were neutral. Therefore, comparing user satisfaction with involvement in user requirements during the configuration of the software can be inferred that they are negatively related. The number of those who were not involved during the requirements process for the software far outnumbers those who participated. In like manner, the number of those who were not satisfied with the use of the software far outnumbers those who were satisfied. With respect to the research questions, it can be deduced that the software users were not very satisfied, and only 15 respondents (37.5%) are likely to recommend the software to others, while 25 respondents (62.5%) will not recommend the software to others.

Tables 2 and 3 show the results of the external measurement (EM) of satisfaction index using QWCOS as indicated in Eq. (1) and FQWCOS as indicated in Eq. (5).

Both QWCOS and FQWCOS yield the same results. However, FQWCOS permits flexibility and a more direct measure of customer satisfaction based on standard error

Table 2 Measuring satisfaction with QWCOS

Questions	N = 40	Freq 4–5	Freq 1–3	Mean (μ)	Rel. Freq	S.E./StD	$O_i = \frac{\mu_i - \mu}{\text{S.E.}}$	$O_i = \frac{\mu_i - \mu}{\text{Std}}$
1. Level of satisfaction with ASAS	40	6	34	2.6000	1.00	0.17097 / 1.08131	31.58	4.99
2. Pretty satisfied with ASAS	40	19	21	2.4250	1.000	0.28168 / 1.7849	19.79	3.13
3. Pretty certain I can use ASAS without assistance	40	14	26	2.2750	1.00	0.26309 / 1.66391	21.76	3.44
4. I have truly enjoyed using ASAS	40	7	33	2.7000	1.000	0.14850 / 0.93918	35.69	5.64
5. I can recommend ASAS to other institutions	40	15	25	3.2308	1.000	0.15351 / 0.95866	31.06	4.97
6. Experience in years in ASAS	40	30	10	3.9250	1.00	0.15766 / 0.99711	25.85	4.09
7. IT competence	40	11	29	1.4750	1.000	0.12904 / 0.81610	50.57	8.00

Table 3 Measuring customer satisfaction with FQWCOS

Questions	$N = 40$	FQWCOS	$O_i = \frac{\mu_i - m}{S.E.}$	$O_i = \frac{\mu_i - m}{Std}$
1. Level of satisfaction with ASAS	EM	$M_{E1}/M_{E2}/M_{E3}/M_{E4}$	31.58	4.99
2. Pretty satisfied with ASAS	EM	$M_{E1}/M_{E2}/M_{E3}/M_{E4}$	19.79	3.13
3. Pretty certain I can use ASAS without assistance	EM	$M_{E1}/M_{E2}/M_{E3}/M_{E4}$	21.76	3.44
4. I have truly enjoyed using ASAS	EM	$M_{E1}/M_{E2}/M_{E3}/M_{E4}$	35.69	5.64
5. I can recommend ASAS to other institutions	EM	$M_{E1}/M_{E2}/M_{E3}/M_{E4}$	31.06	4.97
6. Experience in years in ASAS	EM	$M_{E1}/M_{E2}/M_{E3}/M_{E4}$	25.85	4.09
7. IT competence	EM	$M_{E1}/M_{E2}/M_{E3}/M_{E4}$	50.57	8.00

(S.E.) or standard deviation (StD). However, using S.E. is preferred since the resulting values are higher than the StD values. Moreover, the S.E. values are better expressed as percentages.

5 Summary and Conclusion

The study with the student administration software depicts the likelihood of experience with commercial-off-the-shelf (COTS) software product and bespoke software when there is a gap in requirements engineering (RE) or requirements and reconfiguration engineering (RRE). Therefore, this study proposes a 2D4E framework based on requirements and total quality management (TQM) that would enhance customer satisfaction of software products and services.

5.1 A 2D4E Framework for Achieving Customer Satisfaction on Software-Based Products and Services

D1 Determine what the customer wants using appropriate requirement engineering procedures.

D2 Develop/Configure the software on approved standards/configuration procedures.

E1 Entrench total quality management (TQM) procedures.

E2 Establish appropriate maintenance mechanisms.

E3   Establish evaluation procedures. Need to conduct periodic customer satisfaction
     surveys over appointed timing periods.
E4   Education and train the user.

   E1   Key elements of TQM.
   E1.1   Customer focus—To achieve total customer satisfaction, there is a need
          to:
          Elicit, document, and analyze customers' wants and need.
          Measure and manage customer satisfaction.
   E1.2   Process quality focus—Entrench product quality through process
          Improvement.
   E1.3   Human factor focus—Management is committed to employee partici-
          pation in order to improve quality. This will be made possible through
          employee support and motivation as necessary.
   E1.4   Measurement and evaluation procedures to drive continuous improve-
          ment.

   Using this case study, it is suggested that efforts be made during software requirements engineering (RE) or requirements and reconfiguration engineering (RRE) to
bridge any gaps between designers and users by carrying the users and stakeholders
along throughout the processes of RE or RRE.

References

1. Asher, M.: Measuring customer satisfaction. TQM Mag. **1**(2), 93–97 (1989)
2. Berenbach, B., Kazmeier, J., Paulish, D.J., Rudorfer, A.: Software and Systems Requirements
   Engineering in Practice. Tata McGraw Hill, New Delhi (2009)
3. Boehm, B.: Software risk managements and principles. IEEE Softw. **8**(1), 32–41 (1991)
4. Darcy, P.D., Kemerer, C.F., Slaughter, S.A., Tomayako, J.E.: The structural complexity of
   software: an experimental test. IEEE Trans. Softw. Eng. **32**(1), 54–64 (2005)
5. Duggan, D.: Enterprise Software Architecture and Design: Entities, Services, and Resources.
   Wiley, USA (2012)
6. Fenton, N.E., Pfleedger, S.L.: Software Metrics: A Rigorous & Practical Approach, 2nd edn.
   PWS, London (1997)
7. Gorla, N., Ramakrishnan, R.: The effect of software structure attributes on software development productivity. J. Syst. Softw. **36**(2), 191–199 (1997)
8. http://www.jiludwig.com/requirements_Mgt_Tools. Last accessed 11 Jan 2018
9. http://www.locometrics.com/books.html. Last accessed 11 Jan 2018
10. IEEE: Glossary of software engineering. IEEE, USA (1990)
11. Jacobson, I., Booch, G., Rumbaugh, I.: The Unified Software Development Process. Addison-
    Wesley Longman Publishing Co. Inc., Boston (1999)
12. Kamala, M.I., Tamai, T.: How does requirements quality relate to project success or failure?
    In: Proceedings of the International Requirements Engineering Conference (RE'07) (2007)
13. Kan, S.H., Basili, V.R., Shapiro, L.N.: Software quality: an overview from the perspective of
    total quality management. IBM Syst. J. **33**(1), 4–17 (1994)
14. Okike, E.U.: Measuring class cohesion in object-oriented systems using Chidamber and
    Kemerer metrics and java as case study. Ph.D. thesis. Department of Computer Science, Uni-
    versity of Ibadan (2007)

15. Okike, E.U.: FQWCOS, a flexible model for measuring customer satisfaction on software based products and services. Softw. Eng. **6**(4), 110–115 (2018)
16. Ralph, P.: The illusion of requirements in software development. Requir. Eng. **18**, 293–296 (2013)
17. Smith, S.M.: Customer satisfaction survey questions: 5 sample templates you can use right away. http://www.Qualtric.com. Last accessed 26 July 2018
18. Spacy, J.: Metrics vs measurement. http://www.simplicable.com/new/metrics-vs-measurements. Last accessed 11 Jan 2018
19. Stravrinoudis, D., Xenos, M.: Comparing internal and external software quality measurements. In 8th Joint Conference Proceedings on Knowledge-Based Software Engineering, pp. 115–124. IOS Press, Piraeus Greece, 25–28 Aug 2008
20. Taylor, J.: What is a KPI, metric or measure? http://www.klipfolio/blog/kpi-metric-measure. Last accessed 12 Jan 2018

To Develop a Water Quality Monitoring System for Aquaculture Areas Based on Agent Model

Thai Minh Truong⑩, Cuong Huy Phan⑩, Hoang Van Tran⑩,
Long Nhut Duong⑩, Linh Van Nguyen⑩ and Toan Thanh Ha⑩

Abstract Recently, the environmental pollution has seriously affected the production and life of people in many countries in the world and especially in Vietnam. In particular, aquaculture has been severely affected by the polluted environment, the waste from factories, agricultural production, or even from aquaculture itself. That raises the questions "how to monitor the water quality of aquaculture areas 24/7 and to promptly alert aquaculture farmers and managers to take appropriate response?" Facing this situation, Can Tho University has invested to conduct research and to develop a system to monitor water quality of aquaculture areas based on the actual conditions of the Mekong Delta. In this presentation, we will first describe the agent-based environment monitoring system (AEMS) model, combined triple technologies (sensor, IoT, and agent-based) model. The second introduction is the water quality monitoring system for aquaculture settings based on the AEMS model, we proposed. Thanks to the implement of AEMS model, it not only helps managers and farmers to monitor indicators of water quality at any time, but it also helps to analyze collected data from IoT Agent, send warning messages, and solutions deal with the situation of water environment in each aquaculture farms.

Keywords Water quality monitoring · Agent-based model · IoT · Sensor network

T. M. Truong (✉) · C. H. Phan · H. Van Tran · L. N. Duong · L. Van Nguyen · T. T. Ha
Can Tho University, Can Tho, Viet Nam
e-mail: tmthai@ctu.edu.vn

C. H. Phan
e-mail: phcuong@ctu.edu.vn

H. Van Tran
e-mail: tvhoang@ctu.edu.vn

L. N. Duong
e-mail: dnlong@ctu.edu.vn

L. Van Nguyen
e-mail: nvlinh@ctu.edu.vn

T. T. Ha
e-mail: httoan@ctu.edu.vn

© Springer Nature Singapore Pte Ltd. 2020
X.-S. Yang et al. (eds.), *Fourth International Congress on Information and Communication Technology*, Advances in Intelligent Systems and Computing 1027, https://doi.org/10.1007/978-981-32-9343-4_5

1   Introduction

Aquaculture is an important resource in Vietnam in general and the Mekong Delta in particular
In Vietnam, aquaculture from self-sufficient industry that has become one of the key commodity production industries in all water environments, e.g., fresh, brackish, and seawater in a sustainable way. Moreover, aquaculture in Vietnam is geared toward the development of concentrated production areas. According to data of the General Department of Fisheries cited in [1], the area of aquaculture in Vietnam is continuously increasing each year, for example, in 1981, we only had about 230 thousand hectares of aquaculture, and aquaculture area has reached over 1 million ha.

The area and production of freshwater aquaculture in the Mekong Delta has been increasing continuously in recent years: from 50,000 ha with 200,000 tons in 1999 and 152,000 ha with 600,000 tons in 2005; reaching over 180 thousand hectares of aquaculture with output reaching over 1.6 million tons in 2010 (of which the production of Pangasius is 1.1 million tons). Since 2000, Pangasius and Basa fish has become an important freshwater fish and is the second most important export item after shrimp.

Water pollution in Vietnam affects the productivity of aquaculture
Currently in Vietnam, although management agencies at all levels and sectors have made great efforts in implementing policies and legislation on environmental protection, the situation of water pollution is occurring, affecting the water resources in general and the aquaculture and fishing in Vietnam in particular. In addition to the causes of water pollution from wastewater from industrial parks, waste in agricultural production also affects the quality of water environment. Due to the lack of planning and non-compliance with technical procedures such as incorrect use of chemicals in aquaculture, over-feeding, the aquaculture pond water quality such as pollution, organic matter, increased pH, NH_4, NO_3, develops a number of disease-causing organisms, affecting quality and productivity in aquaculture. We can refer to a case of water pollution on the Buoi River (Thanh Hoa, Vietnam) due to waste discharge from the Hoa Binh sugar factory, which caused fish death in rafts in April— early May 2016 or more than 1,000 tons of fish farming in An Giang and Dong Thap provinces were died in early 2016, according to the An Giang Provincial Department of Natural Resources and Environment. High stocking density, combined with fish waste into water, increases organic matter, leading to the phenomenon decreasing dissolved oxygen in water.

Monitoring the water quality of the aquatic environment for aquaculture is very important
Based on the literature review, and through field surveys at some fish farms in Hau Giang and An Giang, we have found that monitoring of fish pond water quality is a matter of concern and this issue is the great interest to aquaculture companies and organizations today. However, aquaculture farms are applying sampling only

1 or 2 times a day using hand-held meters to measure dissolved oxygen (DO), pH, temperature, and in combination with the test kit use chemicals to measure NH_4/NH_3.

The advantage of the sampling method is that the handset is compatible with the test kit, which is not too expensive for small fish farms. However, the disadvantage of the sampling method is that it does not monitor the water quality of the pond environment in real time, takes a lot of effort in sampling, it is difficult to manage the collected data. It is not also suitable for management of aquaculture ponds as well as to make adapted decisions that are appropriate to the water quality of the pond. Furthermore, it is not suitable for large scale fish farms.

Based on the above analysis, it is clear that the aquaculture industry has brought a great benefit to the whole Vietnam in general and to the Mekong Delta region in particular. However, the environmental pollution caused by the waste from agricultural and industrial activities has had a great impact on the quality and productivity of aquaculture. In order to reduce the damage caused by water pollution, the construction of a water quality monitoring system for aquaculture is essential.

2 Related Work

In modern aquaculture management, water quality monitoring plays an important role. Water quality control is appropriate to keep the concentration of water environment factors within optimum range, which can enhance fish growth and reduce the occurrence of fish disease [1] referring to [2, 3]. And the application of information technology and sensor technology in monitoring pond water quality is becoming a trend in modern aquaculture. With the application of sensor technology, information technology in pond water quality monitoring, it can help managers to monitor the elements of the pond environment in real time from which there are treatments timely to minimize risks and lower costs. There are a number of studies to develop real-time monitoring systems for pond water quality such as: (1) A water environment monitoring model is designed and constructed based on a sensor network consisting of sensor nodes for data acquisition, node data stations from the monitoring node and data transmission to the central node, the remote monitoring center node stores, and displays data received from the data node [4, 5]; (2) An intensive aquaculture pond monitoring system has been developed based on embedded Web-server technology and wireless networking [1]; (3) Design and implementation of wireless sensor networks for fisheries monitoring and control based on ZigBee standard and virtual instruments [6] to monitor and control an aquaculture system in real time. The system includes sensor nodes to measure dissolved oxygen, temperature, pH, and water levels; Coordinate button for receiving information from sensor nodes and sending them to computers and personal computers (PCs) for communicating with users, storing and displaying data.

3   An Agent-Based Environment Monitoring System

In this section, we propose an environment monitoring system that contains monitoring agents. In computer science, an agent is a software or hardware entity that is situated in an environment (virtual or real environment). An agent is able to perceive the environment and other agents and capable of performing autonomous actions in this environment in order to meet its design objective [7, 8].

Major components of an environmental monitoring system
In combination of research and surveys on a number of aquaculture environmental monitoring systems developed on the world, we recognize that an environmental monitoring system consisting of three major components presented in Fig. 1.

In which: (1) Sensor used to collect environmental factors; (2) Communication system, in charge of data packets received from the sensor in predefined format and transmission of information to the processing system; and (3) Data storage and analysis system decodes the received data, stores, analyzes/simulates factors of environment to determine/predict the environmental status and send a warning to the user if the value of environmental factors exceeds the allowable threshold.

An agent-based environmental monitoring system
In this research, we propose an environmental monitoring system (Fig. 2) based on the wireless sensor network (WSN), IoT (Internet of things), data warehousing, and analysis based on agent-based simulation combined with the data warehouse studied in [9]. The deployment system that we build is called "agent-based environment monitoring system" (AEMS). The central component of the AEMS system is the monitoring and warning services server, which provides the following services: (1) Data storage; (2) Analysis model and agent-based simulation model; (3) Warnings services send warning messages and appropriate solutions via Web and mobile applications. Collection of environmental factors was carried out by environmental monitoring stations in charge. The environment monitoring station acts as an agent, which can interact with the environment and other actors in the system. The monitoring station will collect the environmental parameters sent to the data center and simultaneously interact with the user through Web applications or mobile applications. The monitoring stations are built on IoT and agent technologies so we call

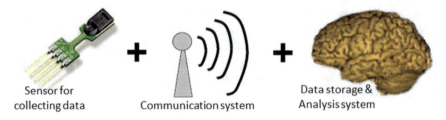

Sensor for collecting data   Communication system   Data storage & Analysis system

Fig. 1   Main components of the environmental monitoring system

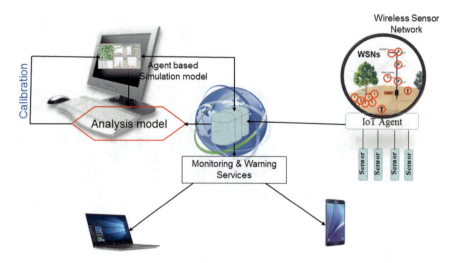

Fig. 2 AEMS—agent-based environment monitoring system

them IoT Agents. IoT Agents connect through wireless networks to create a network of sensors (wireless sensor network).

Depending on the specific implementation:

- *Case 1*: IoT Agents can be placed in a system of ponds that are close to each other, where they are organized into a network of sensor nodes (sensor node). The field and some nodes act as the sink node that communicates with sensor nodes to receive data collected and transmitted to the storage and processing center.
- *Case 2*: IoT Agents are located at some point of isolation (water supply to the farm or to the end of the source, where the discharge from the aquaculture activities takes place) establish and communicate directly with the system's storage and digital processing center.

Basically, the IoT Agent (Fig. 3) is built on the following components: (1) Task scheduling and handling reschedules on demand and performs task based on the set schedule; (2) Open sensor library communicates with the sensors to collect values of environmental factors; and (3) Communication library transfers collected data from the environment to the data center server and receives instructions from the user.

Designing hardware and software for sensor stations

The hardware and software design are based on the model shown in Fig. 3, a sensor station is considered as an agent consisting of two components: the hardware (Fig. 4) and the software (Fig. 6).

The architecture of the hardware components is based on HP's paper slip design concept, in which:

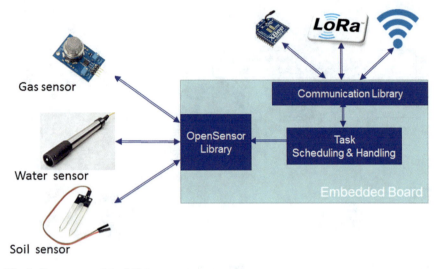

Fig. 3 Components of the IoT Agent

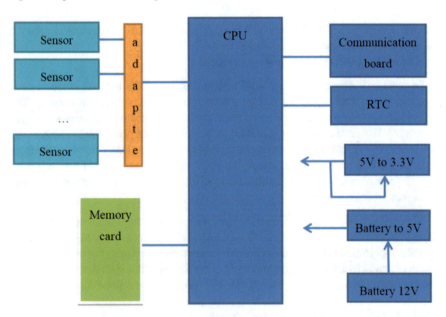

Fig. 4 Hardware diagram of sensor station

- Battery: Provides 12 V power supply.
- 12–5 V converter (Battery 12–5 V): Convert to 5 V power supply for main board and other components used 5 V power.
- 5–3.3 V converter: 5–3.3 V converters used for 3.3 V power supply.
- Central processing unit (CPU): Central processing unit used to execute program commands (software).
- Sensor: An electronic device that senses changes from the external environment and transforms into electrical signals. There may be different types of sensors attached to the substation through adapters such as BTA or DIN connector and so on.
- Memory card: A memory card is used to store data collected from the sensor or the configuration information of the station.
- Real-time clock (RTC): The IC circuit provides real time for sensor stations. RTC have their own battery input, separate from the main source, ensuring accurate timing.
- Communication board: The data transmission circuit that can be a WIFI/Zigbee/LoRa. It is used to transmit data collected from sensor station to data center server, receive commands from the host or contact with other sensor stations.
- Adapter: Depending on the type of sensor connector interface, we can attach different adapters such as BTA, DIN.

In software design, the sensor station was designed to perform three primary tasks group:

- Read data from the sensor (read data sensor);
- Transmission read value from the sensor station to data center server;
- Received the configuration and update configuration.

The program is designed according to the state machine model including three modes (Fig. 5). When the monitoring station in operating mode (RUN), tasks 1, 2, and 3 above are being done. After task 1 is finished then sensors are putted into IDLE mode, after task 2 and 3 finished, CPU and others devices are switched to SLEEP mode for power saving.

In the software architecture, sensor station has two data storages:

Fig. 5 State machine diagram

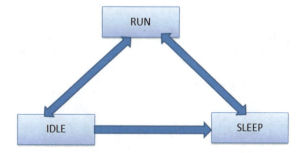

- Configuration data (config system) contains information of the sensor station.
- Data storage store factors of water environment.

The data flow diagram is presented in Fig. 6, in which, the activities of sensor station are scheduled and executed by the task scheduling process. Based on the configuration of the system, the schedule processor:

- Read data from sensor based on the time-based (e.g., reads data from the sensor every 5, 10 min) and then save read data to data log file and deactivate sensor.
- Transmit data from the sensor station to the data center server by opening a connection, send data, and close connection after completing data transfer.
- Receive data from the service server, such as updating the read cycle (time) of the sensor data, the token for the connection, etc.

In the next section, we will present the application of agent-based environmental monitoring system (AEMS) by setting up a water quality monitoring system for aquaculture.

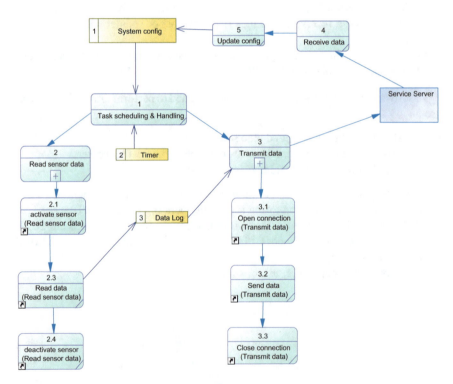

Fig. 6 Data flow diagram

4 Application of AEMS in Monitoring Water Quality of Aquaculture Area

Figure 7 illustrates the aquaculture water environment monitoring system that we have installed based on the "agent-based environment monitoring system" model, which we presented in Sect. 3.

In the aquaculture water quality monitoring system (Fig. 7), the IoT Agents act as environment monitoring stations. These environmental monitoring stations will be fitted with different types of sensors depending on the need to monitor environmental factors at each site. We have now developed environmental monitoring stations that can connect six types of sensors (temperature, dissolved oxygen, pH, ammonium—NH_4, nitrate—NO_3, and salinity). Extends connectivity on other sensors based on actual usage. Monitoring stations can be located at the source of water supplies to the aquaculture area, pond or at the wastewater collection area of the farming area.

Servers in the environment monitoring system are developed on the cloud platform; it is not only helping us easily integrating new services into the system, but also help fish farm companies do not need to invest hardware by themselves. This system provides data transmission, management and analysis services, alerts and response measures that correspond to the state of the environment in each location through Web, mobile applications, and SMS service.

In deployment, we have received the investment from the College of Aquaculture and Fisheries of Cantho University via the fund of project "Research on building models of farming Giant freshwater prawn adaptation to climate change of An Giang province". In this project, three water environment monitoring stations were installed at Bo canal in Phu Thuan commune, Thoai Son district, An Giang province (Fig. 8).

To evaluate the accuracy of the value of environment factors collected from the AEMS system, we collect samples at shrimp farming areas in Thoai Son district

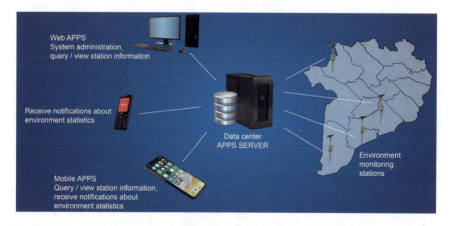

Fig. 7 Deployment model of water environment monitoring system for aquaculture

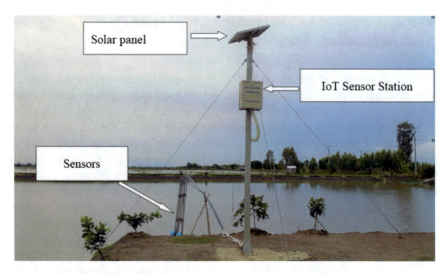

Fig. 8 Water environment monitoring stations were installed at Bo canal in Phu Thuan commune, Thoai Son district, An Giang province

(where our monitoring stations were installed) with three batches by HANNA hand-help meter, collected five samples in each batch, measured continuously with intervals of 5 min. The results were compared with the collected data in our system. RMSE (Root-Mean-Squared Error) [10, 11] has been using to evaluate the deviation between two measurement devices by the formula:

$$\text{RMSE} = \sqrt{\frac{\sum_{i=1}^{n} (x_i - y_i)^2}{n}}$$

where

x_i is the ith measurement value of the Hanna hand-help meter.
y_i is the ith measurement value of the project product.

The comparison results of the three sampling batches on days 1 and October 15, 2017 are reported in Tables 1 and 2.

Comparative results in Tables 1 and 2 show that the RMSE deviation between the two devices (research product and Hanna hand-help meter) for dissolved oxygen does not exceed 0.23, deviation of pH and water temperature does not exceed 0.20. This indicates that the environmental data collection activities of the HANNA instrument and the research product of the project are quite similar.

Table 1 Comparison of results and error estimates at sampling in October 1, 2017

No	Project product			Hanna		
	DO	pH	TMP	DO	pH	TMP
1	3.55	8.76	30.47	4.10	8.40	30.40
2	4.40	8.75	30.41	4.50	8.70	30.30
3	4.92	8.50	30.31	4.60	8.80	30.10
4	4.44	8.45	30.12	4.50	8.40	30.10
5	4.74	8.81	30.16	4.60	8.60	29.90
RMSE (DO)	0.21					
RMSE (pH)	0.16					
RMSE (TMP)	0.11					

Table 2 Comparison of results and error estimates at sampling in October 15, 2017

No	Project product			Hanna		
	DO	pH	TMP	DO	pH	TMP
1	4.62	8.58	31.37	4.50	8.80	30.90
2	4.86	8.66	30.98	4.30	8.40	31.30
3	4.79	8.74	30.80	4.40	8.90	30.70
4	4.78	8.78	30.69	4.90	8.30	30.70
5	4.66	8.80	30.61	4.80	8.60	30.60
RMSE (DO)	0.23					
RMSE (pH)	0.20					
RMSE (TMP)	0.18					

5 Conclusion

In this article, we have developed a model for implementing an agent-based environment monitoring system (AEMS) in combination with technologies such as sensor networks, IoT, and agent-based. An important point in the AEMS system is the environmental monitoring station designed as an operator capable of interacting with other actors and interacting with the environment, e.g., an environmental monitoring station can send water quality factors or receive command from farmer, send command to turn on/off air pump to supply oxygen for fish pond. In installation and actual deployment, we have AEMS application installed a water quality monitoring system for aquaculture, especially systems that help managers and farmers:

- Manage the water quality of the water supply for aquaculture areas and each pond at any time.
- Timely detection of abnormalities caused by production activities, breeding that affect on the quality of water supply from which to provide appropriate responses.

- Information gathered is easily aggregated, processed and shared to multiple people through Web and mobile applications.

Research results show the superiority of using sensor network, IoT, agent-based technologies in monitoring water quality of aquaculture areas. The system is designed to help the farmer monitor water quality indicators 24/7, laboriously gathering more data and collected data managed on the computer. Thereby helping managers and farmers analyze and take action to adapt the water quality of ponds, and helping to minimize the damage caused by water pollution.

Acknowledgements The authors acknowledge that the research was sponsored by Research Fund of Cantho University and financially supported from project "Research on building models of farming Giant freshwater prawn adaptation to climate change of An Giang province".

References

1. Zhu, X., Li, D., He, D., Wang, J., Ma, D., Li, F.: A remote wireless system for water quality online monitoring in intensive fish culture. Comput. Electron. Agric. **71** (2010)
2. Stigebrandt, A., Aure, J., Ervik, A., Hansen, P.K.: Regulating the local environmental impact of intensive marine fish farming. Aquaculture **234**, 239–261 (2004)
3. Simoes Simões, F. dos., Moreira, A.B., Bisinoti, M.C., Gimenez, S.M.N., Santos Yabe, M.J.: Water quality index as a simple indicator of aquaculture effects on aquatic bodies. Ecol. Indic. **8**, 476–484 (2008)
4. Jiang, P., Xia, H., He, Z., Wang, Z.: Design of a water environment monitoring system based on wireless sensor networks. Sensors **9**, 6411–6434 (2009)
5. Luo, H.P., Li, G.L., Peng, W.F., Song, J., Bai, Q.W.: Real-time remote monitoring system for aquaculture water quality. Int. J. Agric. Biol. Eng. **8**, 136–143 (2015)
6. Simbeye, D.S., Zhao, J., Yang, S.: Design and deployment of wireless sensor networks for aquaculture monitoring and control based on virtual instruments. Comput. Electron. Agric. **102**, 31–42 (2014)
7. Wooldridge, M.: An Introduction to Multiagent Systems. John Wiley & Sons (2002)
8. Ferber, J.: Multi-agent concepts and methdologies. In: Phan, D., Amblard, F. (eds.) Agent-Based Modelling and Simulation in the Social and Human Sciences, pp. 7–33. The Bardwell Press, Oxford (2007)
9. Truong, M.T., Amblard, F., Gaudou, B., Truong, M.T., Amblard, F., Gaudou, B., Frame-, C.S.C.A.: CFBM—a framework for data driven approach in agent-based modeling and simulation. In: 2nd EAI International Conference on Nature of Computation and Communication, pp. 264–275. Springer, Cham (2016)
10. Chatfield, C.: A commentary on error measures. Int. J. Forecast. 101–102 (1992)
11. Taylor, S.J.: Comparing forecasts in finance. Int. J. Forecast. **8**, 102–103 (1992)

Identifying Intrusions in Dynamic Environments Using Semantic Trajectories and BIM for Worker Safety

Muhammad Arslan⬡, Christophe Cruz⬡ and Dominique Ginhac⬡

Abstract While there exist many systems in the literature for detecting unsafe behaviors of workers in buildings such as staying-in or stepping into unauthorized locations called intrusions using spatio-temporal data. None of the current approaches offer a mechanism for detecting intrusions from the perspective of a dynamic environment where the building locations evolve over time. A spatio-temporal data model that is required to store worker trajectories should have a capability to track a building evolution and seamlessly handles the enrichment of stored trajectories with the relevant geographical and application-specific information sources for studying the worker behaviors using a building or a construction site context. To address this requirement of maintaining the information, which is generated during the building evolution and for constructing semantically enriched worker trajectories using the stored building information. This work reports a system which offers the ability to perform user profiling for detecting intrusions in dynamic environments using semantic trajectories. Later, Building information modeling (BIM) approach is used for visualizing the intrusions from a standpoint of a building environment so that necessary actions can be performed proactively by the safety managers to avoid unsafe situations in buildings.

Keywords Intrusions · Behaviors · Safety · Construction · BIM

M. Arslan (✉) · C. Cruz · D. Ginhac
University of Bourgogne Franche-Comté (UBFC), 9 rue Alain Savary,
Dijon 21078 Cedex, France
e-mail: muhammad.arslan@u-bourgogne.fr

C. Cruz
e-mail: christophe.cruz@ubfc.fr

D. Ginhac
e-mail: dominique.ginhac@ubfc.fr

© Springer Nature Singapore Pte Ltd. 2020
X.-S. Yang et al. (eds.), *Fourth International Congress on Information and Communication Technology*, Advances in Intelligent Systems and Computing 1027, https://doi.org/10.1007/978-981-32-9343-4_6

1 Introduction

Intrusions are unauthorized stays or walking through the hazardous areas and are very common type of near-miss incidents occur frequently on the construction sites [1, 2]. According to the Health and Safety Executive (HSE), UK, near-miss incidents such as intrusions are one of the major reasons of the construction accidents which often result in falling from heights, electric shocks, caught in or between, and being struck by the construction machinery [3]. Intrusions on sites are often ignored as current construction safety assessment methods are mainly focused on visible consequences such as injuries or deaths [1, 2]. For monitoring the intrusions, real-time safety monitoring solutions containing information and communication technologies (ICTs) are employed on the construction sites for safety management [4]. Spatio-temporal datasets [5] collected from the Global Positioning System (GPS) or Indoor Positioning System (IPS)-based systems have been used extensively to understand the mobility dynamics of moving objects [6]. A raw spatio-temporal dataset consists of ordered sequences of discrete-time triples in the form of <latitude, longitude, timestamp> known as trajectories [5]. Though, additional contextual information [7] is required to make raw trajectories more meaningful for supporting advanced trajectory applications [5]. The additional pieces of contextual and application information are called annotations [7], which are acquired from various external data sources such as Web-based repositories and geographical information systems. The process of tagging annotations to a small segment of a trajectory or a complete trajectory is known as semantic enrichment [5]. Enriching raw trajectories with the semantics of space (e.g., the physical location, where a trajectory was generated), and the semantics of time (e.g., time interval having a start and stop timestamps) give more meaning in discovering behaviors to study movement dynamics of people and moving objects [7]. Eventually, state-of-the-art 3D-visualization approaches such as BIM can be used for visualizing the unsafe behaviors [8] for identifying occurrences of intrusions on sites from a perspective of the building environment for better safety planning and access control.

The rest of the paper is organized as follows: Sect. 2 describes the background of the study. Section 3 is based on the proposed prototype system. Section 4 presents the discussion, a conclusion, and an outlook of the future work.

2 Background

In this section, a brief review [9–15] is presented on the existing solutions developed for tracking construction workers to prevent near-miss incidents such as intrusions on sites for safety management. A study presented by Naticchia et al. [9] is one of the most appropriate examples of detecting intrusions on construction sites using wireless communication technology. As intrusions typically involve such untrained workers who carry out construction activities without having an appropriate awareness on carrying out works inside the hazardous areas on sites such as confined spaces which ultimately pose great threats to the construction safety. A framework [10] based on deep learning method uses the images of the workers obtained from the videos for capturing their identifications and verifies whether an on-going work is being carried by the certified workers or not. A similar study [11] based on virtual fencing logic using ultra-wideband technology is demonstrated for preventing construction workers from accessing predefined hazardous zones. The system also generates alerts in real-time if the safety risk level to access the hazardous zone increases. Existing literature also includes safety monitoring systems, which are developed based on the requirements acquired from the historical records of accidents for further preventing the near-miss incidents. An example of such systems [12] in which information requirements for preventing struck-by-falling-object accidents are collected and a system is built using ZigBee-radio-frequency identification (RFID) sensor network for tracking workers and materials on sites. Moreover, Fang et al. [13] designed a system mainly for indoor localization to monitor construction sites for worker safety. Another similar ZigBee-RFID-based study [14] is present in the literature for tracking identity information of workers. The system is developed after analyzing historic case-studies concerning near-miss accidents. Apart from outdoor and indoor safety systems for reducing near-miss incidents on construction sites, there also exist solutions for preventing accidents in underground construction scenarios. Ding et al. [15] described a study in an underground metro tunnel construction setting based on Fiber Bragg grating (FBG) sensor system and an RFID. Their system determines if workers are working in such locations, which can become hazardous with time as the nature of locations is changing as construction work progresses. Analyzing such risky locations in real-time with environmental monitoring sensors can help to generate timely warnings to safety managers for preventing accidents.

After reviewing the above safety solutions for minimizing the intrusions, it was concluded that the existing systems are intended for recognizing the intrusions, where the building locations are treated as static entities and do not involve over time. However, in case of dynamic environments such as construction sites, locations evolve in geometry (shape and size) or attributes (alphanumeric semantic information) with time as construction works progress. These building evolutions introduce data management challenges for storing worker trajectories and annotating them with the most updated building information. To address this data modeling requirement for storing trajectory data captured from the dynamic environments such as construction sites; initially existing spatio-temporal trajectory data modeling approaches [16, 17] are reviewed. Literature [16] suggested that ontology-based modeling approaches represent semantic trajectories with richer semantic information related to location and time by integrating different types of information enrichment processes as compared to the other approaches. In these models, initially, segmentation approaches are used to output the structured trajectory episodes. Later, these trajectory episodes are mapped on ontologies for annotating them with relevant point of interest (POI) information. In addition, ontology-based models offer many benefits such as tagging of semantic locations and their environmental properties, clustering of similar trajectories based on the common behavior or an activity, reducing trajectories for visualizations, segmenting trajectory episodes into stop and move episodes and determining the transportation mode used in the move episodes of the trajectories. More details on the present semantic trajectory modeling approaches can be found in [16–19]. Therefore, an ontology-based modeling approach is used for developing our prototype system.

3   Intrusion Detection System for Dynamic Environments

To address the research gaps identified in the previous section, an intrusion detection system is presented for dynamic environments using semantic trajectories and an open-sourced BIM platform. The developed system has the ability to keep track of the dynamically changing locations for constructing semantic trajectories to understand the user behaviors. The developed prototype system focuses on the following.

3.1   Data Preprocessing and Semantic Enrichment

To detect intrusions using spatio-temporal trajectories, location data of users is acquired after installing 200 Bluetooth beacons in a building. For collecting the location coordinates (longitude and latitude) of the users, an application is developed using the Android platform. The application selects the best 3 beacons' signals and performs the geo-localization technique for estimating the location coordinates. Using this approach, 1,021 location points are recorded in a day with the sampling interval of 5 s, and later stored in the document database (MongoDB) for further processing. Later, R studio is used for executing the preprocessing task, i.e., data cleaning. For transforming processed trajectories into semantic trajectories, we have used our data model STriDE (spatio-temporal trajectories in dynamic environment) [19] (see Fig. 1 shows the logical connections between different STriDE model entities, timeslices, concepts, and user profiles).

In Fig. 1, an entity denotes real-world entities consist of an identity, spatial, and alphanumeric descriptive properties. The STriDE model is built for the moving objects which change their geometry and semantic information over the evolution of time. To capture the behaviors of moving and changing objects, their movements are represented as a set of trajectories' timeslices. A timeslice consists of an identity, alphanumeric properties, a time component indicating the validity of the timeslice, and a geographical component depicting the spatial representation of the entity [19]. Each timeslice is valid for certain time duration. In case of any change in the timeslice's components, excluding the identity, a new time slice is generated inheriting the components of the last known state of the entity [19]. In addition, the STriDE model uses ontologies offering the conceptualization of a specific domain for showing relationships among different 'concepts' as a hierarchy for representing trajectories [19]. Here, a 'concept' is defined as an object with a set of attributes (see Fig. 2, in which a

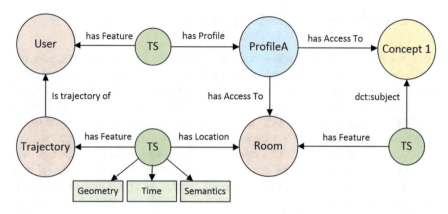

Fig. 1 STRiDE model (user, trajectory, and room are the entities that may change over time. Whereas, TS is the timeslice and concepts are defined using SKOS. Each entity is represented using a TS having the start and end dates)

SPARQL Results (returned in 47 ms)		
userName	**profileName**	**concepts**
User 1	User profile	Building, Floor, Door, Corridor, Footway, Stairs, Room, Office, Meeting room, Amenity, Bathroom
Maintenance 1	Maintenance profile	Building, Floor, Door, Corridor, Footway, Stairs, Room, Office, Meeting room, Amenity, Bathroom, Storage

Fig. 2 Users and profiles of workers

concept defines a building space). In our case, the users, the construction machinery and the workspaces of the building are referred to as dynamic entities which change their properties over time without changing their identity.

To perform the semantic enrichment using the information extracted from semantic data sources (application domain knowledge and geographic databases), the STRiDE model uses an OpenStreetMap (OSM) file of the building and taxonomy along with a set of semantic rules written in the Resource Description Framework (RDF) language. The purpose of an OSM file is to describe the entire building structure in a data model. To label each key-value pair defined in an OSM file, taxonomy is created as a hierarchy of concepts written as RDF triples using Simple Knowledge Organization System (SKOS) vocabulary. Using the constructed taxonomy, OSM's key-value pairs are clustered and maintained using the domain knowledge. In addition, a JavaScript Object Notation (JSON) file is created describing a set of semantic rules to link each OSM object with the taxonomy. Eventually, Java objects are created using an OSM file, semantic rules, and taxonomy as per the semantic definition. These Java objects are later stored in a Stardog (www.stardog.com), a triple store for achieving the complete representation of the building environment.

3.2   Detecting and Visualizing Intrusions

After transforming raw spatio-temporal trajectories using stored contextual information in the STriDE model into semantic trajectories, user profiling is achieved using 'concepts' for detecting intrusions in a building. In the STriDE model, as already described above that we have defined 'concepts' for tagging different locations of a building. These concepts are defined using SKOS vocabulary and stored in the 'concept scheme.' Using the 'concepts,' all the building locations are described in a model. Because the 'concept' of a room or any building location may change over time, it cannot be tagged directly but with its timeslice. By means of 'concepts,' user profiles are created as shown in Fig. 2. Here, a 'profile' is a collection of 'concepts' that shows the building locations to which a user holds an access. For creating a user in the model, a profile needs to be specified. This approach will allow us in detecting any user intrusion according to the allocation of the 'concepts' to different user profiles. This can be verified by checking each timeslice of the user's trajectory for the 'concepts' that it has access to. So, if the user's profile allows him to access the room means that the specific 'concept' is tagged with its profile then everything is good, otherwise we have detected an intrusion (see Fig. 3).

For our study, Autodesk Revit (a BIM software) is used because of its open-sourced application programming interface (API) support and extensive use in the construction industry [20]. The building which is used for the experimentation, its BIM model doesn't exists. For showing a proof-of-concept integration of systems, a sample Revit model file is used. To take advantages from the functionalities offered by the Revit API, a visual scripting tool Dynamo is used as a Revit plug-in [20]. Dynamo graph constructed for the STriDE-BIM integration consists of four main steps which are: (a) All the building locations (tagged as 'rooms') in the BIM model are extracted by defining a category of elements as 'rooms' in the Dynamo. (b) After extracting a list of room names, the details of occurrence of intrusions, i.e., intruder ID, time of occurrence of intrusion and room name where the intrusions are detected are imported into Dynamo. (c) On receiving room's data from an active Revit document and intrusions' data taken from the STriDE model, the room location where an intrusion has occurred is highlighted in 'Red' color on a BIM model using the mapping of room names in both lists (see Fig. 4). (d) As soon as the problematic room is highlighted on a BIM model, this process invokes another user window to ask for details of the intrusion occurred in a specific room location. After inputting the desired room name, time of occurrence of the intrusion in the particular room and its corresponding intruder ID will be displayed to the user in the Revit software.

SPARQL Results (returned in 154 ms)				
userName	froom	roomLabel	sdate	edate
User 1	⌐ stride:W1002	Storage room	2018-01-02T09:36:00	9999-12-31T23:59:59

Fig. 3 Detecting an intrusion from trajectory's timeslices

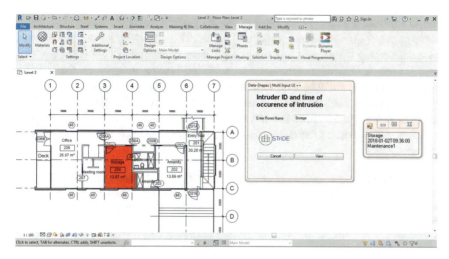

Fig. 4 Visualizing STriDE data in Autodesk Revit software

4 Discussion and Conclusion

The proposed system captures raw spatio-temporal data from low energy Bluetooth beacons and constructs semantic trajectories using the STriDE model. The main feature that distinguishes our STriDE model from the previous trajectory models is that it holds an ability of tracking the relevant semantic information related to the change in geometry and functionality of the building spaces for modeling the seman-tic trajectories of moving objects (workers and machinery) on a construction site. Another important feature that differentiates our developed system from the existing systems is that we have integrated the output of our semantic trajectory data model with a smart city solution, i.e., open-sourced BIM platform for the industrial users. The reason of using BIM is that it's a preferred approach for generating a building visualization in the architectural, engineering, and construction (AEC) industry [21]. It aims to provide reliable and up-to-date information about a facility throughout the building lifecycle to AEC team members for conducting studies, simulations, and operations [21]. In addition, application of BIM enhances the inter-organizational collaboration of information among members because it is based on the Industry Foundation Classes (IFC) standard that is universal and supports easy and fast infor-mation exchanges [21].

The system is developed using semantic trajectories and BIM will help building supervisors and health and safety (H&S) managers in achieving the access control of workers to hazardous locations in building or on construction sites. In case of detection of an intrusion, an intruder can be recognized easily from its trajectory identification using a BIM software, and necessary actions can be performed to avoid unsafe situations. As this proposed system is still in the development phase and further work is required to test a developed system on a real construction site and to evaluate it based on its tagging accuracy of the semantic locations with the corresponding spatio-temporal trajectory points. Though, using the designed system on a construction work site will not compromise its usefulness but will raise concerns in the process of location data acquisition. Placing beacons on the building infrastructure and making sure that beacons remain intact to their original positions during dynamic nature of construction works can be challenging. However, the use of advanced geo-localization technologies can be considered for increasing the accuracy of the acquired spatio-temporal data to study worker behaviors for safety management.

Acknowledgements The authors thank the Conseil Régional de Bourgogne-Franche-Comté, the French government for their funding, SATT Grand-Est, and IUT-Dijon (http://iutdijon.u-bourgogne.fr). The authors also want to thank Orval Touitou for his technical assistance to this research work.

References

1. Heng, L., Shuang, D., Skitmore, M., Qinghua, H., Qin, Y.: Intrusion warning and assessment method for site safety enhancement. Saf. Sci. **84**, 97–107 (2016)
2. Wu, W., Yang, H., Chew, D.A., Yang, S.H., Gibb, A.G., Li., Q.: Towards an autonomous real-time tracking system of near-miss accidents on construction sites. Autom. Constr. **19**(2), 134–141 (2010)
3. HSE, Investigating accidents and incidents. http://www.hse.gov.uk/pubns/hsg245.pdf. Last accessed 01 Oct 2018
4. Williams, T., Bernold, L., Lu, H.: Adoption patterns of advanced information technologies in the construction industries of the United States and Korea. J. Constr. Eng. Manag. **133**(10), 780–790 (2007)
5. Zheng, Y.: Trajectory data mining: an overview. ACM Trans. Intell. Syst. Technol. (TIST) **6**(3), 29 (2015)
6. Teizer, J., Cheng, T.: Proximity hazard indicator for workers-on-foot near miss interactions with construction equipment and geo-referenced hazard areas. Autom. Constr. **60**, 58–73 (2015)
7. Yan, Z.: Semantic trajectories: computing and understanding mobility data. Doctoral dissertation, Lausanne, EPFL (2011)
8. Becerik-Gerber, B., Jazizadeh, F., Li, N., Calis, G.: Application areas and data requirements for BIM-enabled facilities management. J. Constr. Eng. Manag. **138**(3), 431–442 (2011)
9. Naticchia, B., Vaccarini, M., Carbonari, A.: A monitoring system for real-time interference control on large construction sites. Autom. Constr. **29**, 148–160 (2013)
10. Fang, Q., Li, H., Luo, X., Ding, L., Rose, T.M., An, W., Yu, Y.: A deep learning-based method for detecting non-certified work on construction sites. Adv. Eng. Inform. **35**, 56–68 (2018)
11. Carbonari, A., Giretti, A., Naticchia, B.: A proactive system for real-time safety management in construction sites. Autom. Constr. **20**(6), 686–698 (2011)
12. Wu, W., Yang, H., Li, Q., Chew, D.: An integrated information management model for proactive prevention of struck-by-falling-object accidents on construction sites. Autom. Constr. **34**, 67–74 (2013)
13. Fang, Y., Cho, Y.K., Zhang, S., Perez, E.: Case study of BIM and cloud—enabled real-time RFID indoor localization for construction management applications. J. Constr. Eng. Manag. **142**(7), 05016003 (2016)

14. Yang, H., Chew, D.A., Wu, W., Zhou, Z., Li, Q.: Design and implementation of an identification system in construction site safety for proactive accident prevention. Accid. Anal. Prev. **48**, 193–203 (2012)
15. Ding, L.Y., Zhou, C., Deng, Q.X., Luo, H.B., Ye, X.W., Ni, Y.Q., Guo, P.: Real-time safety early warning system for cross passage construction in Yangtze Riverbed Metro Tunnel based on the internet of things. Autom. Constr. **36**, 25–37 (2013)
16. Albanna, B.H., Moawad, I.F., Moussa, S.M., Sakr, M.A.: Semantic trajectories: a survey from modeling to application. In: Information Fusion and Geographic Information Systems (IF&GIS'2015), pp. 59–76. Springer, Cham (2015)
17. Parent, C., Spaccapietra, S., Renso, C., Andrienko, G., Andrienko, N., Bogorny, V., Damiani, M.L., Gkoulalas-Divanis, A., Macedo, J., Pelekis, N., Theodoridis, Y.: Semantic trajectories modeling and analysis. ACM Comput. Surv. (CSUR) **45**(4), 42 (2013)
18. Arslan, M., Cruz, C., Ginhac, D.: Understanding worker mobility within the stay locations using HMMs on semantic trajectories. In: 14th International Conference on Emerging Technologies (ICET), pp. 1–6. IEEE, Islamabad (2018)
19. Cruz, C.: Semantic trajectory modeling for dynamic built environments. In: Data Science and Advanced Analytics (DSAA), pp. 468–476. IEEE, Tokyo (2017)
20. Autodesk. http://paulaubin.com/_downloads/2017_AU/BIM128338-Aubin-AU2017.pdf. Last accessed 01 Oct 2018
21. Azhar, S.: Building information modeling (BIM): trends, benefits, risks, and challenges for the AEC industry. Leadersh. Manag. Eng. **11**(3), 241–252 (2011)

The Ontological Approach in Organic Chemistry Intelligent System Development

Karina A. Gulyaeva⬤ and Irina L. Artemieva

Abstract The amount of knowledge in organic chemistry grows exponentially inducing a need for robust intelligent systems that can promote the process of R&D. Although the methods of intelligent system design vary significantly juxtaposing expert systems, neural networks, genetic algorithms, and fuzzy logic, effective intelligent system development can start only after answering the following essential questions: "How is the application area structured? What is its ontology?". Ever since the DENDRAL Project, the challenge of knowledge representation has been embraced by the scientific community. The notion of ontology has appeared in knowledge engineering delivering a possible solution. As the practice shows, taxonomies provide little expressiveness. Therefore, we suggest that the ontological approach advocates consider applied logic methodology. This framework proposes that complex-structured domains, such as organic chemistry, be represented as interconnected modules of applied logic theories. Employing the described technique, we introduce the model of organic chemistry intelligent system. Most special aspects of this methodology are depicted together with a historical overview of intelligent systems and the roots of knowledge representation models.

Keywords Ontology · Intelligent system · Organic chemistry · Applied logics

1 Introduction

Although the debates concerning genuine origins of the modern organic chemistry continue, the undeniable fact is that synthetical and analytical approaches in organic chemistry have undergone significant development since 1828 when Friedrich Wöhler converted ammonium cyanate into urea [20]. The emergence of synthesis has resulted not only in the proliferation of novel molecular structures but also in better

K. A. Gulyaeva (✉) · I. L. Artemieva
Far Eastern Federal University, Vladivostok, Russia
e-mail: kgulyayeva@gmail.com

I. L. Artemieva
e-mail: artemeva.il@dvfu.ru

© Springer Nature Singapore Pte Ltd. 2020
X.-S. Yang et al. (eds.), *Fourth International Congress on Information
and Communication Technology*, Advances in Intelligent Systems
and Computing 1027, https://doi.org/10.1007/978-981-32-9343-4_7

understanding of reaction mechanisms and structure–activity relationships. Now that the globalization process pushes scientists to work faster and more efficiently, there is a need for robust computer-based intelligent systems that can take the process of research and development to another level.

Since Warren McCulloch and Walter Pitts' "A logical calculus of the ideas immanent in nervous activity" [16] and Alan Turing's momentous "Computing machinery and intelligence" [19], the field of artificial intelligence (AI) has evolved in several directions. From weak methods that used little information about the problem domain to separate areas of expert systems, neural networks, genetic algorithms, and fuzzy logic, the field of AI now, after its as many as 50 years of existence, is moving toward combining the strengths of its distinct methods to create robust intelligent system models.

Regardless of a broad spectrum of methods, we insist that effective intelligent system architecture development can start solely after answering three basic questions:

1. What is the application area? (Where are we? How is the application area structured? What is its *ontology*?)
2. What are the classes of problems solved in the application area?
3. How are these problems solved? (What are the methods to solve them?)

In an extended sense, the classes of problems that are solved in any application field are the subclasses of three major ones: information transfer, storage, and processing. The classes of problems that are solved by means of intelligent systems generally include the problems of classification, diagnostic assessment, monitoring, planning, management, maintenance, project development, interpretation, prediction, and problems that are the mixtures of several classes. The methods for finding possible solutions are elaborated individually in each case predominantly based on efficiency and the amount and quality of information in the field. These methods can incorporate diverse combinations of various mathematical algorithms, expert systems, neural networks, fuzzy logic, and evolutionary strategies. However, the effective choice cannot be made without answering first and foremost, arguably the most challenging question: what is the application area? (How does it function? What are the application field information objects and the relations between them?). The approach in intelligent system design that focuses on the application area *ontology* is called the ontological approach. This approach has been utilized in the current work to develop organic chemistry intelligent system model.

2   Historical Background

Although there exists a broad variety of "artificial intelligence" definitions, one can advert to the definition of Marvin Minsky, one of AI founding fathers. He stated that the machine could be named intelligent "if the task it is carrying out would require intelligence if performed by humans" [4]. Brilliant mathematicians of the past—John von Neumann, Harry Huskey, Herman Goldstine, etc.—together with glorious engineers and physicists, J. Presper Eckert Jr. and John Mauchly, made one of the first attempts to create intelligent machines: "Electronic Numerical Integrator and Calculator" (ENIAC) and "Electronic Discrete Variable Automatic Computer" (EDVAC) [7]. First stored-program machines opened a way to the new era of great expectations.

The 1956 "Dartmouth Summer Research Project on Artificial Intelligence" [14]— Dartmouth College summer workshop—became the cite of attraction for AI enthusiasts, naming a few, Claude Shannon and John McCarthy. The latter, the inventor of "AI" term and LISP programming language, in his "Programs with Common Sense," which is considered the first paper on knowledge representation, introduced the model of program "Advice taker" [15]. The program was supposed to have declarative instructions, reason deductively from the knowledge, and use the calculus of situations, which grounded on the first-order logic [15]. Several major ideas stated by McCarthy were the following: *"In order for a program to be capable of learning something it must first be capable of being told it.* [...] A machine is instructed mainly in the form of a sequence of imperative sentences; while a human is instructed mainly in declarative sentences describing the situation in which action is required together with a few imperatives that say what is wanted"* [15].

Another promising project in the early ages of AI was 1958 "General Problem-Solving Program I" (GPS-I) developed by the RAND Corporation scientists, Allen Newell and John Clifford Shaw, and Herbert Simon—Turing Award and Noble Prize winner from Carnegie Mellon. The major features of GPS-I stated by its developers were

1. "The recursive nature of its problem-solving activity.
2. The separation of problem content from problem-solving technique as a way of increasing the generality of the program.
3. The two general problem-solving techniques [...]: means-ends analysis, and planning.
4. The memory and program organization used to mechanize the program ([...] the computer languages (IPL's) used to code GPS-I)" [17].

Although capable of solving simple problems, such as proving some trigonometric identities, GPS-I failed to solve complex real-world tasks, where the number of possibilities grew exponentially. By 1970, most AI enthusiasts realized that general methods with weak information about the problem domain were condemned to fail.

When elusive dreams of any-task-solver were gone, a major paradigm shift in AI took place: an impactful restriction of the problem domain. Narrowing the problem domain and incorporating human expertise into computer program gave birth to what is now called expert systems. One of the first and most influential expert system frameworks was DENDRAL [13]. The "grandfather of expert systems," as the DENDRAL Project founders—Edward Feigenbaum, Bruce Buchanan, and Joshua Lederberg—call it, started with the task of "the formation of hypotheses of organic molecular structure from mass spectral data" at Stanford in 1965 [6]. Remarkably,

DENDRAL was the first expert system with "outside client" due to Joshua Lederberg's participation in NASA Mariner campaign to search for life on Mars; inter alia, a mass spectrometer and a computer running DENDRAL were planned to be onboard. The project was arguably the first interdisciplinary collaboration between artificial intelligence and organic chemistry fields. Justifying its slogan "Knowledge is Power," DENDRAL drew in the expertise of the eminent scientists: biologist and Nobel prize laureate Lederberg, a world-class expert in mass spectrometry Djerassi, and their collaborators [6]. As the volume of new chemistry-specific knowledge (arranged as LISP code) was almost unsustainable, the use of productions, which had initially been inspired by the work of Newell and Simon for GPS-I, was reengineered into modular situation-action "rules" architecture. Since then, most expert systems have been designed as rule-based.

DENDRAL was originally the acronym for DENDRitic ALgorithm—"a procedure for exhaustively and non-redundantly enumerating all the topologically distinct arrangements of any given set of atoms, consistent with the rules of chemical valence" [12]. Heuristic DENDRAL appeared before Meta-DENDRAL. The basic method of heuristic DENDRAL was "plan-generate-test" [12]. The system was designed as a family of components, each performing a distinct operation: planning (e.g. PLANNER), generating (e.g. CONGEN and GENOA), and testing (e.g. PREDICTOR and MSPRUNE). The glue that connected all components was a uniform representation of chemical graphs, in terms of which all components operated. Meta-DENDRAL, one of the first machine learning systems, was based on heuristic DENDRAL and could automatically infer knowledge from mass spectra.

Obviously, DENDRAL induced a new era in AI, where toy problems and any-task-solver methods were replaced by real-world domain-specific challenges that could only be handled by knowledge-intensive methods. Ever since, a search for effective knowledge representation models has persisted an essential task of knowledge-based, or expert, systems.

3 Methodology

3.1 The Ontological Approach in Its General Form

The ontological approach appeared in the information system and intelligent system models when the word "ontology" propagated its eternal philosophy dwelling to knowledge engineering. "The study of being" acquired its complementary connotation in the work of Stanford researcher Thomas Gruber, who defined it as "an explicit specification of a conceptualization" [8]. The conceptualization implies the existence of some objects, concepts, and the relationships among them in some application area. What can be represented in the information system is said to exist. Gruber accentuated: "When the knowledge of a domain is represented in a declarative formalism, the set of objects that can be represented is called the universe of discourse.

This set of objects, and the describable relationships among them, are reflected in the representational vocabulary with which a knowledge-based program represents knowledge" [8]. Fundamentally, ontology is comprised of terms, term definitions, and attributes.

The ontological approach has been vastly used in the semantic Web and taxonomy design. Web ontology language (OWL) has been utilized by "Open Biological and Biomedical Ontologies" Foundry, which include taxonomies of genes, anatomical formations of several species, tissue types, molecular processes, microbe phenotypes, and the variety of entities [18]. Although the family of OWL languages gradually evolve toward better reasoning capabilities (e.g. OWL DL, which is based on description logics, has been used lately [3, 5]), the major limitation of classifiers remains, which is the deficit of expressive power. Albeit not extensively used, the methodology of representing complex-structured domains as interconnected modules of applied logic theories can be employed to advantage organic chemistry intelligent system development. This approach has been introduced in the following sections. It has been described as meticulously as the format of this paper allowed. Moreover, it is suggested that the reader refer to [1, 9–11] for further details.

3.2 Desideratum for Defining Domain Ontology in Mathematical Terms

The reality of any domain is the set of all situations that happened in the past, happen at present, and will happen in the future. Each situation is confined in space and time. The better the set of supposed situations approximates the reality of a domain, the more accurate the conceptualization is. Any relevant conceptualization includes the application area reality. Using Gruber terms, ontology can be viewed as "an explicit specification" of an *implicit* "conceptualization" [8]. Ontology is not supposed to alter and demonstrates the conformance and shared understanding of domain term denotations among application area experts. Different elements of domain ontology are represented in the ontology model as terms for situations and situation definitions, terms for knowledge and knowledge definitions, ontological agreements, and supplemental and mathematical terms. Relevant conceptualization can be represented by the conceptual framework that contains the terms for expressing the following: concept notations, concept scope, and the relationships among them. Basic and supplementary concepts should be explicitly defined. For this purpose, the language of applied logics has been introduced. This language and the whole mathematical apparatus for domain ontology models have been presented in the series of articles [9–11]. The language is declarative. It uses the elements of predicate calculus and set theory; thus, these elements are uniformly understood. Since mathematical terms have a consistent interpretation and are defined outside the conceptual framework, they are suitable for various application areas.

3.3   Applied Logics

The language of applied logics is employed to describe conceptual framework modules. The kernel of the language is modest. The syntax and semantics of terms, formulas, propositions (each comprised of a prefix and a body), and applied logic theories are meticulously described in [9]. Although minimal, the kernel can be easily extended. The standard extension (ST) defines quantifier constructions, sets, conditional term, set-theoretical relations and operations, constants, tuples, projections, and Cartesian power. Specialized extensions "Intervals" and "Mathematical quantifiers" are described in [11]. Each language of this type is generally comprised of the kernel, standard extension, and various specialized extensions. In fact, it is distinguished by the names of extensions it utilizes. Distinct logical theory in the language of applied logics defines the signature. Each theory is identified by its name, and the theory name parameters are the names of the extensions this theory uses.

Example 1. The applied logic theory form
Module "Molecules" is used in the definition of applied logic theory "Elemental Constituents." "Elemental Constituents" module uses the standard extension of the language, "Mathematical quantifiers" and "Intervals" extensions.

```
Elemental Constituents(ST, Mathematical quantifiers,
Intervals) = <{Molecules}, SS>
Where SS = {propositions of the theory "Elemental
Constituents"}
```

[The applied logic theory "Elemental Constituents" form. This module is to be included in the organic chemistry ontology model.]

This modular approach is particularly useful for complex-structured domain knowledge representation.

3.4   Complex-Structured Application Area

The area of organic chemistry is complex-structured. The notion of a complex-structured domain has been introduced in [1]. The domain is complex-structured if its sections and subsections can be viewed as complex-structured areas, where the professional activity takes place; moreover, each section and subsection is characterized by its own set of tasks. These tasks can be formulated in terms of various ontology and knowledge modules. The difference between domain ontology and domain knowledge is represented in [10] and can be characterized by the statement: "Unenriched logical relationship systems simulate domain ontologies, their enrichments simulate domain knowledge, and enriched logical relationship systems simulate domains themselves" [10].

The methodology described above is used in the development of organic chemistry intelligent system model. The architecture of this system is introduced further.

Notably, the following approach has also been endeavored in the development of "Nanomaterials" ontology model, which is described in [2].

4 Organic Chemistry Intelligent System Model

Science-oriented domains, such as organic chemistry, can undergo not only knowledge alterations, but also the alterations in ontologies themselves, which can provoke the alterations in classes of problems that are solved in the domain and subdomains (note that the domain is complex-structured). Figure 1 introduces the architecture of the organic chemistry intelligent system that meets this challenge. The system is comprised of three subsystems that include information components, program components, and support (or maintenance) component. The separation of information components from program components has become a general principle for intelligent system development since it gives more control over the system.

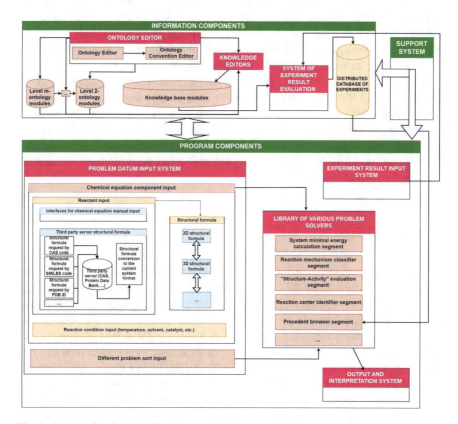

Fig. 1 Organic chemistry intelligent system architecture diagram

Basic information components include the multi-level ontology editor, which controls ontology modules (from the second level to the level *m*), knowledge editors, and knowledge base modules. Ontology editor and ontology convention editor are ontology editor essential parts. Knowledge editors are also the indispensable constituents of information components. In turn, the experimental result evaluation system is controlled by knowledge base modules to make sure the results that scientists have obtained and introduced as an input are consistent with the *current* laws of organic chemistry. Such laws can be named *current* because these laws can change, e.g. given new experimental evidence in the scientific community. These *current* laws of organic chemistry and knowledge tables comprise distinct knowledge base modules. The number of knowledge base modules coincides with the number of ontology model modules. If the experiment results are consistent with the knowledge base, they will be stored in the distributed database of experiments. The data from this database can be used for the precedent search performed by the precedent browser in the program components.

The essential program components of the organic chemistry intelligent system include separate input systems for problem data and experiment results, the library of various problem solvers, and the output and interpretation system. The library of problem solvers is expendable. Should ontology model change, new problem solvers can be added to the system or deleted in case of their obsolescence. In turn, the support subsystem enables the programmer to change program components.

It should be noted that the system has special interfaces for graphical representation of structural formulas, which are widely used in organic chemistry. Moreover, the system can take as an input the data from third-party servers over the Internet, particularly, the data represented as CAS code, SMILES code, or Protein Data Bank (PDB ID). One-to-one correspondence is established between two-dimensional and three-dimensional formulas. Other types of chemical formula representations can be added to the system if needed.

5   Conclusion

Due to the regular appearance of novel molecular structures and better understanding of structure–activity relationships, the amount of knowledge in organic chemistry has been constantly growing for several decades. Evidently, savvy knowledge representation plays a critical role in promoting the process of research and development. Although there exists a broad spectrum of methods for intelligent system design, such as expert systems, neural networks, fuzzy logic, and genetic algorithms, a simple technique has been introduced to start effective intelligent system development. It is based on three essential questions: (1) What is the application area? (2) What are the classes of problems solved in the application area? (3) What are the methods to solve them? As the experience shows, the first question presuming the meticulous description of the application area ontology (especially, if the application area is complex-structured, such as organic chemistry) can be the most challenging. To address this issue, the methodology that differs from traditional taxonomy design is depicted. This methodology grounds on the representation of any complex-structured domain as interconnected modules of applied logic theories. This approach of representing ontology model in mathematical terms has been utilized and organic chemistry intelligent system model that is based upon it has been introduced. As an illustration, applied logic theory form for a separate organic theory module has been introduced. Since the described methodology allows the specialists to extend the number of ontology modules, the information components of this intelligent system are customizable. Moreover, the program components are also expandable because the library of various problem solvers can be edited. Should ontology model change, other problem solvers can be added to the system. The authors hope that the depicted methodology can be employed for intelligent system creation in other application areas as well.

References

1. Artemieva, I.: Domains with complicated structures and their ontologies. In: Int. J. Inf. Theor. Appl. **15**(4), 330–337 (2008)
2. Artemieva, I.L., Ryabchenko, N.V.: Nanomaterials ontology model. Adv. Mater. Res. **905**, 65–69 (2014). https://doi.org/10.4028/www.scientific.net/AMR.905.65
3. Baader, F., Horrocks, I., Sattler U.: Description logics as ontology languages for the semantic web. In: Hutter, D., Stephan, W. (eds.) Mechanizing Mathematical Reasoning. Lecture Notes in Computer Science, vol. 2605. Springer, Berlin (2005)
4. Born, R.: Artificial intelligence: the case against. Routledge (2018). ISBN 9781351141505
5. Cuenca Grau, B., Halaschek-Wiener, C., Kazakov, Y. et al.: Incremental classification of description logics ontologies. J. Autom. Reasoning **44**, 337 (2010) https://doi.org/10.1007/s10817-009-9159-0
6. Feigenbaum, E.A., Buchanan, B.G.: DENDRAL and META-DENDRAL: roots of knowledge systems and expert system applications. Artif. Intell. **59**, 233–240 (1993)
7. Goldstine, H.: The computer from Pascal to von Neumann. Princeton University Press (1993)
8. Gruber, T.R.: A translation approach to portable ontology specifications. Knowl. Acquisition **5**(2), 199–220 (1993)
9. Kleshchev, A., Artemjeva, I.: A mathematical apparatus for domain ontology simulation. An extendable language of applied logic. Int. J. Inf. Theor. Appl. **12**(2), 149–157 (2005)
10. Kleshchev, A., Artemjeva, I.: A mathematical apparatus for domain ontology simulation. Logical relationship systems. Int. J. Inf. Theor. Appl. **12**(4), 343–351 (2005)
11. Kleshchev, A., Artemjeva, I.: A mathematical apparatus for domain ontology simulation. Specialized extensions of the extendable language of applied logic. Int. J. Inf. Theor. Appl. **12**(3), 265–271 (2005)
12. Lindsay, R.K., Buchanan, B.G., Feigenbaum, E.A., Lederberg, J.: DENDRAL: a case study of the first expert system for scientific hypothesis formation. Artif. Intell. **61**, 209–261 (1993)
13. Lindsay, R.K., Buchanan, B.G., Feigenbaum, E.A., Lederberg, J.: Applications of artificial intelligence for organic chemistry. The DENDRAL Project. McGraw-Hill, New York (1980)
14. McCarthy, J., Minsky, M.L., Rochester, N., Shannon, C.E.: A proposal for the Dartmouth summer research project on artificial intelligence. August, 31, 1955. AI Mag. **27**(4) (2006)
15. McCarthy, J.: Programs with common sense. In: Proceedings of the Teddington Conference on the Mechanization of Thought Processes, Her Majesty's Stationery Office, London (1959)

16. McCulloch, W.S., Pitts, W.: A logical calculus of the ideas immanent in nervous activity. Bull. Math. Biophys. **5**, 115–137 (1943)
17. Newell, A., Shaw, J.C., Simon, H.A.: Report on a general problem-solving program. The RAND Corporation, Paper P-1584, December 30 (1958)
18. The OBO Foundry Homepage. http://www.obofoundry.org/. Last accessed 29 Dec 2018
19. Turing, A.M.: Computing machinery and intelligence. Mind **59**, 433–460 (1950)
20. Wöhler, F.: Ueber künstliche Bildung des Harnstoffs. Ann. Phys. **88**, 253–256 (1828). https://doi.org/10.1002/andp.18280880206

Effective Way of Deriving the Context from a Handwritten Image/Object

Komal Teja Mattupalli and Sriraman Kothuri

Abstract This paper is mainly proposed to identify the text given in the input hand-written object format (irregular shape), i.e., recognized/unrecognized patterns by utilizing chain code using SASK algorithm. To find out the output of the written object, we would need to identify the hulls and layer of it using the layered algorithm which is explained in detail in further sections. After identifying the layer, we are supposed to get out the list of outcomes that suits for the output pattern, and hence, that will be provided as the recognized object. It helps us to easily identify the outcome of the text provided in handwritten, where the input is either in recognized or unrecognized format. Another advantage apart from that is by identifying structured layers will be helpful in the traffic to define the traffic time/list of vehicles waiting on the queue to get way further.

Keywords Handwritten text recognition · Unrecognized patterns · Character and hull recognition · Matching patterns · Monograms · Chain code · Hull regions · Hough transform · Layered outline extraction · SASK algorithm

1 Introduction

When we saw any handwritten image then it is not easy to define the text of what they have written as most of the people use different writing styles and different languages. Although there is a challenge to identify text of what it is when the input is an unrecognized pattern, like few people may not write complete word and understand with their own pattern styling. Here, we will be using the machine learning process [1], to maintain the training data set which will be like a library

K. T. Mattupalli (✉)
L&T Infotech, Hyderabad, India
e-mail: komalteja@gmail.com

S. Kothuri
Vel Tech Rangarajan DR. Sagunthala R&D Institute of Science and Technology,
Avadi, Chennai, India
e-mail: sriramankothuri@veltech.edu.in

© Springer Nature Singapore Pte Ltd. 2020
X.-S. Yang et al. (eds.), *Fourth International Congress on Information and Communication Technology*, Advances in Intelligent Systems and Computing 1027, https://doi.org/10.1007/978-981-32-9343-4_8

Fig. 1 Identification of the hull regions in the above images

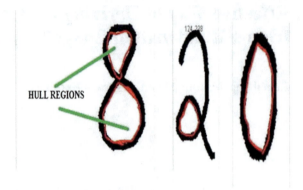

of the information where exactly the patterns with the combinations will be placed. User interacts to the system using GUI enabled for them to draw an input image, which then needs the identification of the context like humans.

From Fig. 1, we can notice the regions identified in the digits like 0, 1, 2, etc. To easily understand the concept of hulls and its identification, here we make use of digits as the best example. It is very helpful in many areas of the image processing; here, we have extended this paper from the temporal pattern mining using SASK algorithm, where this algorithm used to get the layers and associated hulls in the input image. This is now extended to identify the context of the image and output the data given.

Initially, this process is applied for the digits, later it got extended to the characters, and now it is been extended to identifying the text in the original input image. Chain code is implemented to extract the hulls and get the boundary extracted from the different regions of the image. This method is particularly effective for images consisting of a reasonably small chunk of connected components.

In Fig. 1, it is clearly shown the hulls and outline of the digits, and the same will be obtained for the alphabets and for the text written. Here, we used SASK algorithm to get the hull regions in the digits.

2 Previous Work

Many existing vision algorithms are defined to detect the outline/contour of images. But, some of those successfully achieved the same working progress on the image having irregular shape. It is easy to get the contour of the image when it is in the regular shape by using detection algorithms [2] and but still it did not work for the structure with having huge and many subregional objects placed in it. This can be achieved by using the algorithm we have proposed. This kind of the proposed system can be used in many applications for the real-time use.

3 Overall Design of the Proposed System

3.1 Input Data

Input data will be given by interacting with the GUI enabled by the application and the GUI us developed using the JAVA swings. The user draws the input image with the help of mouse. This might be regular/irregular shape. Then that will be passed to the system as the input data set to start proceeding further with the next steps.

Overall design of the new proposed system and steps of process are shown in the below image (Fig. 2).

Fig. 2 Overall proposed design

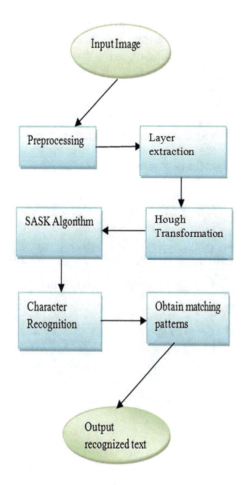

3.2   Preprocessing

After getting the input original image from the above step, this image needs will be saved with .png file extension. This will be considered as the preprocessing [3]. The input data will be saved in the image format with .(PNG) extension. Later, it will process the image to the two-dimensional layer and passes to the next step for process.

3.3   Layer Extraction

In this step, it will try to obtain the boundaries of the original image that got received from the above step. To extract the boundaries of the image, it needs to parse and check every single pixel of the original image. Every pixel property (explained in papers [4]) needs to check both the four-connectivity and eight-connectivity pixel.

Complexity of the algorithm that used here in this process is having $O(n)$. In every input original image, we need to check whether the pixel is from eight-connectivity or with four-connectivity, based on that we can get the outline of the original image; with eight-connectivity, we may not be able to get the layer of the input, and hence, we are making use of four-connectivity as well, which then checks for every pixel to get the original layer to be returned back.

Algorithm

Input: Original .png image
Output: Layer of the original image
Steps to be applied:

E.   Get the original proprocessed image.
F.   Pass the image for checking every pixel of the image needs to check to obtain the layer.
G.   Starts from the top left corner.
H.   Check the total dimensions of the image
I.   Validate every pixel location based on eight/four-connectivity pixel
J.   Remove pixel

3.4 Voting Scheme Using Hough Transform

There are two schemes to be used to get the irregular shapes, edge detection and Hough transformation. Once we detect the layers of the original image from a regular/irregular shape when the image is either in bulk of shape or not, then after, the outline of the image needs to be identified and returned. Before moving to the next step of getting the hulls from an image, here we get the edge of the image extracted and pass that as input to the next step. These edges detected as an individual need to be linked out by making use of Hough transform.

To get better results, they listed out to make use of line detection using an enhanced scheme of voting. This line detection is presented by using an effective way of voting scheme for the transformations [5]. This is mainly aimed to get better performance when the image is in bulk of size. It improves the performance of the voting scheme to give more robust for the detection of lines. The lines are detected based on the clusters which are producing the approximately collinear pixels.

Once the clusters are detected, then this process will be started for every cluster, and associated votes need to be obtained using elliptical-Gaussian kernel. This model shows the best fitting line even though it is having uncertainty. The resultant matrix will be having the accumulator, and then based on the highest proportionate values returned for every pixel, it will be considered and drawn as the line.

Placing votes and getting outline returned:

- From the original image, use the edge detection algorithm first.
- Get the most relevancy of every pixel and by getting them to clusters.
- On every cluster, apply the fitting line voting scheme and get the results from that.

By using elliptical-Gaussian kernel model, all the lines/outline will be located even though the image is having irregular shapes.

3.5 SASK Algorithm

For the binary output, it first checks for the '1' marked pixel from left to right; whenever it appears, mark it as any number; here, we consider it as irregular shape (not exactly the same shape can place any other kind of shape); it checks for the upward pixel; if it is present, it moves the position to that pixel, and if it is not present, it rotates 450 angle clockwise direction; for every 1 pixel, we have to check it from upwards only (no need of exact upward any direction but the same direction follows for every pixel) and checks again, and this process repeats until we reach to the starting pixel; finally, all these 2s becomes the outer outline of the hull, and the 1s are present inside the outer outline which forms a hull region.

Algorithm steps
Input: Outlined image from the voting scheme
Output: Original image with detected hulls.
Step 0: Binary image with the boundaries returned from the voting scheme using hull transformation and edge detection algorithms.
Step 1: Binary file data will be placed
Step 2: Scan the file from top corner and check for every binary value of 1/0. Based on that define the row position as X and column position as Y
Step 3: Check for all eight-connectivity pixels and set the values to either 1 or 2 for the row/column
Step 4: This will continue until the above step returns same position
Step 5: By this step, we got outer layer and now we will need to find out other regions but with different number placed as to get differentiate.
Step 6: Here, we applied two color codes for the original image of the layers extracted. Colors will be either black or red which depends on the regions that need to be identified in the final output of this step.

Overall time complexity for this algorithm to the outer boundary of original image will be less than 'n' and to reset the image to back will take $O(n)$ time complexity. Many edge detection algorithms [6] are designed and used mostly for the regular shapes. SASK algorithm is designed to mainly detect the number of layers present in a layered structure. Mainly focused to separate the layers and intended to do the work on irregular shapes. This algorithm can also apply to the digits and characters also and can possess the better results in it. The characters or digits are having single hulls inside it and where coming to the irregular shapes it may contain number of hulls in it. We can also take the regular shapes that are also applicable to it.

3.6 Character Recognition

There are two different types of patterns like unrecognized and recognized patterns. Recognized patterns are easily identified to get the character recognized. Unrecognized patterns are those which are difficult to identify as they are not completely formed characters with hulls to be recognized. To recognize the text and characters, here we use optical character recognition (OCR) process to identify those.

Recognizing text from the images is useful in many applications such as document analysis, search text in the image. An OCR function [12] in the MATLAB provides an easier way of identifying text by using its text recognition functionality in many applications.

Step 1: Load an image using imread ('ipImage.png')
Step 2: Apply OCR function to that loaded image
Step 3: Display the recognized words as results

At the end, it returns the text that got recognized, and associated confidence value, corresponding location of the input text. By this, we can identify the location of the unrecognized text in the input original image.

Step 1: Identify the confidence values of the characters.
Step 2: Identify the locations of the characters like the boundaries of it.
Step 3: Annotate image with the corresponding character confidences.
Step 4: Display the values of the text to know the text is recognized or not.

For example, if we write something that does not exist and when that is considered as the character and tries to recognize and throw an error, before raising to as the OCR error, we can check the confidence value to identify that is not valid.

OCR gives the best results when the text is located in a uniform background and in a formatted way. When that text appears on a non-uniform background, it needs to follow additional steps to preprocess the image to get the best results. For example, if an image of having non-uniform background with characters written over in that image, then it is quite challenging to get the results. Then, in that case, we need follow below steps to get the background converted and be uniform.

Step 1: Load an image using imread ('ipImage.png').
Step 2: Convert the color image to gray of that input image.
Step 3: Display the image.

Finally, we can run the image using OCR to get the characters/text recognized and returns as result.

3.7 Matching Patterns

- Input image may be part of monograms which needs to be identified by having the proper combination sets.
- In the early 350 BC, monograms are first appeared on the coins. At earliest, there are many examples of the names of cities in Greek which issue the coins, often the first two letters of the city name will be printed on the coin. For example, the

monogram of Achaea consists of the letters Alpha (A) and Chi (X) joined together a created as AX [7].

- It is often being used as the signatures by the artists on the paintings, sculptures and furniture, etc. These are especially used when guilds enforced measures against unauthorized participation in the trade.
- Famous example of a monogram is an artist's signature, i.e., 'AD' is being used by Albrecht Durer (Figs. 3, 4).

After identifying the characters, we need to use the machine learning process [1], i.e., to impart the intelligence to the system by maintaining appropriate training data

Fig. 3 Handwritten text

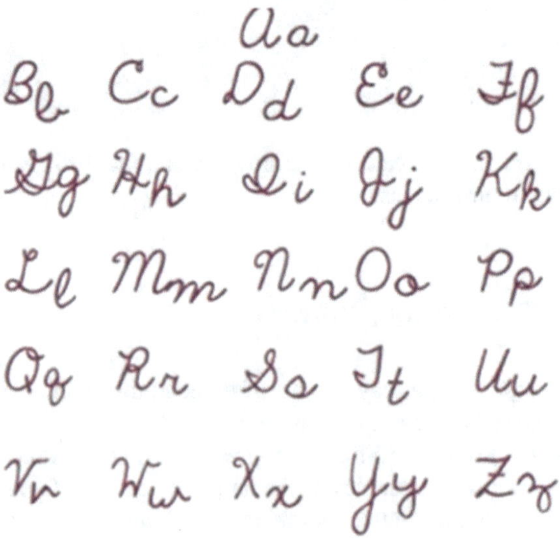

Fig. 4 Alphabets written monograms

set. This data set will have the combinations of the characters to be matched; all the possible patterns will be returned in this step.

3.8 Finding Out Accurate Result

In this final step, we will find out the accuracy of the original image to the list of the matching patterns returned from the training data set. Based on the highest proportionate value, an output text will be returned by the system to the user. By making use of the above steps, we can identify the unrecognized pattern.

4 Conclusion

By this process using SASK with chain code, we can get the layers and hull identified in the image and based on that we can further identify the unrecognized/recognized patterns easily by having different matching sets and pick the highest value of matching pattern. This work embraces the descriptor-based unrecognized known patterns, and further more work shall be done on the same with machine learning algorithms or either. This can be extended further with three-dimensional images processing and passing image without handwritten.

References

1. Baird, H.S.: Anatomy of a page reader. In: IAPR Workshop on Machine Vision Applications, Tokyo (1990)
2. Nagy, G.: Optical character recognition—theory and practice. In: Krishnaiah, P.R., Kanal, L.N. (eds.) Handbook of Statistics, Volume H: Classification, Pattern Recognition and Reduction of Dimensionality. North Holland (1982)
3. Digital Image Processing 2nd Edition by Gonzalez and Woods Pearson Publications
4. Annapurna, P., Kothuri, S., Lukka, S.: Digit recognition using freeman chain code. Int. J. Appl. Innov. Eng. Manag. (IJAIEM) 2(8) (2013)
5. Guo, S., Pridmore, T., Kong, Y., Zhang, X.: An improved Hough transform voting scheme utilizing surround suppression. Pattern Recogn. Lett. 30, 1241–1252 (2009)
6. Thilagavathy, S.K., Indra Gandhi, Dr. R.: Recognition Of distorted character using edge detection algorithm. Int. J. Innovtive Res Comput. Commun. Eng. 1(4) (2013)
7. OCR function from recognize-text-using-optical-character-recognition-ocr under. www.mathworks.com
8. Park, S.C., Choi, B.K.: Boundary Extraction Algorithm for Cutting Area Detection. Computer Aided Design, Elsevier (2000)
9. Debbarma B., Ghoshal, D.: Modified canny edge detection algorithm with variable sigma. Int. J. Emerg. Trends Electr. Electron. (IJETEE) 1(2) (2013)
10. Hyvarinen, A., Oja, E.: Independent component analysis: algorithms and applications. Neural Netw. 13(4–5), 411–430 (2000)

11. de Campos, T.E., Babu, B.R., Varma, M.: Character recognition in natural images. In: Proceedings of the International Conference on Computer Vision Theory and Applications, Lisbon, Portugal, February (2009)

E-Voting System Based on Multiple Ballot Casting

Liudmila Babenko(ID) **and Ilya Pisarev**(ID)

Abstract The use of electronic voting is gradually replacing the traditional one. However, the issue of creating an honest and reliable electronic voting system is still open. Creating a reliable e-voting system that meets all security requirements is a difficult task. The paper presents an electronic voting system based on the principle of blind intermediaries using multiple ballots. The basic requirements for systems are described: eligibility, fairness, individual verifiability, universal verifiability, vote-privacy, receipt-freeness, coercion-resistance, vulnerable software/hardware resistance. Blind intermediaries allow you to exclude the user's vote from communicating with any authentication data. The principle of multiple ballot casting allows verification of the correctness of the ballot without the use of evidence with zero disclosure, as well as ensures compliance with other requirements for electronic voting systems. The system architecture is described, and the components involved in the voting process are described. The voting process is described, consisting of several stages: preparation, voting, counting of results. Cryptographic protocols are used, which are used for data transfer between system components. The compliance with the security requirements of the system is justified.

Keywords Electronic voting · Cryptographic protocols · Cryptography · Encryption · System

L. Babenko (✉) · I. Pisarev
Information Security Department, Southern Federal University, Taganrog,
Russian Federation
e-mail: lkbabenko@sfedu.ru

I. Pisarev
e-mail: ilua.pisar@gmail.com

© Springer Nature Singapore Pte Ltd. 2020
X.-S. Yang et al. (eds.), *Fourth International Congress on Information and Communication Technology*, Advances in Intelligent Systems and Computing 1027, https://doi.org/10.1007/978-981-32-9343-4_9

1 Introduction

The security of e-voting systems depends on compliance with security requirements. The main requirements for the systems are: eligibility, fairness, individual verifiability, universal verifiability, vote-privacy, receipt-freeness, coercion-resistance, vulnerable software/hardware resistance. There are a large number of e-voting systems and schemes [1–16], which justify compliance with certain requirements. However, at the moment there is no system that completely and without any assumptions met all the security requirements. In each system, there are a number of limitations and assumptions, taking into account which one can talk about compliance with these requirements. The main task is to maintain a reasonable level of assumptions and limitations under which the safety properties are achieved.

2 Requirements for E-Voting System

There are many requirements for electronic voting systems. It is worth mentioning the main requirements [17]:

1. Eligibility: Only certain voters listed on the checked list can vote.
2. Fairness: The results of voting cannot be determined until the elections are over.
3. Individual verifiability: A voter can verify that his vote has been counted.
4. Universal verifiability: Any outside observer can verify that votes are correctly calculated.
5. Vote-privacy: It is impossible to prove that a certain voter voted in a certain way.
6. Receipt-freeness: The voter has no information (residual keys, identifiers), which can be used as evidence that he voted in a certain way.
7. Coercion-resistance: The voter will not be able to prove to the attacker, even if he cooperates with him and the attacker provides his data to him that he voted in a certain way.
8. Vulnerable software/hardware resistance: It means that on the client side, any manipulation of data can take place and this should not influence the results of the elections.

3 Preparation

The components of the system are the client application (V), the authentication server (AS) and the voting server (VS). The current version of the system is a modified version of the system, previously developed by the authors of [18], and brings a number of key points. In this paper, voting is considered one of many. During the preparation stage, random *IDs* are generated equal to the number of voters. The public/private key pair *pk*, *sk* is generated for the asymmetric cipher. Each *ID* is signed with a secret key, and the output is *Sign(ID)*. The secret key is divided according to the secret sharing scheme, which is given to authorized people. All security is attached to the fact that if at least one of the authorized people is honest, then the elections will be fair. The *ID* list is stored on the voting server and is not published in clear form. The list of *Sign(ID)* signatures is published.

4 Voting. Multiple Ballot Casting

The principle of multiple ballots is to create L number of ballots, where L is the number of candidates. The server sends the voting *ID* to the authenticated user. Fake identifiers are generated. The user makes his choice, for example, in our case he votes for Petrov. Four ballots are created, and votes are put in random order in them, so that there are not two equally filled ballots. For a ballot with a real vote, the *voteID* is used, and for the rest of the *FakeID*. After that, all ballots are asymmetrically encrypted on public key with OAEP and sent to the BB, and the voter can get 1 ballot with *FakeID*. You can see example of ballot filling in Fig. 1. For receipt-freeness, coercion-resistance, it is necessary to remove any ways to prove that the user voted in a certain way. Therefore, you cannot keep a ballot with a real vote. At the same time, a ballot with *FakeID* will make it possible to verify that all the user's ballots have reached the probability of 1/4 in this case. Since the server does not know which of the ballots is real, it will have to publish everything without a substitution. Otherwise, if at least 1 user discovers that the ballot has not reached BB, it will show that the server has violated the rules by changing the ballot or not publishing it. The client uses the verification that the vote has reached the BB regardless of whether the user wanted to check his vote or not.

In the system, the ballot is a pair: [ID, Candidate Index]. For example above, the list will be as follows:

[ID1, 2], [ID2, 4], [ID3, 1], [ID4, 3]

Only the identifiers are encrypted and in the form shown below, the ballots are sent to the BB.

[Epk(ID1), 2], [Epk(ID2), 4], [Epk(ID3), 1], [Epk(ID4), 3]
[EID1, 2], [EID2, 4], [EID3, 1], [EID4, 3]

The voting server only misses ballots with a uniform distribution of votes, that is, there should not be two equally filled ballots. This is the process of verifying the

Fig. 1 Example of filling out voting ballots type 1 out of 4

ID: ff23ab	ID: abfc52	ID: 414f2c	ID: abaca3
1. Petrov	1. Petrov	1. Petrov	1. Petrov
2. Sidorov	2. Sidorov	2. Sidorov	2. Sidorov
3. Pupkin	3. Pupkin	3. Pupkin	3. Pupkin
4. Serov	4. Serov	4. Serov	4. Serov

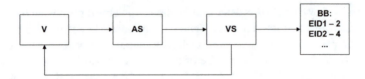

Fig. 2 System architecture and direction of data transfer

vote. In our principle, verification is carried out openly and immediately, unlike the application of the proof with zero-knowledge proof in other schemes. You can see system architecture and direction of data transfer in Fig. 2.

Before starting the protocol, the symmetric keys *vas*, *asvs*, *vvs* are generated using the ECDHE—Diffie-Hellman protocol on elliptical curves using ephemeral keys between the sides of V–AS, AS–VS, V–VS, respectively. In our case, we use a modified version of ECDHE-RSA, where authentication is done using a signature RSA and a server certificate which help to prevent MIMT (man in the middle) attacks. The protocol description is as follows.

ECDHE:

(1) $V \rightarrow S$: *"Hello"*
(2) $S \rightarrow V$: *DHs, Sign$_{SKs}$ (DHs), Certificate*
(3) *S: checks Certificate and sign (DHs)*
(4) $V \rightarrow S$: *DHv*
(5) *Both sides generate a common session key K for further interaction with a symmetric cipher.*

Preparation:

$ECDHE \ (V, AS) = vas$;
$ECDHE \ (V, VS) = vvs$;
$ECDHE \ (AS, VS) = asvs$;

Voting protocol:

1. $VS \rightarrow V$: $E_{vvs} \ (N)$
2. $V \rightarrow AS$: $E_{vas} \ (AuthData, E_{vvs} \ (N))$
3. $AS \rightarrow VS$: $E_{asvs} \ (E_{vvs} \ (N))$
4. $VS \rightarrow V$: $E_{vvs} \ (N, ID, Sign(ID))$
5. V:

   $E_{pk} \ (FakeID1), 2$: $EID1, 2$
   $E_{pk} \ (EakeID2), 4$: $EID2, 4$
   $E_{pk} \ (ID), 1$: $EID3, 1$
   $E_{pk} \ (FakeID3), 3$: $EID4, 3$

6. $V \rightarrow VS$: $E_{vvs} \ (N, EID1, 2, EID2, 4, EID3, 1, EID4, 3)$
7. $VS \rightarrow BB$:

   EID1–2

> *EID2–4*
> *EID3–1*
> *EID4–3*

8. $VS \rightarrow V$: E_{vvs} (N, *Sign(EID1)*, *Sign(EID2)*, *Sign(EID3)*, *Sign(EID4)*)
9. V: *EID1–2*, V check *EID1–2* on BB and *Sign(EID1)*.

The message (1) sends a random number for authentication. In the message (2), using the blind intermediaries principle, a random number N is encrypted on the *vvs* key, *AuthData* user authentication data is applied to it, and all this is encrypted on the vas key and sent to the *AS* side. *AS* can only read *AuthData*. Checks for the presence of these authentication data in the database and, if successful, redirects the other part as a message (3). *VS* checks the value of the number N and sends the user a message (4) with the identifier for voting *ID* and its signature *Sign(ID)*. In step (5), the client makes his choice and the client software generates fake ballots along with the real one, encrypting each identifier with a public key, and then sends the [*EID*, *vote*] pairs to the voting server as a message (6). *VS* sends ballots to BB. *VS* sends back the *EID* signatures after which the user selects a checkup, checks its presence on the BB and its signature *Sign(EID)*.

5 Counting of Votes

The procedure for counting votes is open. Authorized persons restore the secret, which is the private key of asymmetric encryption. This key is published in public access along with all the *IDs*. Thus, the BB has:

1. Public key *pk*
2. The secret key *sk*
3. List *ID*
4. List of *Sign(ID)*
5. The list of votes *EID*.

Any user with the help of the client application can independently calculate the voting results by using the above-described data. The system itself calculates votes as follows.

1. The equality of the number of *IDs* with the number of *Sign(ID)* signatures and their matching is checked, that is, for each *ID* from the list, there is its signature *Sign(ID)*.
2. *EID* votes are deciphered.
3. Only votes with an *ID* from the list participate in the calculation.
4. In the presence of two or more votes with the same *ID*—all votes with this *ID* are not counted.
5. The final results are published.

6   Compliance with the Requirements

6.1   Eligibility

Achieved with the help of an authentication server and the principle of blind inter-mediaries. Only users whose authentication data is present in the database can vote.

6.2   Fairness

It is achieved using the secret sharing scheme when distributing the secret key of asymmetric encryption. If at least 1 of the authorized persons is honest—the results cannot be recognized before the end of the election.

6.3   Individual Verifiability

After voting, the voter receives 1 ballot out of several with the choice of the candidate not selected by him. The *FakeID* ballot will verify that all of the user's ballots have reached a $1/L$ probability (where L is the number of candidates and one type of vote is used). Since the server does not know which of the ballots is real, it will have to publish everything without a substitution. Otherwise, if at least 1 user discovers that the ballot has not reached BB, it will show that the server has violated the rules by changing the ballot or not publishing it. The client uses the verification that the vote has reached the BB regardless of whether the user wanted to check his vote or not.

6.4   Universal Verifiability

Since on BB after the voting is over, there is all information with the help of which it is possible to count, then any unauthenticated user can count the results of voting and make sure of their correctness.

6.5   Vote-Privacy

Due to the principle of blind intermediaries and the creation of many fake ballots, it is impossible to create an authentication data connection with an identifier or a certain ballot.

6.6   Receipt-Freeness, Coercion-Resistance

The user has only 1 ballot with *FakeID*. Thus, the voter will be able to prove that he did not exactly vote for one of the candidates. The system is restricted to a minimum number of candidates 3. Because of what having information about 1 candidate for whom the user did not vote cannot draw a conclusion about who he voted for. However, at the moment there is no provision for re-voting in the system. It is planned to introduce the possibility of fake voting, when the voter sends ballots without a real *ID* to fully satisfy these two requirements.

6.7   Vulnerable Software/Hardware Resistance

If there is malware on the client side, you can only reduce the negative impact on the system as a whole. There are the following attacks and their consequences:

1. Vote substitution is partially protected. It can be detected with a probability of $1/L$ after the publication of the data for the counting of votes if the checklist contains a real vote, which contradicts the installation of the system. It will result in loss of user's vote.
2. Voting for all candidates is protected. It is found at the stage of vote counting, since in this case several ballots with the same *ID* will be for several candidates.
3. Embed after legal authentication—protected. On the part of one user it is impossible, because 1 *ID* for one vote is sent to him.
4. On the client side, all ballots are sent with fake *FakeIDs*. There are two options:

   4.1. The client software sends all the ballots from fake *FakeIDs,* thereby simply clearing the user's vote—partially protected, but will lead to the disruption of the election. When counting votes, the number of voted users on the authentication server component does not match the number of real votes due to temporary lack of a component synchronization mechanism.

   4.2. With the help of collected real *IDs*, the client software can integrate one or several extra votes at the next legal vote, but only for different candidates. That is, it is impossible to throw in a lot of votes for the same candidate. The best option is to look at the user's vote in the current legal vote, and if it does not match the desired vote of the attacker—put a real *ID* in front of another candidate. Thus, as a result of counting, the number of voters will coincide with the number of votes, but the effectiveness of such a throw is not particularly high. This attack is the most critical at the moment, which can be done by the client software.

7 Conclusion

The work presented a system of electronic voting based on blind intermediaries using multiple ballots. The basic requirements for systems are described: eligibility, fairness, individual verifiability, universal verifiability, vote-privacy, receipt-freeness, coercion-resistance, vulnerable software/hardware resistance. The system architecture is described, and the components involved in the voting process are described. The voting process is described, consisting of several stages: preparation, voting, counting of results. The justification for compliance with the security requirements of the system is given. To meet some of the security requirements, a number of improvements are required. In particular, to meet the requirements of receipt-freeness, coercion-resistance, it is required to develop a mechanism for fake voting from the client side which means that the user can use the fake vote as well as the real one and nobody can find out exactly which one he did.

Acknowledgements The work was supported by the Ministry of Education and Science of the Russian Federation, grant No 2.6264.2017/8.9.

References

1. Overview of e-voting systems, NICK Estonia. Estonian National Electoral Commission. Tallinn (2005)
2. Dossogne, J., Lafitte, F.: Blinded additively homomorphic encryption schemes for self-tallying voting. J. Inf. Secur. Appl. (2015)
3. Chaum, D.: Untraceable electronic mail, return addresses, and digital pseudonyms. Commun. ACM **24**(2), 84–90 (1981)
4. Izabachene, M.A.: Homomorphic LWE based e-voting scheme. In: Post-Quantum Cryptography: 7th International Workshop, PQCrypto 2016, Fukuoka, Japan, 24–26 Feb 2016
5. Hirt, M., Sako, K.: Efficient receipt-free voting based on homomorphic encryption. In: International Conference on the Theory and Applications of Cryptographic Techniques, pp. C.539–556, Springer, Berlin (2000)
6. Rivest, L.R. et al.: Lecture notes 15: voting, homomorphic encryption (2002)
7. Adida, B.: Mixnets in Electronic Voting, Cambridge University (2005)
8. Electronic elections: fear of falsification of the results. Kazakhstan today (2004)
9. Lipen, V.Y, Voronetsky, M.A, Lipen, D.V.: Technology and results of testing electronic voting systems. United Institute of Informatics Problems NASB (2002)
10. Ali, S.T., Murray, J.: An overview of end-to-end verifiable voting systems. arXiv preprint arXiv: 1605.08554 (2016)
11. Smart, M., Ritter, E.: True trustworthy elections: remote electronic voting using trusted computing. In: International Conference on Autonomic and Trusted Computing, pp. S.187–202. Springer, Berlin (2011)
12. Bruck, S., Jefferson, D., Rivest, R.L.: A Modular Voting Architecture ("frog voting"). Toward Strustworthy Elections. Springer, Berlin (2010)
13. Jonker, H., Mauw, S., Pang, J.: Privacy and verifiability in voting systems: methods, developments and trends. Computer Science Review (2013)
14. Shinde, S.S., Shukla, S., Chitre, D.K.: Secure E-voting using homomorphic technology. Int. J. Emerg. Technol. Adv. Eng. (2013)

15. Neumann, S., Volkamer, M.: Civitas and the real world: problems and solutions from a prac-tical point of view. In: Availability, Reliability and Security (ARES), 2012 Seventh International Conference on IEEE, pp. S. 180–185 (2012)
16. Yi, X., Okamoto, E.: Practical remote end-to-end voting scheme. In: International Conference on Electronic Government and the Information Systems Perspective, pp. S. 386–400. Springer, Berlin (2011)
17. Delaune, S., Kremer, S., Ryan, M.: Verifying privacy-type properties of electronic voting protocols. J. Comput. Secur. **17**(4), 435–487 (2009)
18. Babenko, L., Pisarev, I., Makarevich, O.: A model of a secure electronic voting system based on blind intermediaries using Russian cryptographic algorithms. In: Proceedings of the 10th International Conference on Security of Information and Networks (SIN '17), pp. 45–50. ACM, New York (2017)

Against Malicious SSL/TLS Encryption: Identify Malicious Traffic Based on Random Forest

Yong Fang, Yijia Xu, Cheng Huang, Liang Liu and Lei Zhang

Abstract It has become a significant research direction to resist cyberattacks through traffic identification technology. Traditional traffic identification technology is often based on network port or feature matching, which has become inefficient in the increasingly complex network environment. Nowadays, the malicious cyberattacks usually encrypt their traffic to escape the traditional traffic identification, and the most common encryption method is the SSL/TLS encryption. In response to this phenomenon, this paper proposes an encrypted malicious traffic identification method based on the random forest, which uses features based on packet information, time, TCP Flags field, and application layer payload information. We designed the technology and application framework to ensure the success of the experiment and collected a large amount of SSL/TLS encrypted traffic as datasets. Benefit from model optimization by parameter adjusting, the experimental results showed that final model had highly accurate and predictive ability.

Keywords Random forest · SSL/TLS encryption · Encrypted malicious traffic · Statistical features · Model optimization

Y. Fang · Y. Xu · C. Huang · L. Liu (✉) · L. Zhang
College of Cybersecurity, Sichuan University, Chengdu, China
e-mail: liangzhai118@163.com

Y. Fang
e-mail: yongfangscu@gmail.com

Y. Xu
e-mail: whiterabbitxyj@gmail.com

C. Huang
e-mail: opcodesec@gmail.com

L. Zhang
e-mail: welch1983@gmail.com

© Springer Nature Singapore Pte Ltd. 2020
X.-S. Yang et al. (eds.), *Fourth International Congress on Information and Communication Technology*, Advances in Intelligent Systems and Computing 1027, https://doi.org/10.1007/978-981-32-9343-4_10

1 Introduction

With the increasing of the network scale and the complexity of the network environment, the network attack becomes difficult to prevent. In the circumstances, many researchers propose a variety of technical methods to resist cyberattacks, and traffic identification is one of technologies to protect users' network security.

The original traffic identification methods were based on the network port, which relied on the port-mapping table stipulated by IANA. However, with the rapid development of the network, a large number of random ports began to be used, and these port-based traffic identification methods became less efficient. In such circumstances, the traffic identification methods based on feature matching were proposed [1]. These methods mainly analyzed the plaintext payload of the network packet application layer and matched the specific features to classify the traffic.

However, with the increasingly complex network environment, a variety of network applications began to encrypt traffic data to protect user privacy or to achieve other goals, which directly caused the reduction of plaintext information extracted from traffic. The method of traffic identification based on feature matching lost the core support and became inefficient and burdensome. To solve this problem, some researchers tried to find a way to decrypt the encrypted traffic. Nevertheless, due to the difficulties of decryption and the problems related to user privacy, this method was bogged down and difficult to carry out.

In such challenging environment, the field of flow identification needed a way. Under the hot condition of machine learning, the method of traffic identification based on machine learning was published in a step-by-step study [2]. This method collects the features that can be extracted from traffic, then uses the sophisticated machine learning algorithm to model, and finally uses the models to identify traffic. This approach, which takes machine learning as its core, should consider about several problems: features, dataset, and machine learning algorithm.

In the past, most of the machine algorithms used were the single classifier, such as C4.5, naive Bayes [3, 4], etc. The single classifier limits the growth of the model [5, 6], which is hard to improve when it reaches a certain degree of accuracy. Feature selection and extraction is a difficult problem when traffic is encrypted. This is why most traffic studies based on machine learning are used to classify unencrypted traffic rather than identify encrypted traffic. Dataset has always been a problem in machine learning, and the general flow classification study uses Moore_set as a dataset, which was used by the Moore and other researchers in the experiment. However, some unique traffic data collection often does not have an unified and authoritative source.

Faced with these problems, this paper proposes a malicious traffic identification method based on the random forest in SSL/TLS encryption. The principal research object is SSL/TLS encrypted network traffic, so we collect malicious and benign traffic data in SSL/TLS encryption as datasets. In this research, the random forest method is used to break through the accuracy limitation of the single classifier. Because the random forest can deal with the high dimensional features, we select multiple features based on packet information, time, TCP Flags field, and application layer payload information. To get better experimental result, we do the model optimization by adjusting random forest parameters.

The content of this paper is structured as follows. We discuss some problems and explained the framework in Sect. 2. In Sect. 3, we introduced the algorithm and evaluation method used in the paper. Section 4 details our observations of malicious traffic and the selection of features. Finally, the experiment is shown in Sect. 5.

2 Preliminary

In this paper, the "flow" is defined as a set of network data packets: $f = \{p_1, p_2, ..., p_n\}$, and the "traffic" is defined as a set of flows in a time interval t: $\{T_t = f_1, f_2, ..., f_m\}$. If a flow of T_t starts and ends in t, this flow is called a complete session flow, and the other is called a non-complete session flow. Traffic is composed of complete session flows and non-complete session flows.

2.1 Problems

How to describe the relationship between flow and traffic? We consider the scenario where users browse the web page. When requesting a web server resource, a TCP session connection is established, and all packets will pass through this session connection. If the website does not deploy the TCP keep-alive mode, every time the resource is requested, the TCP connection will be re-established. What's more, the third-party resources are likely to exist on the webpage. Therefore, each time you request the third-party resources, new TCP connections need to be established. When the web page is closed, the set of data packets transmitted in each TCP connection is a flow. Since the protocol, IP address, and port number remain unchanged during the session, these flows can be easily extracted by tool such as *wireshark*. If we set time interval between user requesting and closing the web page, the set of these TCP flows is called traffic.

How to define malicious flows generated by malware? During a time interval t from malware start running to finish, the traffic generated by the host is T_t. T_t contains a set of flows, some of which are generated by malware, and others are generated by benign software. We believe that the flow generated by malware is either a direct malicious behavior or a malicious behavior assistance. Therefore, we considered they are all the malicious flow. Since time t covers the beginning to the end of malware running, these malicious flows are all the complete session flow.

Why pay attention to SSL/TLS encryption? Today, ever-increasing malware begins to use encryption to hide malicious intentions. However, designing a encryption algorithms is complex and costly. The malware designers are more likely to use existing encryption algorithms directly, and the SSL/TLS is a common and mature encryption method in network communication. Actually, we have found this phenomenon and trend in our analysis of malware.

How to organize data? Our malicious data comes from open Internet resources. The provider collected the data in the simplest environment (the basic Linux system, only the specific malware and traffic capture tool *tcpdump* is installed), which allows us to simulate same environment locally and collect benign data. At the same time, in order to ensure the anonymity and consistency of the data, we processed the source IP address of all flows into 127.0.0.1.

Our raw data contains a set of traffic $C = \{T_1, T_2, ..., T_k\}$, and for each T_i ($1 \leq i \leq k$), it contains all the flows from the host between the time malware starts running to the end. We extracted the flow generated by the malware from each T_i through the port and destination ip. After that, we filtered out the SSL/TLS encrypted flows and filtered $T_i' = \{e_1, e_2, \ldots e_q\}$, where e_j ($1 \leq j \leq q$) is the malicious SSL/TLS encrypted flow. The final set of malicious encrypted flows is:

$$C_{\text{mal}} = \bigcap_1^k T_i' \tag{1}$$

Similarly, we can get a benign encrypted flow set. The features and labels will be extracted from the encrypted flows in the sets as records, which will participate in subsequent machine learning.

Others? For the convenience of description, in this paper, the set of malicious flows will be called as "malicious traffic", and the set of benign flows will be called as "benign traffic". Our goal is to identify" malicious traffic" in traffic.

2.2 Framework

This paper uses the SSL/TLS encrypted malicious traffic identification technology based on the random forest and divides the technical frame into five levels: data layer, feature layer, model layer, identification layer, and backup layer. The complete technical framework is shown in Fig. 1. The data layer is primarily responsible for processing the original network data flow, which includes filtered the data flow, and the filtering rules are as follows:

(1) It is an encrypted flow.
(2) The number of packets in the stream is higher than ten.
(3) The flow carries the payload information.
(4) It is a complete session flow.

When finishing filtering of the traffic, the data layer will sort all data conforming to the standard into a dataset and transfer it to the feature layer. The feature layer will extract the statistical features of each data as a record. After obtaining all the feature data, the feature layer will sort the data into the next layer as a modeling dataset.

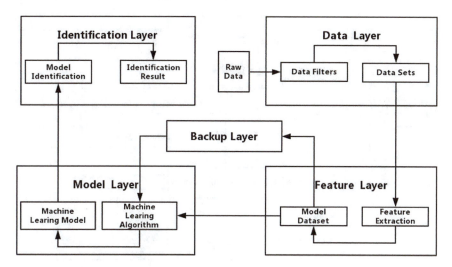

Fig. 1 Technical system framework

The model layer read the modeling dataset assembly in the upper layer; then, the machine learning algorithm is used to train the dataset for a machine learning model. In the end, the machine learning model will be transferred to the identification layer to identify the malicious encrypted traffic. The backup layer is a particular layer that automatically saves the modeling dataset from the feature layer and provides modeling data directly for the model layer.

In the practical application system, the connection of each level module is closer. This paper designs an application system according to the technical frame and divides it into four functional modules: file input/output module, statistical features extraction module, machine learning modeling module, and model identification module. The system modules cooperate with each other and complete the holonomic technology realization process together. Figure 2 shows the collaboration and relationship of application system modules. Each of the numbers in Fig. 2 represents a complete functional process. The number 1 represents the feature extraction and modeling process from the original data flow; the number 2 represents the modeling process starting from the backup data; the number 3 describes the flow identification process beginning with the identified data, which also includes feature data extraction.

3   Methodology

3.1   Algorithm

Random forest (RF) is an integrated classifier proposed by Breiman [7]. The principle of RF is using many decision trees to train and predict the samples provided. RF uses

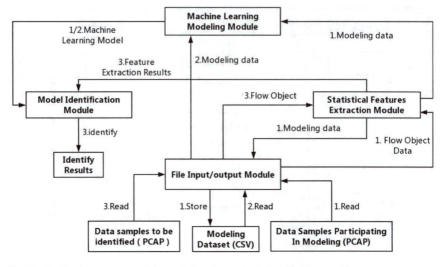

Fig. 2 Application system modules collaboration and relationship

algorithms to generate a decision tree set $\{h(X, \theta), K = 1, 2, ..., k\}$ (k is the number of decision trees). The $\{\theta\}$ is a random vector that is independent of each other and obeys distribution, and it determines the growth process of a single tree. X is the input vector of the unknown label, and all decision trees in the random forest will work together to determine its label.

Every decision tree in the random forest is relatively weak but is often proficient at solving problems in one domain. So when predicting a new input, 99% of the trees are useless. The predictions between them cancel out, and the results of individual excellent trees prediction will be the result of the final prediction. In random forests, the final classification results are:

$$H(x) = \arg\max_{y} \sum_{i=1}^{k} I(h_i(x) = Y) \tag{2}$$

The $H(x)$ in the formula is the combination classifier model, and h_i is a single decision tree classification model, and Y is the target variable.

Random forests are established with the following rules:

(1) When the size of the training set is S, the training sample is randomly selected with the put back. Assuming that N data samples are selected from the training set, the set of N samples is used as the training set for this decision tree. Random extraction ensures the difference of each training set, although it contains a repeating part.

(2) When the characteristic dimension of the training set is M, a value m is specified [$m \ll M$, in general circumstances $m = \mathrm{sqrt}(M)$]. Randomly select the number of M from the feature set and combine it into a feature subset. The feature subset

and the training assembly will be used as input to the decision tree algorithm, and then a decision tree is generated.

(3) Every decision tree grows as much as possible and does not contain pruning behavior.

3.2 Evaluation Method

The receiver operating characteristic (ROC) curve is the relationship between true positive rate and false positive rate, which is often used to evaluate the merits of a binary classifier. The true positive rate defines the situation where the true positive case is classified correctly, which can be expressed as:

$$TPR = \frac{TP}{TP + FN} \tag{3}$$

The false positive rate defines the situation where the true negative case is misclassified into a positive case, which can be expressed as:

$$FPT = \frac{FP}{TN + FP} \tag{4}$$

For the overall evaluation of a classifier, researchers often use the area under the ROC curve (AUC). If a classifier is perfect, its AUC value is 1. If a classifier always does a random classification, its AUC value is 0.5. Therefore, the closer the AUC value of a classifier is to 1, the better the overall performance of the classifier.

4 Features

To make better training effects, we decided to conduct deeper research and feature selection. Because the ports and IP addresses used by malware are variable, we decide to identify malicious traffic by statistical features. After a period of analysis of the malware, we use features based on packet information, time, TCP Flags field, and application layer payload information. A complete data flow is usually a bidirectional flow, and the bidirectional flow can be separated into uplink flow and downlink flow, which can be understood as uploading and downloading. In order to extend the feature set, we extract features from the uplink flow, downlink flow, and bidirectional flow.

4.1   Feature Type

Packet-information-based features. In the long-term experimental analysis of malicious traffic, we found that the malicious flow in the length of some particular packets has a fixed, and this packet is often the largest packet in the flow, occasionally the smallest packet. The average packet length of the malicious data flow is usually larger, and the fluctuation interval is stable. Therefore, we finally choose the total number of packets, the total length of bytes, the maximum and minimum packet length, the average packet length, and the standard deviation of the packet length as traffic features.

Time-based features. At the time level, the frequency of sending packets of malicious flows has the certain regularity, and especially compared with benign traffic, the time interval of transmitting packets of malicious traffic fluctuates more. Therefore, we choose the maximum and minimum time interval between packets, the average time interval, the standard deviation of the time interval, and the number of packets and bytes arriving per second as the traffic features.

TCP Flags field-based features. The TCP Flags field includes several types: SYN, FIN, ACK, PSH, RST, URG. The URG is the emergency pointer which tells the system that there are urgent data in this segment and must process this data first. When a large number of packets containing the URG arrive at the system, the standard functionality of the system will be suppressed, which sounds more like malicious traffic would do. In the actual data sample observation, we did find in the malicious traffic that it used URG more frequently than benign flow. The PSH is the push pointer. When the system receives a message with a PSH value of 1, it immediately empties and submits the data in the buffer. When studying the data samples, we found that benign traffic does not use PSH at most time, but a few malicious traffic use PSH to encroach on buffer space as a pre-attack preparation forcibly. So, we choose the number of PSH and URG in the data flow as the traffic features.

Application layer payload-information-based features. As for the application layer payload information in encryption, we can still extract features at the bit and byte level to replace the plaintext features, as shown in Fig. 3. We choose "Byte_Equality_Meter", "Bit_Positions_Meter", "Bit_Value_Meter",

```
Encrypted Application Data: 000000000000000113b18f9501aeb9b80e2f2dce879
```

```
00010111 00000011 00000011 00001111 01110111 00000000 00000000 00000000
00000000 00000000 00000000 00000000 00000001 00010011 10110001 10001111
10010101 00000001 10101110 10111001 10111000 00001110 00101111 00101101
11001110 10000111 10011001 01111111 00001101 11100010 11001101 10101000
00010010 10110010 10101010 01011111 10100000 00010010 11001011 11111001
00001110 10100010 01100110 00100111 11100101 01011110 01011000 11011000
10011101 11000011 11000101 11000000 01101101 11000011 11000111 00001010
```

Fig. 3   Bit- and byte-level information for encryption data

"Byte_Frequency_Meter" as the traffic features, which are four kinds of application layer statistical features in SPID theory proposed by Hjelmvik [8]:

(1) Bit_Positions_Meter. It represents the regulation of bit values (0 and 1) distribution. This feature shows whether the distribution of the same bit value is sparse or tight.
(2) Bit_Value_Meter. It represents the occurrence frequency of bit values (0 and 1). This feature shows whether the same bit value appears frequently or not.
(3) Byte_Equality_Meter. It represents the variation regularity of data packet payload information in the same direction data packets. This feature shows the byte values change rate in the same direction packets.
(4) Byte_Frequency_Meter. It represents the statistical analysis of all 256 possible values for each byte of the packet payload section. This feature shows the probability that various byte values appear in the packets.

4.2 Flow Direction Features

According to the features of flow direction, the all features can be classified into uplink-flow-features, downlink-flow-features, and bidirectional-flow-features. The uplink-flow-features represent the feature extracted from the uplink flow; the downlink-flow-features represent the feature extracted from the downlink flow; bidirectional-flow-features represent the feature extracted from the complete bidirectional flow.

After reference to a large number of literature and experimental research reports, this paper selected 33 traffic features (13 uplink-flow-features, 13 downlink-flow-features and 7 bidirectional-flow-features) as feature set for random forest features. Tables 1 and 2 show the detail of this features.

The suitable feature set has a significant influence on the efficiency and accuracy of machine learning algorithms [9]. Therefore, the selected features of this paper are the quality features confirmed by the relevant experimental research, and the features quantity is kept in a range which does not affect the efficiency of the algorithm.

5 Experiments and Evaluation

5.1 Dataset

Malicious traffic data Before collecting the malicious traffic data, we need to evaluate the worth of the malicious traffic type. After consulting a lot of literature and the 2017 network report [10–15], we believe that ransomware, phishing, trojan, botnet, and malware are the common types of cyberattack that companies and organizations

Table 1 Features of uplink/downlink flow

Feature type	Feature name	Description
Packet information based	Total_up(down)_packets	Total number of uplink (downlink) packets
Packet information based	Total_up(down)_bytes	Total number of uplink (downlink) bytes
Packet information based	Min_up(down)_packet_length	Minimum uplink (downlink) packet length
Packet information based	Max_up(down)_packet_length	Maximum uplink (downlink) packet length
Packet information based	Mean_up(down)_packet_length	Average uplink (downlink) packet length
Packet information based	Std_up(down)_packet_length	Standard deviation of uplink (downlink) packet length
Time based	Min_up(down)_packet_interval	Minimum uplink (downlink) packet time interval
Time based	Max_up(down)_packet_interval	Maximum uplink (downlink) packet time interval
Time based	Mean_up(down)_packet_interval	Average uplink (downlink) packet time interval
Time based	Std_up(down)_packet_interval	Standard deviation of uplink (downlink) packet time interval
TCP Flags field based	Up(Down)_push_flag	Number of PSH tags in uplink (downlink) packet
TCP Flags field based	Up(Down)_urgent_flag	Number of URG tags in uplink (downlink) packet
Application layer payload information based	Up(Down)_Byte_Equality_Meter	Uplink (Downlink) packet byte comparison value

Table 2 Features of bidirectional flow

Feature type	Feature name	Description
Time based	Total_packets_per_second	Number of packets reached per second
Time based	Total_Byte_per_second	Number of bytes reached per second
Time based	Mean_total_packet_interval	Average packet arrival time interval
Time based	Std_total_packet_interval	Average time interval standard deviation
Application layer payload information based	Bit_Positions_Meter	Bit value distribution
Application layer payload information based	Bit_Value_Meter	Bit value frequency ratio
Application layer payload information based	Byte_Frequency_Meter	Byte frequency statistics

face over the past year. Given the prevalence of Bitcoin and its frequent use by ransomware, we also consider malicious bitcoin mining as a valuable type of malicious attack.

Therefore, the traffic generated by the aforesaid malicious network attack type will be regarded as data with modeling value. Most of the flow classification experiments chose *Moore set* as the experimental dataset. However, the data contained in this dataset was too complicated and extensive, which did not apply to the research of directional SSL/TLS encryption traffic.

So, this paper used the malicious encrypted traffic data provided by the website named *malware-traffic-analysis.net* [16]. The site is dedicated to providing the latest malicious traffic data, which has been continuously updated from 2013 onward to August 2018. Data content contains ransomware, mail-fishing, phishing, trojans, malicious bitcoin mining, malware, botnets, and other types of malicious traffic, which are all the original data encrypted by SSL/TLS.

Benign traffic data The benign traffic data originated from the data crawled from the local simulated environment. The collection of benign flow data adopted the method of cyclic grasping, which caught in a 24-h cycle for a total of 7 days. To make the benign traffic data sample more representative, we designed the local environment to visit the Top 500 website published by Alexa in turn and simulate the use of global classic software to capture the raw data generated by it.

When completing the collection of raw data, we needed to extract the experimental dataset from it. At first, the data flow containing SSL/TLS encryption should be filtered out from the raw data flow including HTTP, SMTP, FTP, DNS, and HTTPS. We implemented this step by Lua language script in conjunction with T-shark. After that, to ensure the quality of the experiment, we had to select high-quality traffic data from the extracted data flow. This step refers to the data-layer work of the framework in Chap. 2.

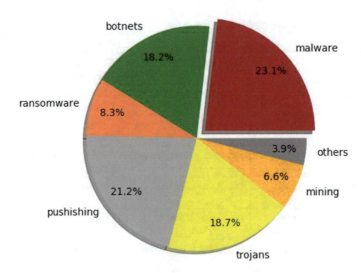

Fig. 4 Proportion of various types of traffic in malicious encrypted dataset

Dataset and analysis We collected two different datasets, and we referred to them as *current dataset* and *future dataset*. The *current dataset* contained 1000 malicious encrypted traffic and 5000 benign encrypted traffic with random extraction. The traffic data in *current dataset* was all occurring before 2018, and this dataset would participate in modeling as modeling data. The *future dataset* contained 600 malicious encrypted traffic and 600 benign encrypted traffic with random extraction. The traffic data in *future dataset* was all occurring in 2018, and this dataset would be used as a test set to evaluate the predictability of the model. It is worth mentioning that the data in the *future dataset* was collected from January to June in 2018, in which 100 malicious data and 100 benign data were collected each month.

We analyzed the malicious encrypted traffic data in the *current dataset*, and Fig. 4 shows the proportion of various types of malicious traffic attack. Our analysis found that ransomware had fewer samples than other common attack types because many ransomware had self-destruction, which made samples difficult to be collected. But the samples' quantity of the malicious bitcoin mining was less because it was difficult to be detected.

We enumerated the number of malicious attack samples and the number of encrypted malicious traffic that can be extracted each year in Table 3. After analysis, we thought the number of malicious attack samples in 2013 was too small to be referenced. Since not every attack used encryption, some attack samples did not carry encrypted malicious traffic, so we counted the average to evaluate the frequency of malicious attacks using encryption. From the statistics, the frequency of malicious attacks using encryption in 2014–2016 was relatively stable. But in 2017, the frequency soared to double, and the total number of encrypted malicious traffic was almost the sum of the previous three years. In the first half year of 2018, the frequency of malicious attacks using encryption increased again, and the total number

Table 3 Annual malicious data sample statistics from source

	2013	2014	2015	2016	2017	2018 (half year)
Malicious encrypted traffic number	188	501	390	402	1308	3621
Malicious attack samples number	19	267	172	192	249	118
Average	9.89	1.87	2.26	2.09	5.25	30.68

of encrypted malicious traffic was almost three times of 2017. It can be seen that the use of encryption for malicious attacks has become frequent in the past two years, and it is increasingly important to accurately identify encrypted malicious traffic.

5.2 Model Optimization

In order to get better experimental results, we optimized the random forest model by adjusting the model parameters. Because there is no dependency between each weak classifier in the random forest model, parameter adjusting is meaningful. This paper used *current dataset* as training dataset to establish random forest model, and evaluated the advantages and disadvantages of the model by using the out-of-bag score (oob_score), which reflected the generalization ability of a model fitting.

Parameters of random forest are divided into random forest (RF) parameters and RF decision tree parameters. RF parameters are few, and the primary core is the maximum number of the decision tree (n_estimators), while the RF decision parameters are mainly related to decision tree parameters. According to the repeated test, we found that the parameter adjusting of the RF decision tree had little effect on the overall model performance and almost adverse effect. Therefore, this paper only adjusted the maximum number of decision trees in the RF parameters and took the out-of-bag score as the model evaluation standard. We experimented many times and recorded the results. The results of the parameters adjusting were shown in Fig. 5.

We could learn from Fig. 5 that the average value of the out-of-bag score rose to the maximum when the maximum decision tree number increased to 110. When the number of decision trees continued to grow, there was an over-fitting phenomenon, and the out-of-bag score began to decrease. The standard deviation from the out-of-bag score showed that the stability of the model was highest when the maximum decision tree number was 110. Therefore, the experiment set the RF parameter value to 110.

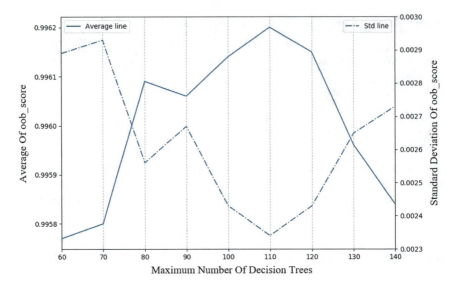

Fig. 5 Generalization ability/stability of model with maximum decision tree quantity

5.3 Evaluation and Result

Model evaluation. The experiment built a random forest model in *current dataset* and used tenfold cross-validation to assess the accuracy of the model. We plotted the ROC curve to represent the tenfold cross-validation test results, as shown in Fig. 6.

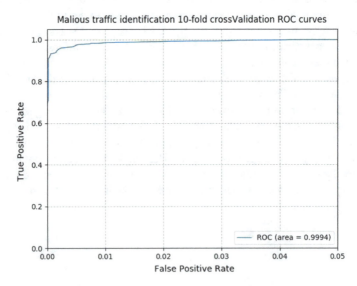

Fig. 6 ROC curve for tenfold cross-validation of random forest model

Table 4 Feature importance of random forest model

Feature type	Feature name	Importance
Packet information based	Max_down_packet_length	0.11352482
Packet information based	Min_down_packet_length	0.08373189
Time based	Min_down_packet_interval	0.07465366
TCP Flags field based	Down_push_flag	0.06962413
Time based	Max_down_packet_interval	0.06446696
Application layer payload information based	Byte_Frequency_Meter	0.06434366
Packet information based	Total_down_packets	0.05018637
Time based	Max_up_packet_interval	0.04172961
Packet information based	Mean_down_packet_length	0.03212983
Application layer payload information based	Up_Byte_Equality_Meter	0.03152412

The AUC of the ROC curve could be seen to reach 0.9994, which showed that the model had a high accuracy rate.

Feature importance. We evaluated the importance of the feature in the experiment. In Table 4, the Top 10 features with the highest importance weight in the classification process were listed. Analyzing based on the feature of flow direction, it was found that the downstream-traffic-features were more important, which accounted for 7 of the Top 10. We analyzed the feature types of Top 10, and find that there were four packet-information-based features, three time-based features, one TCP Flags field-based feature, and two payload-information-based features. It was showed that the importance of feature types in modeling is relatively balanced.

Model prediction. Over time, some malicious attacks can mutate or evolve, or even appear new malicious attacks. These malicious attacks generate entirely new malicious traffic, so we need to evaluate the predictive power of the model. We would use data generated before 2018 to model, then predict the data generated in 2018 to reach the prediction effect. This paper used *current dataset* as the modeling dataset to build a random forest model, and *future dataset* would be used as a test set to evaluate the model's predictive ability. The data in the *future dataset* was distributed evenly in 2018 from January to June, with a total of 600 encrypted malicious traffic and 600 encrypted benign traffic. Figure 7 showed the ROC curve of model prediction. The final experimental results showed that the AUC value of ROC curve reaches 0.9709, which indicated that the model had excellent predictive ability and could identify the malicious traffic generated by the new malicious attack.

6 Limitation

Although the model we designed showed excellent accuracy and predictive power in the experiment, the operation of the system framework was not optimistic in terms of

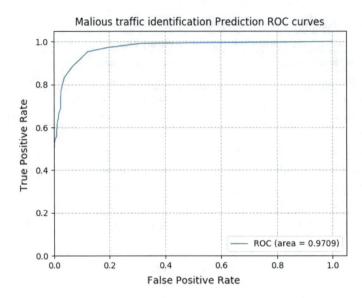

Fig. 7 ROC curve for prediction of random forest model

time consumption, especially the feature extraction module. Due to the large number of features we used, the time cost of the feature extraction module was terrible when the analysis and calculation process of some features were complicated (such as the application layer payload-information-based features).

According to our statistics, when the packets number in a flow increased to 2000, the extraction of features had taken 10 s. It can be expected that when a malicious attack constructs many traffic with large number of packets, the identification power of the model will be severely weakened. We considered blocking the flow whose packet number more than the certain threshold, but this method reduced the model's ability to recognize malicious attacks. Therefore, how to balance model performance and predictive ability is our next issue.

7   Conclusion

In this paper, the research of malicious encrypted traffic identification based on the random forest is carried out. This paper discusses some problems that deserve attention, then introduces the technical framework and practical application system, random forest algorithm and evaluation method. After a long period of research, the paper finally determined 33 features to conduct experiments. In the experimental part, firstly, the paper introduces the data collection and processing, and analyzes the data in the final dataset; then, the paper uses experiments to obtain the optimal parameters of random forest algorithm; finally, the paper evaluated the accuracy, the feature importance and the predictive ability of the model. The experiment showed that the model had high accuracy under ten cross-validation; the downlink-flow-features were more important for the model, but different types of features still have some impacts on the model; the model had excellent predictive ability and could identify the malicious traffic generated by the new malicious attack.

References

1. Moore, A.W., Papagiannaki, K.: Toward the accurate identification of network application. In: Proceedings of Passive and Active Networks Measurement Workshop, pp. 41–54 (2005)
2. Ding, L., Yu, F., Peng, S., et al.: A classification algorithm for network traffic based on improved support vector machine. J. Comput. **8**(4), 1090–1096 (2013)
3. Kalaiselvi, T., Shanmugaraja, P.: Hybrid algorithm for the traffic flows. J. Comput. **8**(4), 340–343 (2016)
4. Wei, L., Marco, C., Moore, et al.: Efficient application identification and the temporal and spatial stability of classification schema. Comput. Netw. **53**(6), 790–809 (2009)
5. Arthur, C., Judith, K., Djamel, S., et al.: Better network traffic identification through the independent combination of techniques. J. Netw. Comput. Appl. **33**(4), 433–446 (2010)
6. Alberto, D., Antonio, P., Kimberly, C.C.: Issues and future directions in traffic classification. IEEE Netw. **26**(1), 35–40 (2012)
7. Breiman, L.: Random forests. Mach. Learn. **45**(1), 5–32 (2001)
8. Hjelmvik, E., John, W.: Statistical protocol identification with SPID: preliminary results. In: Swedish National Computer Networking Workshop (2009)
9. Jnguyen, T.T., Grenville, A.: A survey of techniques for internet traffic classification using machine learning. IEEE Commun. Surv. Tutor. **10**(4), 56–76 (2008)
10. Cybercrime tactics and techniques Q1 2017. https://www.malwarebytes.com/pdf/labs/ybercrime-Tactics-and-Techniques-Q1-2017.pdf
11. Three-Quarters of Organizations Experienced Phishing Attacks in 2017, Report Uncovers. https://www.tripwire.com/state-of-security/security-data-protection/three-quarters-organizations-experienced-phishing-attacks-2017-report-uncovers/
12. Banking trojans, not ransomware, are the biggest threat to the enterprise now. https://www.techrepublic.com/article/banking-trojans-not-ransomware-are-the-biggest-threat-to-the-enterprise-now/
13. Top 10 Banking Trojans for 2017: What you need to know. https://blog.barkly.com/top-banking-trojans-2017
14. Spamhaus Botnet Threat Report 2017. https://www.spamhaus.org/news/article/772/
15. 2017 State of Malware Report. https://www.malwarebytes.com/pdf/white-papers/stateofmalware.pdf
16. Malware-Traffic-Analysis.net. http://www.malware-traffic-analysis.net/

Automatically Determining a Network Reconnaissance Scope Using Passive Scanning Techniques

Stefan Marksteiner, Bernhard Jandl-Scherf and Harald Lernbeiß

Abstract The starting point of securing a network is having a concise overview of it. As networks are becoming more and more complex both in general and with the introduction of IoT technology and their topological peculiarities in particular, this is increasingly difficult to achieve. Especially, in cyber-physical environments, such as smart factories, gaining a reliable picture of the network can be, due to intertwining of a vast amount of devices and different protocols, a tedious task. Nevertheless, this work is necessary to conduct security audits, compare documentation with actual conditions or find vulnerabilities using an attacker's view, for all of which a reliable topology overview is pivotal. For security auditors; however, there might not much information, such as asset management access, be available beforehand, which is why this paper assumes network to audit as a complete black box. The goal is, therefore, to set security auditors in a condition of, without having any *a priori* knowledge at all, automatically gaining a topology oversight. This paper describes, in the context of a bigger system that uses active scanning to determine the network topology, an approach to automate the first steps of this procedure: passively scanning the network and determining the network's scope, as well as gaining a valid address to perform the active scanning. This allows for bootstrapping an automatic network discovery process without prior knowledge.

Keywords Network scanning · Security · Network mapping · Networks

1 Introduction

As today's networks' complexity, especially with the introduction of *Internet of things (IoT)*-related technologies and, protocols and architectures with its snares [3,

S. Marksteiner (✉) · B. Jandl-Scherf · H. Lernbeiß
Joanneum Research, Graz, Austria
e-mail: stefan.marksteiner@avl.com

B. Jandl-Scherf
e-mail: bernhard.jandl-scherf@joanneum.at

H. Lernbeiß
e-mail: harald.lernbeiss@joanneum.at

© Springer Nature Singapore Pte Ltd. 2020
X.-S. Yang et al. (eds.), *Fourth International Congress on Information and Communication Technology*, Advances in Intelligent Systems and Computing 1027, https://doi.org/10.1007/978-981-32-9343-4_11

12], is becoming higher, the possibility of structural security vulnerabilities also rises. In order to discover these vulnerabilities, it is beneficial not only to use documentation and verify known countermeasures, which would only lead to already paved roads, the usage of black-box techniques, resembling an attack could be beneficial [17]. To do so, a network's structure must be uncovered first to reveal weak spots [14]. With said network complexity, however, this can be a tedious task. It is, therefore, crucial to automate as much of this process as possible in order to yield suitable results with workable effort. There is already work on automated toolchain-based scanning that is dedicated to achieve this task [13] . Based on this work, this paper shows an approach to automatically determining a network's scope to prepare the ground for automated network scanning without any given information. This way, a (possibly external) auditor might be able to black-box map an unfamiliar network by hitting a single button, and not using any *a priori* knowledge.

2   Related Work

More than three decades ago, it became apparent that the original two-level hierarchy of the Internet was no longer adequate for technical as well as organizational reasons [15]. In 1985, the Internet Engineering Task Force (IETF) specified the rules on how to integrate the new third (*subnet*) layer [16]. Since then, a host that gets attached to a subnet of the Internet needs to know two additional parameters besides its Internet address to be able to communicate with hosts outside the own subnet: the subnet mask and the address of a gateway that connects the subnet with other parts of the Internet. The earlier version [15] already envisioned two major ways for providing the subnet mask. One is *hardwired* (*a priori*) knowledge (such as reading from persistent configuration), the other is an extension to the ICMP protocol that allows for dynamic determination. This extension was then specified in the form of the *Address Mask Request/Response* message pair in [16]. Together with the *Information Request/Reply* [19] and the *Router Discovery* ICMP messages [9], the scene could be perfectly set for dynamic discovery of the three required communication interface parameters.

Today, the *Information Request/Reply* and the *Address Mask Request/Response* messages are deprecated [10] and the *Router Discovery* messages were never widely implemented [1]; the DHCP protocol [8] is used instead. In scenarios lacking services for dynamic configuration (like no DHCP server present or the DHCP server not willing to lease addresses to unknown hosts), procedures for automatic configuration [11] must be used. The IETF Zeroconf Working Group addressed automatic IP interface configuration as one of their requirements [23]. The working group defined how to obtain a so-called *link-local* IPv4 address [7]. But link-local addresses are not suitable for communication with devices not directly connected to the same physical (or logical) link; thus, they cannot be used for network mapping beyond that link.

If obtaining a routable address is difficult, why not just passively listen to network traffic? In [18], it is shown that a wealth of information can be gathered by observing mDNS messages. However, approaches like this require services that announce themselves and cannot reach beyond their vLAN on their own. Several drafts up to number 15 [1] of the standard to detect network attachment [2] suggested (in an appendix) to listen for the network traffic caused by several protocols to make an *educated guess* as to which network a device has moved to. Although used in a different scenario, the idea of listening passively for packets of relevant protocols points into the right direction.

3   Methodology

The implemented network reconnaissance procedure can be divided into three main phases. Within the first phase, passive scanner modules observe the network traffic and gather information about hosts in the network. The second phase consists of an analyzer module processing the results from the passive scanners and determining network ranges that will serve as input for subsequent active scanners, which form the third phase (that is outside this paper's scope). The ultimate scope is to black-box detect a network where the analyst, e.g., an auditor performing a security audit or a penetration-testing consultant discovers the given network without any *a priori* knowledge. The point of origin of the scan is, therefore, some point in the local area network (e.g., a given network port to plug in). Figure 1 shows an overview of this process. In case the network device configuration does not fit to any determined network range, the analyzer module will additionally determine all elements needed for a device reconfiguration before start of the active scanners. The described procedure was implemented as a module for the *Tactical Network Mapping* (*TNM*) plug-in framework. This framework provides functionality to generate a network graph with only a given target network range as input by augmenting data from a configurable toolchain consisting of existing scanning tools such as *nmap, amap, Dmitry, zmap*, and *snmpwalk* to gather... from a network, with genuine analytics that try to assess the network's structure [13]. These analytics work mainly through determining the parent (or gateway) of each host using *traceroute* information, operating system guessing information and looking for *usual* addresses (such as 0.1 or 0.254). Combination of passive scanning and network range determination presented in this paper enhances this framework to be independent of a target range input.

Fig. 1 Network
reconnaissance procedure

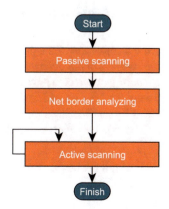

3.1 Passive Scanning Phase

The starting point is formed by a couple of parallel executing passive scanners that sniff the network traffic and gather information about communicating hosts. In order to be able to control the termination of the passive scanning, two parameters were introduced. The first parameter specifies a duration timeout and the second a threshold of detected hosts. After reaching one of these limits a network border analyzing module will be started. The execution of scan and analyzing modules occurs within the *TNM* framework. Therefore, plug-ins were implemented for some passive scanners that are available on Linux: *Netdiscover*, *P0f* and *CDPSnarf*. It became evident that the control possibilities and delivered host information were not sufficient for the need of net border analyzing. The reason is that *CDPSnarf* is restricted to messages via the *Cisco Discovery Protocol (CDP)* only (restricting the senders to networks devices and omitting hosts), *P0f* is dedicated to analyzing own active outbound and inbound connections (which will not happen with passive scanning only) and *Netdiscover* is restricted to ARP messages only (leaving out using IP packets). An additional reason is that none of these scanners yield information about the *time-to-live (TTL)* of the packets, which can be used to be compared against a list of standard TTL values in operating systems (e.g., [21]).[1] Hence, an own *Hostdiscover* module was implemented based on the *libpcap*-library. The *Hostdiscover* module is able to sniff ARP and IP packages. While the ARP approach is straightforward in the sense that it basically sets a *permit all ARP packets* filter to pcap, the IP method set an IP filter and additionally filters to let only packets of hop counts of 1, 64 or 128 through. These values are configurable but set as default, for most operating systems have according default TTL values [22], so that received non-filtered packets are likely to come from the same segment and, thus, network, which eventually poses the primary scanning target. For ARP, such a filter is not necessary, as the protocol is non-routable anyway [6]. Using these filters (capturing ARP, IP or both), passing network traffic is captured in promiscuous mode. As a result, the number of ARP requests and replies and the number of IP packages are stored together with the host address and can be accessed by subsequent analyzer modules.

3.2 Network Range Determination Algorithm

After termination of the passive scanning phase, the network border analyzing module starts with the determination of preliminary network ranges. Three variants for clustering of detected host addresses are implemented, one of which can be chosen by a setting an option within the software implementation. The first variant is the clustering by considering a maximum network size (again configurable, with a default of 256 hosts). Two hosts will fall into different clusters if the host address range would

---

[1]This way, it can be prevented to assume a remote network (with more than one hop away) as current network.

need a network size greater than a given value. Additionally, two networks will be separated, if the addresses span over private [20] or the dynamic link-local space [7]. The second variant makes the clustering against a presumed network prefix. The third variant tries to put all hosts into one big network, except the special ranges mentioned above. Each variant yields one or more clusters of network addresses ranging from a lowest one as starting point to a highest one as an end point determined by the lowest and highest observed addresses that fit into a given range. If, for instance, the lowest observed address is 192.168.0.2 and the highest is 192.168.1.17 with a maximum size of 256, the first variant would split the observed range into two clusters.

The process step compares the current network device configuration with the determined network ranges. If the configured IP address fits to any of the determined network ranges, the subsequent steps needed for reconfiguration of the network device will be skipped and the process will proceed with the definition of the final ranges.

If the network configuration does not fit (or no IP address is configured at all at the chosen scanner interface), the first step is to choose one preliminary range to assume as own network (which also determines the subnet mask). For this purpose, the all determined ranges are ordered by the following criteria:

1. Network type (globally reachable [5], private , or dynamic link-local);
2. The number of detected hosts for the range in descending order;
3. The respective starting address in ascending order.

The first element of the ordered network range list will be assumed as own network.

To be able to configure the network device for active scanning, a valid IP address is needed. The basis for the selection of an address is given by the list of detected host addresses for the previously chosen own network. The first undetected address that is located between the first and last address of detected hosts will be taken. If no address can be found within the host addresses, an address before the first or behind the last host will be taken. The overall procedure is aborted if no free address could be found.

Further, a default gateway address is determined, primarily by analyzing ARP statistics originating from modules of the starting passive scanning phase. As a first approach, the most often found sender address of ARP replies will be used. If this does not succeed, the most often found target address of ARP requests will be selected. If no ARP statistics are available very commonly used candidates for default gateway addresses are used, which is the first or last address of the determined network address range. Figure 2 illustrates the whole process.

After the found configuration is set to the network interface, the selection of final network ranges that will be used as input by active modules follows. This phase should give the possibility to expand or condense the preliminary determined ranges. The currently only available implementation at this stage is the adaption of network ranges based on the subnet mask of fitting device configurations.

Fig. 2 Network border
analyzing in detail

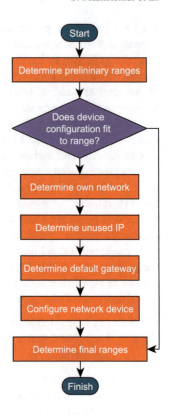

3.3   Active Scanning Phase

When the final range of the current network is determined, this estimated range or
ranges is used as target network for active scanning. Analogously, the found free
IP is configured as the host's IP address (see above), in order to obtain a valid one
to allow for scanning the network without prior knowledge. The active scanning
itself is conducted via an iterative toolchain, containing a variety of scanning tools
(such as *nmap*) and analytics modules (such as an algorithm to determine the default
gateway of hosts using traceroute, operating system and other information) and yields
a topology graph of the target network (i.e., the network, which the host currently
resides in). This process, including the gateway determination, is described in detail
in [13].

The system was tested in a live, productive network setting. During three test
runs, six different networks were found: network X containing the anonymized range
$\alpha.\beta.x.0/24$, network Y containing $\alpha.\beta.y.0/24$, network A containing $\alpha.\beta.a.0/24$,
network M containing $\alpha.\beta.m.0/24$, network N containing $\alpha.\beta.n.0/24$ and network
Q with the only known address being $\alpha.\beta.q.206$, whereby X and Y, as well as M and
N are direct neighbors, while A and Q have a range that is disjunct from the other
networks. Each test used a toolchain that consisted of the *Hostdiscover* module (see

Sect. 3.1), the network range determination (described in Sect. 3.2) and an *Nmap* scanner actively scanning the determined range plus a default gateway determination algorithm (see Sect. 3.3).

A first test run, with the scanning host residing in network X, used a threshold of 10 found hosts and 5 minutes time for the passive phase. After detection of 10 hosts, which resided in the networks X and A, the network range determination correctly identified the ranges for X and A that formed the target for the active phase and was able to discover 244 hosts by determining the network and running an active scan on it (using *Nmap*'s traceroute und OS detection options), 73 from network X and 172 from network A. Furthermore, this method automatically yields, next to the network structure, port scanning, and operating system detection results for further analysis (see Fig. 3 for an anonymized graph of the scan result produced by the *TNM* tool).

A second test with a threshold of 500 hosts and 5 minutes of time, with the scanning host inside network Y yielded in 93 passively detected hosts of which 1 is from network X, 51 are from network Y, and 41 are from network A. As with the first test run, the network ranges were determined correctly using the scanning host's own network mask (see Sect. 3.2), for its IP address fits in one of the network ranges. Knowing its own home network (Y), it is evident that X and Y are indeed two separate /24 networks and not a single /23 range. The active scanning over the

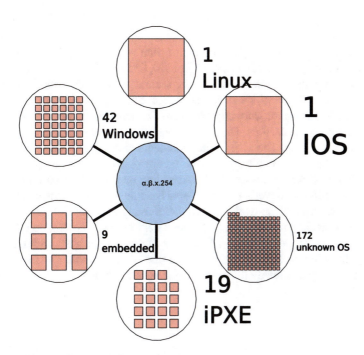

Fig. 3 Resulting topology graph from a test run

Fig. 4 Comparison graph
from another test run,
highlighting the hosts found
in active phase (green)

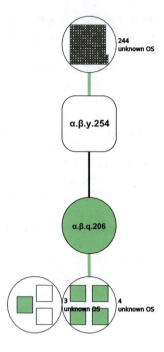

three networks X, Y, and A (for the sake of time without using *Nmap*'s operating
system detection) yielded 309 hosts; 75 from X, 63 from Y, and 171 from A.

A third run with the same setting as the second, found 105 hosts in the passive
phase; 50 from network Y, 54 from network A, and 1 from network M. The network range determination correctly identified the networks Y and A, but assumed
a combined network $M + N$. The separate nature latter, however, became evident
in the active scanning. The active phase yielded a total of 253^2 hosts: 63 from Y,
171 from A, 13 from M, 5 from N, and 1 from Q. The host from Q is a special
case, as it only occurs in traceroute data (residing on the path toward network N).
Through the different traceroute data, it also became evident that networks M and
N were in fact separate networks (the former directly adjacent to Y, the latter having an additional hop residing in Q). Figure 4 depicts the discovered topology with
a comparison graph out of the *TNM* tool that highlights the hosts additional hosts
found in the active phase in green, while the uncolored (white for nodes and black
for edges) ones were already found during the passive phase.

---

[2]3 of which were detected additionally through the still active-passive scanners (represented by the
disjunct bubble in Fig. 4). This indicates that the hosts were not online at the very moment of the
Nmap scan.

Table 1 Results of the test runs

Test run no.	1	2	3
Passive results	10	93	108
Number of networks	2	3	5
Active results (including passive)	244	309	253
Network X	73	75	–
Network Y	–	63	63
Network A	172	171	171
Network M	–	–	13
Network N	–	–	5
Network Q	–	–	1
Detection gain ratio	24.4	3.32	2.38

4 Results

The three test runs showed, apart from the knowledge gain regarding the network's structure, a significant increase in discovered hosts by the active phase, compared to simple passive sniffing alone (see Table 1).

5 Conclusion and Discussion

The work outlined in this paper has shown a path toward automating network scanning without any *a priori* knowledge. It demonstrated how a network range can determined automatically through network sniffing and automated analysis and be used as a target for an active scanning toolchain. It does so by sniffing the ARP and/or IP traffic on a given network segment and cluster the result into likely segments. Out of these, the algorithm tries to choose a free address and configure it automatically to the used network interface. This equips the scanner automatically with a valid IP address and a target (the discovered ranges) for active network scanning. From this point onward, it is possible to do automatic active scanning as described in a different paper [13]. The overall method showed a significant increase in device detection compared to sole passive sniffing methods, as well as the ability to retrieve topology and other information (such as device types, operating systems, etc.) automatically.

This work has also uncovered some shortcomings in detecting more segmented (sub)networks. This means that the current process is vital in the sense that a preliminary estimation on the basis of Sect. 3 is necessary, but might the final determination be enhanced by using an algorithm resembling a binary search [24] that compares traceroute information[3] information of portions[4] of the preliminary ranges, once they exist and the scanning hosts has acquired a valid IP address, refining the results' accuracy. This improvement is subject to continuative works. Furthermore, continuative research will elaborate approaches to handle IPv6 networks.

5.1 Improvements for Gateway Determination During the Active Phase

When a router's internal (that is non-outside) interface residing inside the target range is directly addressed, it yields the same hop count as the external interface. The rationale is that a router must reduce the initially set *time-to-live (TTL)* value of each packet and must discard it if its decreased to zero, except if is destined to the router itself, where it has to act as a host [4]. In this case, the TTL has no hop count function (except in the rare case of source routing usage) [6]. That means that the internal interface of a network, assuming that, like any other scanned, it is reachable from the outside, will display a hop count that is one less than the hop counts of the other hosts of the scanned network, when viewed from the outside. The missing address, representing this subtracted hop count, will naturally be the one of the outside interface of this very router.

That allows for determining the internal address of the connecting router closest to the target, which is probably the standard gateway.

Acknowledgements This work was partly supported by the Austrian Research Promotion Agency (FFG) within the *ICT of the future* grants program, grant nb. 863129 (project *IoT4CPS*), of the Federal Ministry for Transport, Innovation and Technology (BMVIT) and by the Federal Ministry of Defence (BMLV).

References

1. Aboba, B.: Detecting Network Attachment (DNA) in IPv4. RFC Draft, Internet Engineering Task Force. https://tools.ietf.org/html/draft-ietf-dhc-dna-ipv4-15 (2005)
2. Aboba, B., Carlson, J., Cheshire, S.: Detecting Network Attachment in IPv4 (DNAv4). RFC 4436, Internet Engineering Task Force (2006)
3. Adat, V., Gupta, B.B.: A DDoS attack mitigation framework for internet of things. In: 2017 International Conference on Communication and Signal Processing (ICCSP), pp. 2036–2041 (2017). https://doi.org/10.1109/ICCSP.2017.8286761
4. Baker, F.: Requirements for IP Version 4 Routers. RFC 1812, Internet Engineering Task Force (1995)

---

[3]For instance with *nmap -sn -Pn [network_portion]*. This ensures traceroutes to be carried out, even when no active host resides in the network to be examined.

[4]A possibility would be to compare the most distant addresses and, by binary splitting the set, keep comparing until the traceroutes are equal to yield actual subnetworks.

5. Bonica, R., Cotton, M., Haberman, B., Vegoda, L.: Updates to the Special-Purpose IP Address Registries. RFC 8190, Internet Engineering Task Force (2017)
6. Braden, R.: Requirements for Internet Hosts—Communication Layers. RFC 1122, Internet Engineering Task Force (1989)
7. Cheshire, S., Aboba, B., Guttman, E.: Dynamic Configuration of IPv4 Link-Local Addresses. RFC 3927, Internet Engineering Task Force (2005)
8. Droms, R.: Dynamic Host Configuration Protocol. RFC 2131, Internet Engineering Task Force (1997)
9. Eastlake, D.E. (eds.): ICMP Router Discovery Messages. RFC 1256, Internet Engineering Task Force (1991)
10. Gont, F., Pignataro, C.: Formally Deprecating Some ICMPv4 Message Types. RFC 6918, Internet Engineering Task Force (2013)
11. Guttman, E.: Zero configuration networking. In: INET 2000 Proceedings. Yokohama, Japan. https://www.isoc.org/inet2000/cdproceedings/3c/3c_3.htm (2000)
12. Marksteiner, S., Expósito Jiménez, V.J., Vallant, H., Zeiner, H.: An overview of wireless iot protocol security in the smart home domain. In: Proceedings of 2017 Internet of Things Business Models, Users, and Networks Conference (CTTE), pp. 1–8. IEEE, New York, NY, USA (2017). https://doi.org/10.1109/CTTE.2017.8260940
13. Marksteiner, S., Lernbeiß, H., Jandl-Scherf, B.: An iterative and toolchain-based approach to automate scanning and mapping computer networks. In: Proceedings of the 2016 ACM Workshop on Automated Decision Making for Active Cyber Defense, SafeConfig'16, pp. 37–43. ACM, New York, NY, USA (2016). https://doi.org/10.1145/2994475.2994479. https://doi.org/10.1145/2994475.2994479
14. Mirkovic, J., Reiher, P.: A taxonomy of DDoS attack and DDoS defense mechanisms. ACM SIGCOMM Comput. Commun. Rev. **34**(2), 39–53 (2004)
15. Mogul, J.: Internet Subnets. RFC 917, Internet Engineering Task Force (1984)
16. Mogul, J., Postel, J.: Internet Standard Subnetting Procedure. RFC 950, Internet Engineering Task Force (1985)
17. Muelder, C., Ma, K.L., Bartoletti, T.: Interactive visualization for network and port scan detection. In: Recent Advances in Intrusion Detection, pp. 265–283. Springer (2005)
18. Pickett, G.: Port scanning without sending packets. Presentation at DEF CON 19. https://defcon.org/images/defcon-19/dc-19-presentations/Pickett/DEFCON-19-Pickett-Port-Scanning-Without-Packets.pdf (2011)
19. Postel, J.: Internet Control Message Protocol. RFC 792, Internet Engineering Task Force (1981)
20. Rekhter, Y., Moskowitz, B., Karrenberg, D., de Groot, G.J., Lear, E.: Address Allocation for Private Internets. RFC 1918, Internet Engineering Task Force (1996)
21. Siby, S.: Default TTL (time to live) values of different OS. http://subinsb.com/default-device-ttl-values/ (2014). Retrieved 06 Nov 2018
22. Straka, K., Manes, G.: Passive detection of nat routers and client counting. In: Olivier, M.S., Shenoi, S. (eds.) Advances in Digital Forensics II, pp. 239–246. Springer, US, Boston, MA (2006)
23. Williams, A.: Requirements for Automatic Configuration of IP Hosts. Internet-Draft, IETF Zeroconf Working Group. http://files.zeroconf.org/draft-ietf-zeroconf-reqts-12.txt (2002)
24. Williams L.F., Jr.: A modification to the half-interval search (binary search) method. In: Proceedings of the 14th Annual Southeast Regional Conference, ACM-SE 14, pp. 95–101. ACM, New York, NY, USA. https://doi.org/10.1145/503561.503582 (1976). https://doi.org/10.1145/503561.503582

Using a Temporal-Causal Network Model for Computational Analysis of the Effect of Social Media Influencers on the Worldwide Interest in Veganism

Manon Lisa Sijm, Chelsea Rome Exel and Jan Treur

Abstract Over the years, a clear and steady rise can be seen in the interest in veganism. Although research has been conducted to determine the reasons why veganism has grown, ultimately there is still a necessity for further research on how social networks affect its growth. This paper aims to provide a possible explanation for the rise in interest, using computational analysis based on a temporal-causal network model focussing on social contagion. This model portrays a simulation of a sample size population on Instagram, showing how a social influencer can influence the opinions of people directly (influencers' followers) and indirectly (followers of the influencers' followers), and how this compares to a situation in which this influencer is not there.

Keywords Social contagion · Social media · Veganism · Network-oriental modelling approach · Temporal-causal network

1 Introduction

Over the last years, the number of vegans has noticeably increased. From 2014 to 2018, this number has quadrupled in the UK [13]. While there has been research studies towards the reasons why people became vegan, it has not yet been established how exactly this rise came to be. The main reasons why people become vegan are health and ethical reasons [11]. McDonald [9] attempted to research how people learn to become vegetarian or vegan, by conducting a qualitative study. According to this study, most participants already felt affection for nonhuman animals prior to

M. L. Sijm · C. R. Exel (✉) · J. Treur
Behavioural Informatics Group, Vrije Universiteit Amsterdam, Amsterdam,
The Netherlands
e-mail: chelseaexel@gmail.com

M. L. Sijm
e-mail: manonsijm@hotmail.com

J. Treur
e-mail: j.treur@vu.nl

© Springer Nature Singapore Pte Ltd. 2020
X.-S. Yang et al. (eds.), *Fourth International Congress on Information and Communication Technology*, Advances in Intelligent Systems and Computing 1027, https://doi.org/10.1007/978-981-32-9343-4_12

129

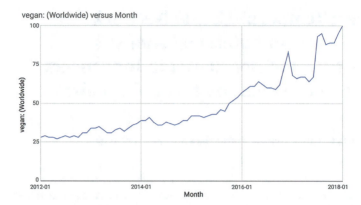

vegan: (Worldwide) versus Month

Fig. 1 A Google search regarding the interest in vegan spread out over a period of six years, where 100% equals the highest number of searches for this term

becoming vegan, but they became vegan after experiencing one or more catalytic experiences. These experiences involved information about animal cruelty that was presented to the participant, which led to further action. After learning more about animal cruelty, participants eventually made the decision to give up animal products in their entirety. McDonald argued that openness and the willingness to learn were salient factors into the decision of becoming vegan. After becoming vegan, the participants in this study stated that their vegan lifestyle included the desire of educating others about animal cruelty [9]. Even though this research provides more insights into how people become vegan, ultimately it does not explain the substantial rise of interest in veganism in the last 10 years. As Fig. 1 shows, in 2018 the interest in 'vegan' shows a monotonically increasing trend: it has been strongly increasing since 2012. The assumption is that people are becoming more aware and learning more about veganism, which could contribute to the rise in interest of it entirely.

With the rise of the Internet and social media, people have obtained more access to all types of information compared to twenty years ago. Instagram, for example, has been a popular platform to show and sell products to people, proving to have an impact on buyers. Posting a picture next to a sales item appears to boost the sale conversions with a factor of seven [18]. This impact is not only true for sales but also appears to work for lifestyle changes. Nine out of ten experiments conducted by Maher et al. [8] showed significant improvements in health behaviour, and it is argued that behaviour changed because of social network sites. Vaterlaus [17] confirmed this, by showing that at least 38% of participants showed that their food choices were influenced by social media. This provided more support for the social ecological model, which indicates that several factors, including social media, appear to have an influence on health behaviours [5]. Social influencers especially appear to contribute to the fact that social media seem to have an impact on health behaviours. Social influencers are perceived as more appealing because they are arguably considered more popular than others. This perception of popularity even increased the perceived opinion leadership of some influencers [4]. This effect was also the case for pictures.

Pictures with more 'likes' were liked even more by participants than other pictures. This was also detectable within the neural responses in the brain, namely the nucleus accumbens, where popular pictures showed a greater response in this area [12]. This could possibly be explained by a persuasion principle of Cialdini; the number of 'likes' on a picture provides the opportunity of social proof, which means 'when a lot of people are doing something, it is the right thing to do' [2, 3].

2 The Temporal-Causal Network Model

According to the above-mentioned research, it appears that social media and social influencers can affect the lifestyle and buying behaviours of their followers. This can be categorized under social contagion, which can be explained as the spread of belief, affect or behaviour, where people influence on another, e.g. [1]. To simulate and analyse this computationally, the network-oriented modelling approach, presented in [15], was used; see also [16]. This approach can be considered as a branch in the causal modelling area which has a long tradition in AI; e.g., see [6, 7, 10]. It distinguishes itself by exerting a dynamic perspective on causal relations, according to which causal relations manifest effects over time. These causal relations themselves can also change over time. The type of network models that are used as a basis for this is called a *temporal-causal network model*. These network models are widely applicable, varying from biological and mental networks to social networks and beyond [16]. This also includes the social contagion principle.

To analyse the effect of an influencer on the spread of veganism computationally, an agent-based social contagion model was designed. The model consists of different nodes (agents) interacting with each other. The nodes only interact with the nodes that they are directly connected to. This connection can be bidirectional, which means that a node can express their opinion to a connected node, but also receive the opinion from that same node. On the other hand, it can be unidirectional, where an opinion will either be received or expressed. The nodes can still be influenced indirectly via some intermediate steps by other nodes that they are not connected to. The opinion that is communicated in this particular case will regard the nodes' attitudes towards veganism. These opinions can differ in weight, where 0 is the lowest weight a node could have and 1 the highest. In this case, a value of 0 would mean no interest at all, whereas a value of 1 would mean high interest in veganism. Each node has an activation value, which varies over time. Based on the temporal-causal network that has been defined, this activation value depends on the interaction between the agents according to the following three elements: **connection weight** $\omega_{X,Y}$, which represents the strength of the connection from state X to state Y, **speed factor** η_Y, which represents how fast state Y is changing upon causal impact and **combination function** $c_Y(..)$, which combines the causal impacts of other states on state Y. As there is not much specifically known about how specific agents are linked to an influencer, a scale-free network approach based on Tapan [14] was used to represent the population. This is a connected graph, where the majority

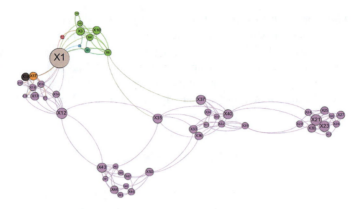

Fig. 2 Conceptual representation of the temporal-causal network model

of the nodes have one or two connections and only a few nodes have a plethora of connections. A sample size of 50 nodes has been chosen, where one node was chosen as the influencer (X_1). This influencer represents a popular person on social media who is actively posting about vegan. While looking at the real world, it is evident that not everyone is directly connected to each other. Therefore, this model divided all nodes into five subgroups, representing different clusters of the population. Subsequently, it is unlikely that all clusters in a population would be influenced by the same influencer(s). For this reason, the influencing node X_1 only impacts the first two clusters. This way, indirect effects are also presentable in the model. The conceptual representation of this model can be seen in Fig. 2, where the size of the nodes represents the number of outgoing nodes (influence) and the colours represent strongly connected nodes. The conceptual representation of a temporal-causal network model can be transformed into a numerical representation as shown in Table 1; see also [15, 16].

The following *difference* and *differential equation* for each state Y are obtained:

$$Y(t + \Delta t) = Y(t) + \eta_Y[\mathbf{c}_Y(\omega_{X_1,Y}X_1(t), \ldots, \omega_{X_k,Y}X_k(t)) - Y(t)]\Delta t$$
$$\mathbf{d}Y(t)/\mathbf{d}t = \eta_Y[\mathbf{c}_Y(\omega_{X_1,Y}X_1(t), \ldots, \omega_{X_k,Y}X_k(t)) - Y(t)] \tag{1}$$

The combination functions that are used to obtain a realistic simulation for the influence of a social media influencer on the overall interest in veganism are the *identity function* **id**(.) for states with a single impact and the *advanced logistic function* **alogistic**$_{\sigma,\tau}$(..) for states with multiple impacts, where σ is a parameter for steepness and τ a parameter for threshold.

$$\mathbf{id}(V) = V$$

$$\mathbf{alogistic}_{\sigma,\tau}(V_1, \ldots, V_k) = \left[\frac{1}{1 + e^{-\sigma(V_1 + \cdots + V_{k-\tau})}} - \frac{1}{1 + e^{\sigma\tau}}\right](1 + e^{-\sigma\tau})$$

$$\tag{2}$$

Table 1 From conceptual representation to numerical representation of a temporal-causal network model; adopted from Treur [16]

Concept	Representation	Explanation
State values over time t	$Y(t)$	At each time point t, each state Y in the model has a real number value in [0, 1]
Single causal impact	$\mathbf{impact}_{X,Y}(t)$ $= \omega_{X,Y} X(t)$	At t state X with connection to state Y has an impact on Y, using connection weight $\omega_{X,Y}$
Aggregating multiple impacts	$\mathbf{aggimpact}_Y(t)$ $= \mathbf{c}_Y \left(\mathbf{impact}_{X_1,Y}(t), \ldots, \mathbf{impact}_{X_k,Y}(t) \right)$ $= \mathbf{c}_Y \left(\omega_{X_1,Y} X_1(t), \ldots, \omega_{X_k,Y} X_k(t) \right)$	The aggregated causal impact of multiple states X_i on Y at t is determined using combination function $\mathbf{c}_Y(..)$
Timing of the causal effect	$Y(t + \Delta t) = Y(t) + \eta_Y \left[\mathbf{aggimpact}_Y(t) - Y(t) \right] \Delta t$ $= Y(t) + \eta_Y \left[\mathbf{c}_Y (\omega_{X_1,Y} X_1(t), \ldots, \omega_{X_k,Y} X_k(t)) - Y(t) \right] \Delta t$	The causal impact on Y is exerted over time gradually, using speed factor η_Y; here, the X_i are all states with connections to state Y

An example of a numerical representation for agent state X_{14} in difference and differential equation format is, respectively:

$$X_{14}(t + \Delta t) = X_{14}(t) + \eta_{X_{14}}[\mathbf{c}_{X_{14}}(\omega_{X_1,X_{14}} X_1(t), \omega_{X_{12},X_{14}} X_{12}(t)) - X_{14}(t)]\Delta t$$
$$\mathbf{d}X_{14}(t)/\mathbf{d}t = \eta_{X_{14}}[\mathbf{c}_{X_{14}}(\omega_{X_1,X_{14}} X_1(t), \omega_{X_{12},X_{14}} X_{12}(t)) - X_{14}(t)] \tag{3}$$

with $\mathbf{c}_{X_{14}}(..) = \mathbf{alogistic}_{100,1}(..)$.

3   Simulation Results

The influence of a social media influencer on the overall interest in veganism of a population is analysed by using two scenarios. For each scenario, the timescale of the model is *time* 70 = 1 *year* with *time* 0 = 2012 and $\Delta t = 1$. Based on Google search data, we know that the interest in veganism in January 2012 was only 29% of the population compared to the interest six years later in 2018 (100%). Therefore, 15 people (14.5 rounded up) of the 50 people in this network are starting with an interest in veganism at *time* = 0 (2012) and are given an initial state value of $X_i(0)$ = 1. The remaining people (35) are not interested in veganism at *time* = 0 and are given an initial value of $X_i(0) = 0$. The connection weights in this network are either 0 (state X is not influencing state Y directly), 1 (state X has a direct positive effect on state Y) or -1 (state X has a direct negative effect on state Y). The speed factor η_Y differs for each agent and varies between 0 and 0.1. For each agent Y with a single incoming connection, the *identity combination function* is used to estimate the activation value. For each agent Y with multiple incoming connections, given the activation values at time t, the *advanced logistic combination function* is used to calculate the activation value at time $t + \Delta t$ with *steepness* (σ) between 10 and 100, and *threshold* (τ) between 0 and 1.

Scenario 1 models the development when the influence of a social media influencer on the population is present. In this scenario, agent state X_1 with an *initial value* of 1 has 13 outgoing connections and 0 incoming connections and is called an 'influencer'. The rest of the states have a maximum of 6 outgoing influences. All states together have an average of 3.32 outgoing influences. The other settings are constructed according to the description above. The simulation results of the model can be found on the left side in Fig. 3. In the simulation of the first scenario, it is apparent that the interest in veganism grows over time. After 6 years (*time* = 420), every agent state is interested in veganism, but some agent states rise faster and sooner than others. This proves to be a realistic process due to personal differences and the strength of connections between agents.

To get a better view of the overall interest in veganism, the average is derived from the simulation and the result can be seen on the left side in Fig. 4. This shows that the average interest in veganism increases from 30 to almost 100% in 6 years. These results are in line with the expectations we had. If we compare this trend to

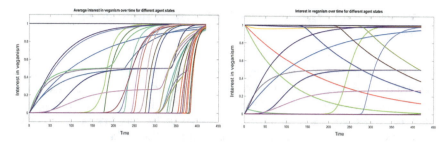

Fig. 3 Simulation of the temporal-causal network model for scenarios 1 (left, with influencer) and 2 (right, without influencer), where each line represents one agent state

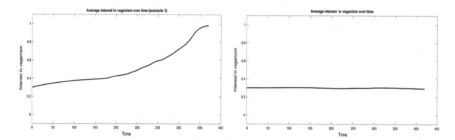

Fig. 4 Average interest in veganism of all agents combined over time for scenarios 1 (left, with influencer) and 2 (right, without influencer)

the worldwide interest in veganism derived from the Google search data and the literature, we can perceive approximately the same pattern.

Scenario 2 models the same process (with the same settings), but without the influencer X_1. This means that the outgoing connections of X_1 are 0 to each state Y. The simulation results of this adjusted model can be found in Fig. 3 on the right. Looking at the simulation results of the second scenario, we see that over six years, most of the people stay either interested in veganism or not interested in veganism. Only seven people that were initially uninterested in veganism became more interested over time. There were also seven people that became uninterested in veganism after previously harbouring interest. The reduced interest in veganism of some people arises since they are no longer influenced by the influencer, which leaves room for an increased influence of other (maybe) non-vegan people in their network. To obtain a better view of the overall interest in veganism, the average is derived from the simulation and the result can be seen on the right in Fig. 4. This shows that the interest in veganism is stationary over the time when the network is not influenced by the vegan influencer. If we compare this trend with the trend we have gathered in the first scenario, the influential power of a social influencer becomes visible. These findings are in accordance with our expectations since we expected the influencer to have a substantial influence on the overall interest due to social contagion and social proof.

4   Verification of the Network Model by Mathematical Analysis

In order to verify whether the implemented model does what is expected from the model specification, a mathematical analysis of stationary points was carried out. A state Y has a *stationary point* at some time point t if $dY(t)/dt = 0$. For temporal-causal networks, there is a simple criterion to check whether there is a stationary point at t for state Y: a state Y in a temporal-causal network has a stationary point at time point t only if the speed factor of Y is 0 or $\mathbf{aggimpact}_Y(t) = Y(t)$, where $\mathbf{aggimpact}_Y(t) = \mathbf{c}_Y(\omega_{X_1,Y} X_1(t), \ldots, \omega_{X_k,Y} X_k(t))$ (with $X_1, \ldots, X_k$, which are the states with outgoing connections to Y).

From scenario 1 shown in Sect. 3, eight stationary points with their time points t and their state values $X_i(t)$ were identified. To verify the model, these state values were compared to the values at the same time point t calculated using the right side of the equation $Y(t) = \mathbf{c}_Y(\omega_{X_1,Y} X_1(t), \ldots, \omega_{X_k,Y} X_k(t))$ in the above criterion. To explain the mathematical analysis, the observed state value of agent state X_{22} at a stationary point at $t = 324$ is compared to the value $\mathbf{aggimpact}_Y(t)$ expressed in the equation above. The agent state X_{22} has incoming connections of X_{21}, X_{23} and X_{26}. The equation for the agent state X_{22} at $t = 324$ with combination function *advanced logistic function* with $\sigma = 20$ and $\tau = 0.5$ is as follows: $\mathbf{aggimpact}_Y(t) = \mathbf{alogistic}_{\sigma,\tau}(V_1, \ldots, V_k)$ with $V_i = \mathbf{impact}_{X,Y}(t) = \omega_{X,Y} X(t)$. Here for the case of agent state X_{22} it holds

$$V_1 = \mathbf{impact}_{X_{21},X_{22}}(324) = 1 \times 1 = 1$$
$$V_2 = \mathbf{impact}_{X_{23},X_{22}}(324) = -1 \times 1 = -1$$
$$V_3 = \mathbf{impact}_{X_{26},X_{22}}(324) = 1 \times 0.5 = 0.5$$

Then

$$\mathbf{aggimpact}_{X_{22}}(324) = \mathbf{alogistic}_{20,0.5}(V_1, V_2, V_3)$$

$$= \left[\frac{1}{1 + e^{-20}((1 - 1 + 0.5) - 0.5)} - \frac{1}{1 + e^{20*0.5}} \right](1 + e^{-20*0.5})$$

$$= 0.500 \tag{4}$$

The difference between the simulation value for state Y and the value $\mathbf{aggimpact}_Y(t)$ is called the *deviation*, and this portrays the accuracy of the model. If we compare this state value 0.500 for X_{22} at $t = 324$ with the value of $\mathbf{aggimpact}_Y(t)$ derived from the other state values at $t = 324$ in the simulation, which is 0.449, the deviation is $0.500 - 0.449 = 0.001$. The state values found in the simulation and the equations for $\mathbf{aggimpact}_Y(t)$ for X_{22} and other agent states with a stationary point can be found in Table 2. The stationary point equations all contain an accuracy <0.01, which contributes to confidence that the model was implemented in a correct manner.

Table 2 Stationary point equation outcomes

State Y_i	X_2	X_{16}	X_{20}	X_{22}	X_{24}	X_{26}	X_{28}	X_{42}
Time point t	406	401	152	324	283	192	397	411
State value $Y_i(t)$	1.000	0.996	0.495	0.499	0.262	0.496	0.996	0.999
aggimpact$_{Y_i}(t)$	1.000	1.000	0.500	0.500	0.267	0.500	1.000	1.000
Deviation	0.000	0.004	0.005	0.001	0.005	0.004	0.004	0.001

5 Validation Using Empirical Data and Parameter Tuning

Lastly, comparing the model to the empirical data provided a validation of the model. This empirical data were retrieved from the Google search data mentioned in the beginning. To create an accurate model, time points were compared to the empirical data and the model was adjusted by tuning the parameters. This was achieved by comparing the average state of the agents in the model that was retrieved from the output created in MATLAB with the data provided by the Google search. The number of time points used in the proposed model was $n = 420$ (420 iterations with $\Delta t = 1$). Since the time point dimensions did not match yet with each other, the empirical data were converted into the same number of time points as the proposed model has. This provided a basis for a detailed and uniform way of comparison.

First, the empirical data were scaled to the dimensions of the simulation model (0–420). This means that the initial and final points from the empirical data, which were 2012-01 and 2018-01, were changed into 0 and 420, respectively. Each month that lies between 2012 and 2018 is converted to the scale of the simulation model by the formula: **timepoint**$(t) =$ **timepoint**$(t - 1) + 5.83$, where t is a specific time point from the empirical data. The data for the time points that were not known yet were estimated by interpolation using a third-order polynomial trend line, which provided a formula to calculate the value for each time point of the model. This third-degree polynomial formula was as follows (with $t =$ **timepoint**):

$$(0.000001\, t^3 - 0.0002\, t^2 + 0.0718\, t + 27.048)/100 \qquad (5)$$

Following this, the proposed model was compared to the empirical data, which was done in MATLAB. A sum of squares error SSR of 0.031953 was computed, which leads to root mean square, RMS = SQRT (0.031953/420) = 0.0087. These results indicate that the differences between the simulated data and the empirical data are quite small. However, there are still some differences, which can also be seen on the left in Fig. 5; in particular everywhere in the time interval the average of the simulation values is higher than the value of the empirical data, which indicates that there is room for improvement.

To improve the model and decrease the error, parameter tuning was used to find more optimal speed factor values. The speed factor of the average state X_{51} was 1.722 in the proposed model, which is based on the total speed factor values for all

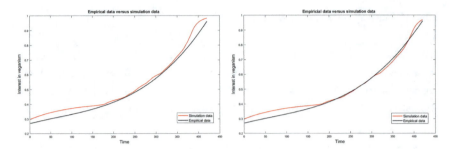

Fig. 5 Empirical data compared to the data of the proposed model before (left) and after (right) parameter tuning

Table 3 Results of parameter tuning by exhaustive search

$\eta_{X_{51}}$	1.722	1.700	1.682	1.665	1.648	1.631
SSR	0.0320	0.0279	0.0250	0.0240	0.0251	0.0282
RMS	0.0087	0.0082	0.0077	0.0076	0.0077	0.0081

states. An exhaustive search was used to find the speed factors that best represent the empirical data, which was executed by lowering the speed factor step by step for each state with 1% of its value at a time. The total speed factors $\eta_{X_{51}}$ and their SSR and RMS values for the different options in the search space can be found in Table 3. Table 3 shows that a total speed factor of 1.665 provides the lowest sum of squares error (0.0240) and root mean square (0.0076). This means that when the original value of each speed factor is multiplied by 0.97, this results in the best speed factor for the proposed model. The right side in Fig. 5 shows the proposed model after parameter tuning.

6 Discussion

In this paper, a temporal-causal network model concerning the influence of social media influencers on the overall interest in veganism is introduced. The model uses the network-oriented modelling approach described in [15, 16] and is based on the principle of social contagion and findings in the literature regarding social media and veganism. To verify the model, a mathematical analysis has been performed. To validate the model, parameter tuning has been performed by comparing it with empirical data.

   The computational analysis presented in this paper provides more insights into possible reasons why people became vegan and proposed that this could be due to the rise of social media and vegan social media influencers. There are, however, some limitations that need to be taken into account while interpreting this model. The quantitative data that were found to validate the model is the amount of Google searches on the word 'vegan' over time, interpreted as the interest in veganism over time. There was no quantitative data available concerning the influence of social media influencers on their followers, or information on the composition of the network of an influencer. Therefore, this model represents a possible way of how the growing interest in veganism could have arisen, but it should be noted that this is one of many possible ways, maybe including ways where social media does not play an important role. Further research may be needed to determine in more detail the extent the influence social media influencers have on their followers, and particularly about the relationships among vegans on social media.

References

1. Bosse, T., Hoogendoorn, M., Klein, M.C.A., Treur, J., van der Wal, C.N., van Wissen, A.: Modelling collective decision making in groups and crowds: integrating social contagion and interacting emotions, beliefs and intentions. Auton. Agent. Multi-Agent Syst. **27**(1), 52–84 (2013)
2. Cialdini, R.B.: Influence: Science and Practice. Pearson, Boston (2009)
3. Cialdini, R.B.: Influence, vol. 3. A. Michel, Port Harcourt (1987)
4. De Veirman, M., Cauberghe, V., Hudders, L.: Marketing through Instagram influencers: the impact of number of followers and product divergence on brand attitude. Int. J. Advert. **36**(5), 798–828 (2017)
5. Freeland-Graves, J.H., Nitzke, S.: Position of the academy of nutrition and dietetics: total diet approach to healthy eating. J. Acad. Nutr. Diet. **113**(2), 307–317 (2013)
6. Kuipers, B.J.: Commonsense reasoning about causality: deriving behavior from structure. Artif. Intell. **24**, 169–203 (1984)
7. Kuipers, B.J., Kassirer, J.P.: How to discover a knowledge representation for causal reasoning by studying an expert physician. In: Proceedings Eighth International Joint Conference on Artificial Intelligence, IJCAI'83. William Kaufman, Los Altos, CA (1983)
8. Maher, C.A., Lewis, L.K., Ferrar, K., Marshall, S., De Bourdeaudhuij, I., Vandelanotte, C.: Are health behavior change interventions that use online social networks effective? A systematic review. J. Med. Internet Res. **16**(2), e40 (2014)
9. McDonald, B.: "Once you know something, you can't not know it": an empirical look at becoming vegan. Soc. Anim. **8**(1), 1–23 (2000)
10. Pearl, J.: Causality. Cambridge University Press (2000)
11. Radnitz, C., Beezhold, B., DiMatteo, J.: Investigation of lifestyle choices of individuals following a vegan diet for health and ethical reasons. Appetite **90**, 31–36 (2015)
12. Sherman, L.E., Greenfield, P.M., Hernandez, L.M., Dapretto, M.: Peer influence via Instagram: effects on brain and behavior in adolescence and young adulthood. Child Dev. **89**(1), 37–47 (2018)
13. Statistics (2018). https://www.vegansociety.com/news/media/statistics
14. Tapan: Scale-Free Network Using B-A Algorithm. https://nl.mathworks.com/matlabcentral/fileexchange/49356-scale-free-network-using-b-a-algorithm. Accessed 28 Jan 2015
15. Treur, J.: Dynamic modeling based on a temporal-causal network modeling approach. Biol. Inspired Cognit. Archit. **16**, 131–168 (2016)
16. Treur, J.: The ins and outs of network-oriented modeling: from biological networks and mental networks to social networks and beyond. Trans. Comput. Collect. Intell. **32**, 120–139 (2019). Keynote Lecture at the 10th International Conference on Computational Collective Intelligence, ICCCI'18

17. Vaterlaus, J.M., Patten, E.V., Roche, C., Young, J.A.: #Gettinghealthy: the perceived influence of social media on young adult health behaviors. Comput. Hum. Behav. **45**, 151–157 (2015)
18. Zaryouni, H.: Instagram Solves the Commerce Problem. L2 Daily. https://www.l2inc.com/instagram-solves-the-commerce-problem/2015/blog. Accessed 4 Mar 2015

Detecting Drivers' Fatigue in Different Conditions Using Real-Time Non-intrusive System

Ann Nosseir⊙, Ahmed Hamad⊙ and Abdelrahman Wahdan

Abstract Driver's fatigue causes fatal road crashes and disrupts transportation systems. Specially in developing countries, drivers take more working hours and drive longer distances with short breaks to gain more money. This paper develops a new real time, low cost, and non-intrusive system that detects the features of the drivers' fatigue. More specifically, the system detects fatigue from the eye closer and yawning. First, the face landmarks are extracted using the histogram of oriented gradients (HOG). Then, the support vector machine (SVM) model classifies the fatigue state from the non-fatigue. The accuracy of the SVM model presented by the area under the curve (AUC) is 95%. The system is evaluated with 10 participants in conditions that can affect the detection of the face. These conditions are different light conditions, gender, age groups, people wearing reading glasses, and males with beard and moustache around their mouth. The results are very promising, and it is 100% accurate.

Keywords Driver fatigue · Driver face detection · Image processing · Blinks · Eyeglasses · Eyelid · Yawning

1 Introduction

Ten to thirty per cent of the car crashes are due to drivers' fatigue [1]. In Egypt, one of the main reasons of trucks' accidents is drivers' fatigue. Truck drivers drive

A. Nosseir (✉) · A. Hamad · A. Wahdan
British University in Egypt, Cairo, Egypt
e-mail: ann.nosseir@bue.edu.eg

A. Hamad
e-mail: ahmed.hamad@bue.edu.eg

A. Wahdan
e-mail: abdelrahman122203@bue.edu.eg

A. Nosseir · A. Wahdan
Institute of National Planning, Cairo, Egypt

© Springer Nature Singapore Pte Ltd. 2020
X.-S. Yang et al. (eds.), *Fourth International Congress on Information and Communication Technology*, Advances in Intelligent Systems and Computing 1027, https://doi.org/10.1007/978-981-32-9343-4_13

continuously for long hours [2] and because of the lack of sleep, they lose track of the road while driving. Unfortunately, there is no clear approach for the policy to control this behaviour rather than the control points in different places. There is also an absence of definitive criteria for establishing the level of fatigue [2].

Recently, there are embedded systems in cars to assess the drivers' fatigue automatically. Lexus Toyota [3] car has a driver monitor assist safety feature in. The car has an infrared camera attached in the steering wheel. Its beams monitor the drivers' eye attention. This is limited to only one car brand and it is not in trucks [3].

They are attachable products to the car that detect the size and direction of the pupils such as MR688 [4] and RVS-350 [5]. These products are quite expensive, and the quality of these systems was not tested with the public yet. The challenge is to develop a low-cost real-time system that detects the drivers' fatigue features. Having more than one measure such as eye closure and yawning can increase the accuracy of the system. Additionally, it needs to work in different light conditions and recognizes features of different faces for different age groups. It can as well identify the eyes of the drivers who are wearing reading glasses and the mouth even if it is surrounded with hair of a beard and a moustache.

This paper presents a non-intrusive system that utilizes the image processing techniques for video processing. The system tracks the driver's current state. It first recognizes the driver face features and then infers the symptoms of fatigue and classifies the fatigue state from non-fatigue using SVM. The paper starts by discussing the related work that detects fatigue drivers. This is followed by the proposed system and its evaluation in different conditions. It ends by discussing the results and conclusions.

2 Related Work

2.1 Current Systems

The increase in the number of accidents [1] raises the importance of developing systems that detect drivers' fatigue. Work in this area uses either intrusive or non-intrusive devices or a hybrid prototype of both categories to increase the accuracy. The former uses sensors attached to the human body to detect physiological fatigue signals from the breathing rate, heart rate, brain activity, muscles, body temperature, and eye movement. 'The physiological signals start to change in earlier stages of drowsiness' [6].

Devices such as electroencephalogram (EEG), electrocardiogram (ECG), and electrooculogram (EOG) are attached to the body that reads these signals. 'Electrocardiogram (ECG) signals of the heart vary significantly between the different stages of drowsiness such as alertness and fatigue' [7]. Electroencephalography (EEG) signals of the brain waves are categorized as delta, theta, alpha, beta, and gamma. A decrease in the alpha frequency band and an increase in the theta frequency band indicate drowsiness [6, 7].

'Electrooculography (EOG) detects the electric difference between the cornea and the retina that reflects the orientation of the eyes. It identifies the rapid eye movements (REM) which occur when a subject is in a drowsy state' [6]. These systems need to be attached to the human body to get reading, and this is main challenge to implement these systems. It causes discomfort to the drivers, especially if they are attached for a long time.

The later uses devices or sensors that are attached to the car. For example, a camera captures eyes' closer, head nodding, head orientation, and yawning of the drivers [7], or sensors are attached on the side of the car to detect drivers' deviation from the driving lane [7].

Some of these systems are embedded in cars. For example, Ford [8], Volkswagen [9], and BMW [10] have an alert system that works based on the driving behaviour. Ford, for instance, has sensors attached on both sides, i.e. right and left of the car to detect road lanes. It alerts the driver if there is a sudden change and a diversion from the current lane. Volkswagen system senses sudden steering to other lanes and how fast drivers steer. BMW system has more features like forward collision warning, pedestrian warning, and city collision mitigation. These systems are limited for certain car brands and types. They are not widely used yet to assess their accuracy and performance.

2.2 Image Processing Techniques

Research works on improving the accuracy of these systems. In the domain of image processing, more efforts are exerted to improve detecting the facial features of fatigue and selecting the best classifiers to differentiate between fatigue and non-fatigue states. A learning-based algorithms such as Haar-like [11], Eigen-/Fisher face [12], LBP map [13], Gabor-based template [14], HOG [15] are implemented to extract the eyes or the mouth.

Based on the algorithm used to extract the eyes and the mouth, an algorithm is developed to identify the eye closure or mouth yawning. Percentage of eyelid closure (PERCLOS) is one of algorithms used to count the eye closure. 'It counts the number of video frames in which there was no eye pupil detected and dividing this by the total number of frames for a specific time interval' [16].

The primary difficulties of real-time estimation of PERCLOS are the algorithm that should provide good accuracy with different lighting conditions, changing backgrounds of the image, and camera shake due to the car motion, for instance. Then, a classifier such as support vector machine (SVM), convolutional neural network (CNN), artificial neural network (ANN), or nearest neighbour is selected to differentiate between two states of fatigue or not.

The selection of the right learning-based algorithm is not quite easy. Each one has its pros and cons. For example, Haar-like features [11] recognize the darker regions than the skins. That is why it cannot be implemented if the face is not straight or with a change of a light conditions. It can miss the eyes or the mouth regions.

The histogram of oriented gradients (HOG) [15] uses a feature descriptor to extract certain feature and ignores other unimportant information. According to Dalal and Triggs [17], in their research using HOG for human detection, HOG detected human with minimum false-positive results.

The accuracy of these algorithms was tested with either images of faces that have different states of fatigue or with videos. A few were tested with real-time streaming videos. The evaluations as well were done in controlled environment like labs with driving cars' simulators. A few have been tested outdoors.

To gain more confident of the techniques, the accuracy of these techniques needs to be tested in different conditions such as in different lighting conditions of day or night and with subjects wearing reading glasses. These conditions affect the accuracy of extracting the face features and the measures of eye blinks. The mouth surrounded with hair of a beard and a moustache can influence the accuracy of measuring yawning.

This work develops a system that detects the face features, extracts a fatigue measures, and tests this system in different conditions.

3   Proposed System

Figure 1 shows the system activities diagram. It starts with the face detection and recognition of landmark, i.e. eyes and mouth. Then, it calculates the eyes blinks using number of eyelid closure and mouth opening for yawning.

3.1   Face Landmarks Detection

The driver fatigue detection system detects the face feature in each frame. Second, the eyes blinks and the yawning are calculated every 1 min.

The development of these steps is done in a number of trials. To detect the face features in real time, first, the Haar-like algorithm is tested with JAVA and MATLAB. The algorithm is affected by different lighting conditions and did not detect tilted faces, i.e. face movements.

Second, another feature-tracking algorithm, which is Kanade–Lucas–Tomasi (KLT) [17], is implemented with MATLAB and JAVA. The KLT tracks a set of points and detects the corners within the face bounding box using the minimum eigenvalue algorithm. The MATLAB and JAVA implementation didn't solve the lighting variations problem. Additionally, the JAVA was too slow to process each frame.

Third, the histogram of oriented gradients (HOG) [15] is a feature descriptor. 'It counts the occurrences of gradient orientation in localized portions of an image' [15].

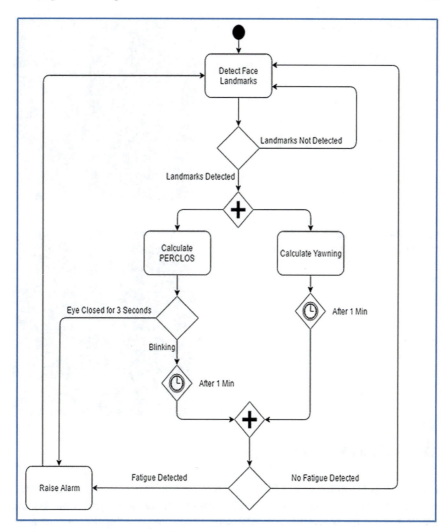

Fig. 1 The system activity diagram of the system

Figures 2 and 3 show the implementation of the HOG algorithm with MATLAB. The implementation was slow, and identifying the eyes and the lips was not very accurate. This was tested with five participants.

Fourth, the D-lib [18] is a general-purpose cross-platform software library. It uses the 'HOG along with linear support vector machine to train face images to get 68 face landmarks points' [18]. The system used the D-lib with C++. It detects the eyes and the mouths lips accurately.

Fig. 2 HOG input image

Fig. 3 HOG descriptor
example

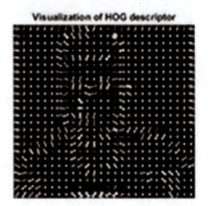

3.2 Fatigue Detection

The fatigue is identified by eye blinking, i.e. eye closure frequency in a frame time, and the yawning is measured by the mouth opening frequency in a frame time. For the eye closure, the Euclidean distance between the upper and lower eyelids or eye aspect ratio (EAR) is calculated. EAR measures the eye closure. For the mouth opening, the Euclidean distance between the upper and lower lip is calculated. The fatigue detection is recognized by the eye closure and the yawing (mouth open), defined as in Eqs. (1) and (2), respectively:

$$E(f) = \sum_{\text{open}=0}^{\text{closed}=1} \tag{1}$$

$$Y(f) = \sum_{\text{not yawning}=0}^{\text{yawning}=1} \tag{2}$$

where $E(f)$ is the total number of eye closure in each frame f. The values are given as follows. Closed eye takes 1 and open is 0. $Y(f)$ is the number of yawning in each frame f and yawning has the value of 1 and not yawning is 0. These measures are calculated in a frame window that has a sequence of frames x_i in Eq. (3).

$$fw(t) = xi \ (t \geq i \geq t - 1) \tag{3}$$

The fatigue is calculated by the following formula in Eq. (4).

$$FA(t) = \left(\sum_{x \in fw(t)} E(x) \geq 21 \right) \wedge \left(\sum_{x \in fw(t)} Y(x) \geq 2 \right) \tag{4}$$

Where the $FA(t)$ is the fatigue FA in a time frame t. It is calculated with the number of eye close equals or more than 21 and there is at least 2 yawning or more. To build a support vector machine (SVM) model that classifies the fatigue from the non-fatigue states, 831 records of fatigue measures, i.e. eye closure and yawning were collected. The model is trained and tested, and the results are presented in the receiver operating characteristic curve (ROC) curve and the area under the curve (AUC) which are 95% accuracy (see Fig. 4).

Fig. 4 ROC curve and AUC

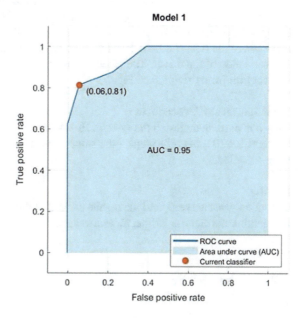

Fig. 5 System installed in
the car

4 Evaluation

The evaluation was guided by two research questions. RQ1. Will the system work in
different light conditions? RQ2. Will the system be able to work with different age
groups, gender, people wearing reading glasses or with male having hair of a beard
and a moustache? The evaluation was done in two phases.

4.1 First Study

Equipment
The system uses the laptop's camera and External HD 1080 camera attached to the
laptop for a live video streaming input to the system. The laptop is HP Intel Core i5
7th generation with 4 GB RAM and 500 GB hard disk. The system was installed in
a car where the laptop was on the right side of the driver, and the webcam is on the
left-hand side of the driver. The camera was fixed in the far left of the driver not to
distract him from driving (see Fig. 5).

Participants and Procedures
Only one participant tested the system. He was asked to drive midday, where the sun
is bright for 10 min and to repeat the same drive at night for 10 min. The two trips
were recorded.

Results
The video was analysed, and the results show that the system managed to detect the
face landmarks and the fatigue measures accurately.

4.2 Second Study

Equipment
The system was tested in a lab, in the morning with the laptop camera. The camera
was on to top of the laptop and facing the participants' eyes.

Fig. 6 The participants' evaluation

Fig. 7 The system working with reading glasses and hair around the mouth

Participants and Procedures

The system is evaluated with ten participants. They are three ladies and seven males. The average age of five of the participants is between 20 and 25, four is between 30 and 35 and one male is above 70. Three of the participants were wearing reading glasses. Four of the males had moustaches and beards (see Fig. 6).

They were told that there are two short sessions of 5 min. In the first session, they will look into the camera, blink, and yawn. In the second, they will just look to the camera.

Results

The system was promptly showing the results of fatigue or not (see Fig. 7). For the ten participants, the system detects all fatigue measures accurately.

5 Conclusions and Future Work

This work presents a novel non-intrusive system that detects drivers' fatigue using a simple webcam. To develop this system, different algorithms and implementation tools have tested to select the optimal. The system detects the facial features using HOC algorithm. Then, it calculates the fatigue symptoms of eye blink and yawning. A SVM classifier model is created to differentiate between fatigue and non-fatigue states. The AUC results of this model are 95%. This system is evaluated in different light conditions, with different age groups and attachments to the face such as reading glasses and hair around the mouth. The evaluation gave encouraging results. The system recognized all the fatigue symptoms and showed the fatigue and non-fatigue states immediately to the participants. To gain confidence in the results, in future, this system will be tested in different context of fatigues.

Acknowledgements This work is partially supported by the 'Science and Technology Development Fund' (STDF) and the 'German Egyptian Research Fund' (GERF) Project Ref. No. 23059.

Ethical Approval All procedures performed in studies involving human participants were in accordance with the ethical standards of the institutional and/or national research committee and with the 1964 Helsinki declaration and its later amendments or comparable ethical standards.

Informed Consent Additional informed consent was obtained from all individual participants for whom identifying information is included in this chapter.

References

1. Yoassry, E.: Fatal and injury fatigue-related crashes on Ontario's roads: a 5-year review why do we need an operational definition of fatigue-related crashes. In: Driver Fatigue Symposium 2007, vol. March, Canada, Toronto (2007). https://www.arrivealive.mobi/facts-about-drowsy-driving-internationally
2. Elshamly, A., Abd El-Hakim, F., Afify, H.: Factors affecting accidents risks among truck drivers in Egypt. In: MATEC Web of Conferences 2017, vol. January 94, Braşov, Romania (2017)
3. Car News Home Page. https://web.archive.org/web/20070927004541/http://www.newcarnet.co.uk/Lexus_news.html?id=5787. Accessed 5 Sept 2017
4. MR688 Homepage. http://www.care-drive.com/product/driver-fatigue-monitor-mr688. Accessed 25 Oct 2018
5. RVS-DC161-V4 Homepage. http://www.rvssystems.com/our-products/safety-accessories/professional-driver-fatigue-monitoring-system-rvs-dc161-v4.html. Accessed 25 Oct 2018
6. Sigari, M., Pourshahabi, M., Soryani, Fathy, M.: A review on driver face monitoring systems for fatigue and distraction detection. Int. J. Adv. Sci. Technol. **64**, 73–100 (2014)
7. Saini, V., Saini, R.: Driver drowsiness detection system and techniques: a review. Int. J. Comput. Sci. Inf. Technol. **5**(3), 4245–4424 (2014)
8. FORD Homepage. https://owner.ford.com/how-tos/vehicle-features/safety/driver-alert-system.html. Accessed 25 Oct 2018
9. VOLKSWAGEN Homepage. http://en.volkswagen.com/en/innovation-and-technology/technical-glossary/fahrerassistenzsysteme.html. Accessed 25 Oct 2018
10. BMW Homepage. https://www.bmw.ca/en/topics/experience/connected-drive/BMW%20ConnectedDrive:%20Driver%20Assistance%20.html. Accessed 25 Oct 2018
11. Viola, P., Jones, M.: Rapid object detection using a boosted cascade of simple features. In: Computer Vision and Pattern Recognition 2001, 8–14 Dec, pp. 511–518. IEEE Computer Society, Kauai, HI, USA (2001)
12. Turk, M., Pentland, A.: Face recognition using Eigen faces. In: Computer Vision and Pattern Recognition 1991, pp. 586–591. IEEE, Lahaina, Maui, Hawaii, USA (1991)
13. Wang, L., He, D.: Texture classification using texture spectrum. Pattern Recogn. **23**(8), 905–910 (1990)
14. Daugman, J.: Uncertainty relation for resolution in space, spatial frequency, and orientation optimized by two-dimensional visual cortical filters. J. Opt. Soc. Am. **2**(7), 1160–1169 (1985)
15. Dalal, N., Triggs, B.; Histograms of oriented gradients for human detection. In: Computer Vision and Pattern Recognition 2005, pp. 886–893. IEEE, San Diego, CA, USA, USA (2005)
16. Wierwille, W.: Historical perspective on slow eyelid closure: whence PERCLOS? In: Ocular Measures of Driver Alertness Conference 1999, MC-99-136, pp. 31–53. Federal Highway Administration Herndon, Washington, DC (1999)

17. Lucas, B., Kanade, T.: An iterative image registration technique with an application to stereo vision. In: Imaging Understanding Workshop, IJCAI, Canada, pp. 121–130 (1981)
18. King, D.: Dlib-ml: a machine learning toolkit. J. Mach. Learn. Res. Arch. **10**, 1755–1758 (2009)

Decentralized Autonomous Corporations

Craig S. Wright

Abstract This document is intended to provide an overview on decentralized autonomous corporations (DACs) based on the blockchain technology. We provide definitions of (1) DACs, (2) secure multiparty computation—a secure multiparty protocol for authorizing transactions, (3) autonomous agents—a set of computer programs that carry out some set of operations on behalf of users. We conclude the document with an application for financial-portfolio management.

Keywords Bitcoin · ECDSA · Security · Blockchain · Distributed corporation · DAC · Autonomous agents

1 Decentralized Autonomous Corporation

A decentralized autonomous corporation (DAC) is a virtual corporation without any central point of control, with a certain agenda, business plan, and protocol.

DACs run without any human involvement and are defined by code and software. They aim to be autonomous in the sense that they implement mechanisms for self-regulation. Once they are fully operating, a DAC may act independently of its creators.

DACs are distributed. There is no failure point that can be attacked, and DACs cannot be shut down or even modified to make them send all their money to an attacker's account.

DACs necessitate a network of autonomous agents which perform functions within an environment to achieve goals, without being directed to do so. One example of an autonomous agent is a computer virus. Malware such as a worm virtually "survives" through replication. A worm will copy itself from host to host in a network and does not require further human action.

DACs generate Bitcoin addresses. DACs construct and sign transactions. Thus, a system of signing transactions that can be computed in a decentralized way is needed.

C. S. Wright (✉)
CNAM, Paris, France
e-mail: c.wright@nchain.com

© Springer Nature Singapore Pte Ltd. 2020
X.-S. Yang et al. (eds.), *Fourth International Congress on Information and Communication Technology*, Advances in Intelligent Systems and Computing 1027, https://doi.org/10.1007/978-981-32-9343-4_14

The first solution is multisignature addresses; given a set autonomous agents, for example 1000, we generate a 501-of-1000 multisignature address between them. The problem is that the transaction would be too large. The maximum size of a standard transaction is 10,000 bytes [1]. Each signature is about 70 bytes [2], so 501 of 1000 signatures would make a 35,000-byte transaction. The second solution is secure multiparty computation.

2   Secure Multiparty Computation—Key Sharing

Secure multiparty computation (SMC) protocols allow a group of mutually distrusting parties to compute a joint function on their private inputs. SMC is a method that allows a threshold group to safeguard confidential information. The secret is split into components—parts or splits. These splits are subsequently circulated between the parties. For the threshold group to recreate the secret, a predefined and specified set of slices—the threshold k—needs to be blended to form the original. The attainment of fewer than the threshold of separate slices provides no knowledge regarding the secret.

Let s be a secret value and n share s_i. From this, we may create a k-out-of-n secret sharing scheme, if the subsequent requirements are met:

- **Correctness**; information disclosing any k or more splits of s allows the secret to be effortlessly calculable.
- **Privacy**; information disclosing any $k - 1$ or fewer splits of s results in the secret utterly undetermined (which is the idea that all its possible values are equally likely).

We consider the case where many autonomous agents wish to securely execute certain intricate distributed calculation on a set of inputs values. This could include signing Bitcoin transactions. In this case, we use Shamir's secret sharing scheme (SSSS) [3, and references therein]. SSSS is formulated through the concept that k points are required to distinctively identify and solve a polynomial of degree $k - 1$.

Shares in SSSS have two values—an index and the evaluation of the randomly produced polynomial associated with the index—$(i, f(i))$. The indices need to be exclusive to each party (no two parties can use the same value). The value zero may not be used. Using zero will disclose the secret $f(0) = s$. To generate the shares, start by selecting n random points that will lie on a polynomial $(x, f(x))$. The points on the function are then allocated to each of the parties. The coefficients are to be randomly selected. We ensure that the free member is equivalent to the secret (Fig. 1). Shamir's scheme uses a random 1-degree polynomial.

A characteristic of elliptic curves is where a (k, n) threshold scheme with polynomial interpolation is set with the public key; the private key can be recovered from k of the n pieces exactly in the same way as the public key.

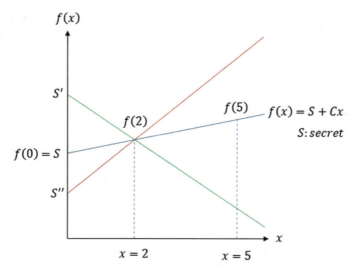

Fig. 1 Categorizing a secret value using Shamir's secret sharing scheme. The green and red lines are failed attempt to reconstruct S knowing $k - 1$ slices. A party with a single point for a 1-degree polynomial, we can draw all the possible values (S, S', S'', etc.). It is infeasible for anybody to know which one is correct. Nevertheless, where a party has knowledge of at least 2 shares, that party may recreate the polynomial $f(X)$ using Lagrange interpolation

3   Autonomous Agents

An autonomous agent is a computer system that is situated in some environment (e.g., the Internet) and that is capable of autonomous action (e.g., buying, selling) in this environment in order to meet its design objectives.

Autonomous agents are purely software. They receive and execute instructions coded in a Practical Agent ProgrAmming Language (PAPAL). PAPAL is the agent programming language. Autonomous agents do not have the capabilities to manufacture a product, write code, and develop hardware. They require actors in the physical world for this purpose, called contractors.

Herein, we define different classes of autonomous agents: voting autonomous agents and task autonomous agents.

3.1   Autonomous Interface Agents (AIA)

Autonomous interface agents can make real-time suggestions to users based on a simple keyword-frequency knowledge recovery value.

For example, an autonomous interface agent may observe where a user is interested in or hold shares in the bank sector, and it may suggest to him that he invests in the metal sectors (if both sectors are independent). The AIA enquires another autonomous agent to look for an asset that is not highly correlated to the bank sector; then, a pop-up window appears showing the correlation and other indicators between the gold price and HSBC share price.

3.2   Voting Autonomous Agents

Voting autonomous agents are agents who are empowered to vote. At the deployment phase, the DAC issues N tokens to raise money (via dominant assurance contract?). The tokens represent the company shares. The tokens grant its holder ownership and voting rights.

The DAC issues N shares to raise money; if the goal is not reached, repayment follows (dominant assurance contract):

1. Tokens are used to represent company shares and necessitate a smart contract for revenue sharing [the revenue may be automatically shared as it is earned in real time. It is also possible that the share/token holds a piece of DAC private/public keypair, i.e., $(x, f(x)) \rightarrow$ if someone holds k pieces, it is equivalent to a 51% attack]
2. The share price can be time-dependent, like airline tickets.

The colored coin transaction contains a pair or multiple pairs of $(x, f(x))$. Tokens can be exchanged, and new share owners request new pairs $(x_{new}, f(x_{new}))$. Previous $(x_{old} f(x_{nld}))$ should be removed from the DAC database, and $(x_{old} f(x_{nld}))$ cannot be accepted as a valid share.

3.3   Task Autonomous Agents

Task autonomous agents search for unfilled tasks piloted by its own predilection orders over existing tasks. It selects a contractor and submits the contractor's proposal for voting either to the voting autonomous agents or to the DAC members. DAC members can review the contractor's proposal and send their share $(x, f(x))$ to the task autonomous agent [Prior to deployment, the entity distributes the shares between the autonomous agents]. If the task autonomous agent collects k shares, it submits a smart contract on the blockchain representing the proposed project. There is a set time frame to vote on any given proposal. After this time, the proposal is closed.

Voting for Proposals

Shareholders of a DAC cast votes weighted by the number of tokens they control. Within the contracts, the individual actions of members cannot be directly determined. There is a set time t_p to debate and vote on any given proposal. The time frame is set by the function autonomous agent. After t_p has passed, any token holder can verify that the threshold value k has been reached; i.e., k shares have been collected.

Table 1 The table shows as an example of a contractor profile and a checklist for assessment. The score for reliability and reputation reflects the commitment of the contractor to established quality principles. For contractor quality, online ratings sites can be used to provide trustworthy information

Attributes	
Reliability and Reputation[a]	●●●●○
Contractor Quality[b]	90%
Using subcontractor	No
Cost comparison, if applicable (cheapest, relative, expensive, most expensive)	Most expensive

[a]Online ratings sites can be used to provide trustworthy information
[b]This table shows as an example a checklist for assessing subcontractor quality. The score reflects the commitment of the contractor to established quality principles

If this is not the case, the proposal is closed. A minimum number of shareholders can also be required for the vote to be valid.

In order to ensure transparency and before a vote can take place, a contractor profile should be presented to the voting autonomous agents and to the shareholders. Table 1 shows an example of a contractor/company profile.

Shareholders can vote using the platform MyVote, or they can delegate their voting power to the voting autonomous agents. In the latter case, shareholders provide information to voting autonomous agents using the platform BlockID.

We propose two voting methods:

- using a voting platform,
- relying on the voting autonomous agents.

MyVote—A Blockchain-Based Voting Platform

MyVote is a platform which utilizes blockchain technology to allow voters to vote at home and in their own time. Through a process of vote casting using transactions, a party may produce a blockchain that can be used to track and record all vote counts and related records. With this system, shareholders can form a consensus concerning the final voting results and this may be automated and auditable by all parties. As an immutable evidence source, the blockchain allows the parties to ensure the integrity of the votes. They can ensure that no invalid votes have occurred and votes will not be able to be added, removed, or altered.

BlockID—A Blockchain-Based Platform for Digital Identity.

BlockID is a blockchain-based identity service that allows shareholders signing up for an account and the DAC submitting questionnaires; see Table 2.

An electronic identity record will consist of attributes, or data characteristics. Such may include:

- Date of birth,
- Social security number,
- Medical history.

Table 2 A questionnaire example received by a shareholder

Questionnaire for assessing shareholder vote	
Question 1	Which of these statements is closer to your view in respect of the time for completion of works? • On-time delivery is crucial • I accept reasonable delay. • N/A
Question 2	Which of these statements is closer to your view on using subcontractors? • Unacceptable • Acceptable • N/A
Question 3	Will you hire a contractor that is not a member of any health and safety organizations? • Yes • No

Identity theft is widespread. The ability to easily access and copy records leaves the Web vulnerable. Consequently, electronic identity authentication and validation procedures are essential to safeguarding the WWW and associated network infrastructure. It is important that the security of both public and private sectors is maintained.

BlockID can be used as a driver license, passport, credit card, apartment keys etc.

Decision Making and Voting autonomous agents
If shareholders decide to not be involved in the decision making, they can rely on their voting autonomous agent. For voting autonomous agents to achieve their function, they require access to the questionnaire completed by the shareholder, and the decision of the voting autonomous agent is handled by one or more classification algorithms. In the following, we propose a simple solution using a decision tree algorithm. But, any suitable machine learning algorithm, e.g., neural network or SVM, can also be used.

Option decision tree is like regular decision tree. The distinction between these is that an ODT can include option nodes as well as the regular decision nodes and leaf nodes on a standard tree. Option nodes allow multiple branching instead of a single branching per node.

Figure 2 illustrates an element of a tangible option decision tree.

4 Application: Digital Asset Management (DAM)

Digital asset management (DAM) is an application of a multi-autonomous-agent architecture to the problem of financial portfolio management. The approach of

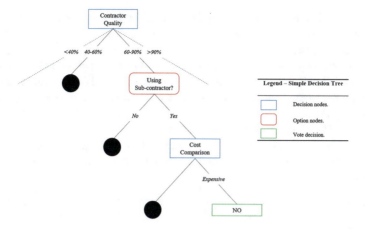

Fig. 2 A simple example of an option decision tree

the multi-autonomous-agent system (i) requires the coordination of several component tasks (gathering information, interpretation, prediction, and more). The system
should be reliable; if one or more agents fail, the system will continue to function.
Autonomous agents are designed with redundant capabilities and/or appropriate coordination mechanisms. Thus, an unresponsive agent is not a problem.

The DAM system deploys several different task autonomous agents:

- Breaking news autonomous agents gather information from breaking news articles
  from the Web. These agents track and filter news articles and decide if they are
  sufficiently valuable that the user through the interface autonomous agent and/or
  other autonomous agents need to know about them immediately.
- Stock tracker autonomous agents gather stock prices in real time. These agents
  express actions that are instigated by a user or an autonomous agent. These are instigated using queries or via a process of observation where various data resources
  are examined seeking the occurrence of a defined pattern; e.g., a stock price has
  exceeded a predefined threshold.
- Model autonomous agents build mathematical models using standard approaches
  (e.g., ARIMA), data mining techniques (e.g., neural network), and complex
  stochastic models to attempt to forecast the immediate prospect for share values
  in an exchange or market.
- Risk autonomous agents evaluate portfolios for financial risk using a stock market
  risk measure.
- Interface autonomous agents interrelate with the user by collecting user requirements and producing solutions. Agents employ user specifications to define and
  align intersystem coordination. Interface autonomous agents display a comprehensive summary of the user's portfolio.
- Trader autonomous agents vote based on their trading strategies.
- Etc....

A simplified representation of the DAM architecture is shown in Fig. 3.

DAM has three primary purposes. The primary and principal purpose is the pro-ducing trading decisions. Next, the transmission and posting of a completed order to market exchanges must be achieved. Lastly, the supervision and monitoring of each order following the transmission or submission must be maintained [portfolio management is not the core of DAM, and it can be outsourced to another system that integrates DAM].

Purpose #1—Complete trading decisions.

Trading decisions are created corresponding to a defined strategy (see Fig. 4). Trading strategies are commonly established using one or more of the subsequent methods:

- Quantitative analysis—created using fundamental indicators and arithmetic basis of rate movements and past/current information and records.
- Event-based—created through the assessment of past and current occurrences. These are frequently linked to outlines obtained using written news sources.
- Sentiment-based—composed using indicators that catch market sentiments.

A strategy of such form may be derived using machine learning algorithms.

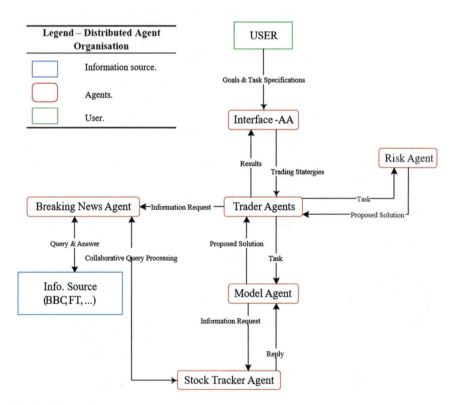

Fig. 3 DAM architecture

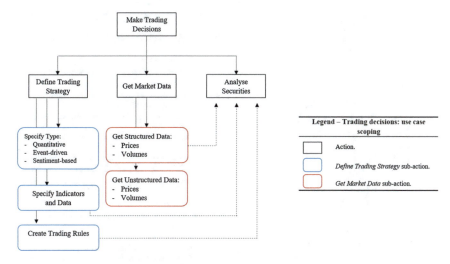

Fig. 4 Make trading decisions: use-case scoping

Trade orders are calls to trade securities (which are a buy or sell process). The purpose of DAM is to present suitable trading orders where such meet the requirements of accuracy, timeliness, and visibility. A non-exhaustive list of types of trading includes:

- Long orders
- Short orders
- Etc….

Purpose #2—Submit orders.
Purpose #3—Manage orders following submissions.
The third goal of DAM is to administer the orders subsequent to a submission ensuring that they remain within configured bounds. The process includes:

- Recording all profits and losses received by the agent
- Logging and recording the transactions for later audit or analysis
- Monitoring any unfulfilled orders.

In the following, we will focus on the interaction between agents and the decision-making process.
We consider a set of autonomous trading agents:

- Moving average autonomous agent (MA Autonomous Agent) uses moving average strategies to predicate on the relationship between a security's price and its moving average trendline.
- Stochastic oscillation autonomous agent (Sto Autonomous Agent) implements a stochastic oscillation trading strategy to follow security price momentum.
- Risk autonomous agent (Rk Autonomous Agent) assesses the risk that is inherent to all trading strategies.

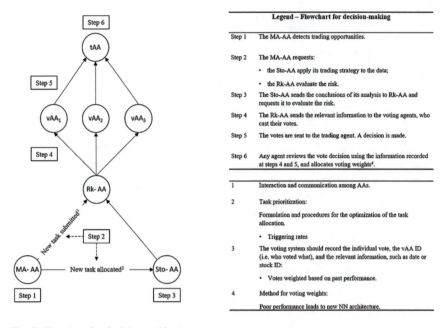

Fig. 5 Flowchart for decision-making

- Voting autonomous agent (vAutonomous Agent).
- Trading autonomous agent (tAutonomous Agent).

Figure 5 shows the sequence of steps in decision making.

Decision-Making Using Human Profiles

In this section, an evaluation protocol is described in which the user, here a human trader, reviews historical data sets and answers questions related to trading strategies. The financial time series and the trader's answers are stored on a database (DHT + blockchain), and are later used for decision making. The experiment aims at assessing the similarity between past and actual observations, and to take future decisions based on past experiences. For the sake of reproducibility, a large sample should be used.

Table 3 shows a small sample of questions to assess trading strategies.

MyProfile is a dedicated interface that uses the blockchain to store historical time series and their quantitative indicators (moving average, stochastic oscillator, etc), and the answers provided by the human traders.

The questionnaire phase occurs after the deployment phase and before the running phase. During the running phase, the DAC will apply feature extraction (MA, Stoch.) and similarity search for time series data.

Finding the degree to which a given time series resembles another one has been an active area of research for a number of years, and a great variety of techniques currently exist for measuring the similarity between time series datasets.

Table 3 A questionnaire example received by a human trader

Question 1	You took a large long position in the S&P500 futures at 1400. Subsequently you watched the futures climb to 1450 over one week. Now, you have substantial paper profits. There is no new market news or information expected, and your charts continue to predict a bull trend. Which of the following do you do now?
	• Sell 50% of your position • Sell you entire position • Do nothing • Increase you position 50% • Double you position
Question 2	An investor has sold 100 shares of FBN stock short at 62.00 and subsequently buys one FBN Jan $65.00 call at 2 If FBN stock rises to $70.00 and the investor exercises the call, what is the gain or loss in this position?
	• $2.00/share • $5.00/share • $8.00/share • Unbounded
Question 3	The chart in Fig. 7 below represents the temporal evolution of EUR/USD exchange rate. Using the moving average indicator (middle plot) and the stochastic oscillator (bottom plot), what is your trading strategy at time T?
	• Sell your position • Increase your position • Do nothing

Our objective is not to present an exhaustive review of the several dissimilarity measures that have been proposed. Classical metric parameters include:

- Euclidean distance,
- Dynamic time warping (DTW)—DTW works by optimally aligning the time series in the temporal dimensions so that the accumulated cost of this alignment is minimal.
- Etc....

Each of these metric parameters will give an insight into the degree of similarity between the time series. Those indicators/statements can sometimes be contradictory; e.g., Euclidean distance may indicate that the signal resembles template 1, while DTW indicates that correlation is higher with template 3.

The size of the phase space—the set of the metric parameters—may contain a large number of dimensions; thus, a clustering algorithm (as illustrated in Fig. 6), such as k-means, may be better suited to assign a cluster to the data point. In addition, if the distance of the data to the centroid exceeds a threshold value, the DAC can take the decision to not cast the autonomous agent vote and to request human votes.[1]

---

[1]Because of the size of the phase space, that is the number of metric parameters, a clustering algorithm, such as k-mean, may be use to assign a cluster to the data point. If the distance of the data to the centroid exceeds a threshold value, the DAC can request human traders to vote.

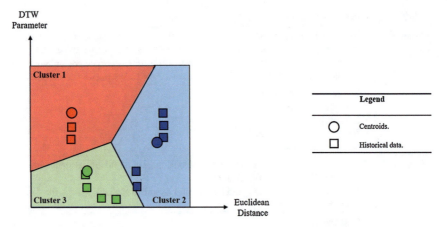

Fig. 6 The figure displays what a *k*-means clustering algorithm would yield using three clusters

Fig. 7 The chart represents the temporal evolution of the EUR/USD exchange rate

Decision Making with Neural Networks

In this section, an innovative approach is presented to decision making using a deep learning algorithm. Instead of using deduction or a collection of rules, neural networks (NNs) rely on the ability to recognize patterns through experience. Although NNs are considered essentially adaptive pattern recognition, several researchers have applied NNs to business decision-making situations in the past.

The available settings on NN architectures enable the user (human or artificial) to design sophisticated structures having incredibly rich and intelligent behaviors. For example, the user can:

- standardize the input variables or the target variables;

- specify the error function, the type of transfer, and the activation functions;
- add hidden layers;
- etc....

For the present problem, the input data vector that feeds the NN may consist of

- Boolean variables (True/False) corresponding to MA signals (Sell/Buy);
- numeric values (momentum or price velocity);
- stochastic oscillator indicators (Overbought/Oversold);
- etc....

The architecture of the network proposed here is very simple. It consists of a single layer of input nodes and one output node (see Fig. 8).

The first phase will consist of training the DAC. A trend-oriented training in parallel to a forecast-oriented training should be developed to enhance the forecasts of stock market movements. Predefined trend targets based on moving average, stochastic oscillator, etc., are used for machine learning, and metric indicators such as Bias and MACD and the input permutations are selected as input signals to the node system (as illustrated in Fig. 9). The potential output may be the categories {buy,

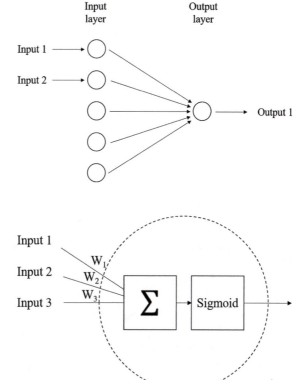

Fig. 8 A simple neural network architecture. The input layer nodes are passive making no action other than transmitting the input data. In comparison, the output layer is dynamic, revising the signals in accordance with ones in Fig. 9. The action of this neural network is defined because of the weights utilized by the output node

Fig. 9 Neural network active node. Individually, an input is multiplied by a weight. These are defined as the wn values. With all of the inputs received. The node sums the values to obtain the output. This generates a single result that is delivered through an 's' shaped non-linear function known as a sigmoid

hold, sell}. Such trend-oriented training provides the node with the ability to deliver consistent learning targets to capture dynamic price movements.

Working with Contract—Pay for Performance

Example: Someone submits a proposal to build Martian space stations. We are going to need a range of technology, skills, and services to implement such a proposal:

- Web designers and Web development specialists to get a creative crowd-funding Web site
- Project managers, architects, and developers to design and implement the project
- Either acquire existing Martian space stations from an existing provider or build them our own
- Use cryptocurrency for payments of services and goods
- Etc....

Appendix 1

See Table 4.

Table 4 Checklist for assessing subcontractor quality. The responses to the questions are rated, as is the score

Checklist for assessing subcontractor quality	
Customer satisfaction?	*Responses*
Will you assure an exclusive relationship with us and agree to terms of non-disclosure?	Yes
What is your history of on-time delivery?	95%
Has any independent quality assessment been performed for your company and what were the results?	No
Are you financially stable? How are you capitalized? Etc...	

References

1. bitcoin/src/script/interpreter.cpp. GitHub Webpage, https://github.com/bitcoin/bitcoin/blob/fcf646c9b08e7f846d6c99314f937ace50809d7a/src/script/interpreter.cpp#L256, last accessed 2019/03/12
2. ECDSA r, s encoding as a signature. StackExchange Webpage, http://bitcoin.stackexchange.com/questions/2376/ecdsa-r-s-encoding-as-a-signature, last accessed 2019/03/12
3. secmultipartycomp/slides/sec_multi_party_comp.pdf. GitHub Webpage, https://github.com/czielinski/secmultipartycomp/blob/master/slides/sec_multi_party_comp.pdf, last accessed 2019/03/12

A Distribution Protocol for Dealerless Secret Distribution

Craig S. Wright

Abstract As the value of bitcoin increases, more incidents such as those involving Mt Gox and Bitfinex will occur in standard centralised systems. The addition of group-based threshold cryptography with the ability to be deployed without a dealer and which supports the non-interactive signing of messages provides for the division of private keys into shares that can be distributed to individuals and groups to provide additional security. This scheme creates a distributed key generation system for bitcoin that removes the necessity for any centralised control list minimising any threat of fraud or attack. In the application of threshold-based solutions for DSA to ECDSA, we have created an entirely distributive signature system for bitcoin that mitigates against any single point of failure. When coupled with retrieval schemes involving CLTV and multisig wallets, our solution provides an infinitely extensible and secure means of deploying bitcoin. Using group and ring-based systems, we can implement blind signatures against issued transactions.

Keywords Bitcoin · ECDSA · Elliptic-curve cryptography · Threshold cryptography

1 Introduction

In this paper, we present a threshold-based dealerless private key distribution system that is fully compatible with bitcoin. The system builds on a group signature scheme that departs from the traditional individual signing systems deployed within bitcoin wallets. As deployed, the system is both extensible and robust tolerating errors and malicious adversaries. The system is supportive of both dealer and dealerless systems and deployment in an infinitely flexible combination of distributions.

C. S. Wright (✉)
CNAM, Paris, France
e-mail: c.wright@nchain.com

© Springer Nature Singapore Pte Ltd. 2020
X.-S. Yang et al. (eds.), *Fourth International Congress on Information and Communication Technology*, Advances in Intelligent Systems and Computing 1027, https://doi.org/10.1007/978-981-32-9343-4_15

Individual parties can act as single participants or in combination as a dealer distributing slices of their protected key slice across machines for security and recoverability or in groups for the vote threshold-based deployment of roles and access control lists.

There is no limit to the depth of how far a slice can be divided. Rather, the issue of complexity needs to be weighed against the distinct deployment. Using the same principle and methodology, as the record of signing and transactions can be hidden from outside participants—all with extensions to be presented in a subsequent paper—and even from those within the groups, we have introduced a level of anonymity and plausible deniability into bitcoin transactions, increasing the standard of pseudonymous protection to the users.

Ibrahim et al. [11] developed an initial robust threshold ECDSA scheme. The following protocol is a further extension of what forms the elliptic curve form of the threshold DSS introduced by Gennaro et al. [10].

The use of group mathematics allows to create a verifiable secret sharing scheme (VSS) that extends the work of Shamir [20] in secret hiding and that of Feldman [9] and Pedersen [18] from RSA and DSA schemes so that we can use it within ECC and ECDSA-based signature systems such as bitcoin [15]. Our system is tolerant against malicious adversaries, halting and is robust against eavesdropping.

In this paper, we start by presenting a method to allow for the cooperative signing of ECDSA signatures where no one-party ever knows the private key. Moreover, we allow for the private key pairs to be updated and refreshed without the necessity of changing the private key. In a subsequent paper following this one, we will detail the mathematics needed to add and remove members to a group while maintaining the secrecy and privacy of the private key.

The process will also be extended in sister papers to demonstrate how cooperative computations can be performed, while the output is dependent on the secure input of separate entities in a manner that does not require trust. Existing solutions all require a trusted party. Utilising the solution presented in this paper will also allow us to extend the work of Chaum [6] from a centralised system into a truly distributed manner of issuing electronic notes that can be settled directly on the bitcoin blockchain making the requirements for alternate blockchains or sidechains obsolete.

Matters of trust: All existing systems require some level of trust. Until this time, bitcoin has needed the protection of a private key using a secure system that is isolated from the world, which has proven difficult to achieve. Of note, systems, where bitcoin can be exchanged or stored, require trust in a centralised authority. Our work changes this requirement entirely distributing and decentralising the key creation and message signing processes within bitcoin while not changing any of the core requirements of the protocol. The methodologies noted in this paper may be implemented without modifying the bitcoin protocol, and in fact, there is no way to determine whether this process has been deployed through the analysing of a signed message.

In creating a distributed signature scheme for bitcoin, we allow for a group of people or systems to securely hold a key in a way that leaves no individual capable of generating a signature on their own. When extended, this scheme also allows for the secure recovery of each of the shares as well as the bitcoin private key itself. The group-generated signature is indistinguishable from that generated from the existing protocol. As such, signature verification remains as if it was enacted through a single-person signer using a standard transaction.

Such increase in trust is achieved as the secret key is shared by a group of n participants or m groups of participants. A threshold number of participants are required for the signing of a transaction, and any coalition of participants or groups of participants that meet the minimum threshold can perform the signature operation. Importantly, this protocol can be enacted synchronously or as a batched process where individuals or groups can attempt to create a coalition of participants.

2 Prior Work

Shamir [20] first introduced a dealer-based secret sharing scheme that allowed for a distributed management of keys. The problems associated with such a scheme come from the necessity of trusting a dealer who cannot be verified. Such a form of the scheme is fully compatible with the current system presented in this paper and can be used for group distribution of individual key slices that are created through the process noted herein.

The said aim of the procedure in joint random secret sharing (JRSS) [18] is to create a method where a group of participants may collectively share a secret without any participant having knowledge of the secret. Each participant selects a random value as their local secret and distributes a value derived from this using $SSSS$ with the group. Each participant then adds all the shares received from the participants, including its own. This sum is the joint random secret share. The randomness offered by a single honest participant is sufficient to maintain the confidentiality of the combined secret value. This state remains true even if all $(n - 1)$ other participants intentionally select non-random secret values. Joint zero secret sharing (JZSS) [4] is like JRSS, with the difference that each participant shares 0 as an alternative to the random value. The shares produced using this technique aid in removing any potential weak points in the JRSS algorithm.

Desmedt [8] introduces the concept of group-oriented cryptography. This process allowed a participant to send a message to a group of people in a manner that only allowed a selected subset of the participants to decrypt the message. In the system, the members were said to be known if the sender is required to know them using a public key, and the group is anonymous if there is a single public key for the group that is held independently of the members. Our system integrates both methodologies, and allows for known and anonymous senders and signers to exist within a group simultaneously.

3 Methods

For any elliptic curve (CURVE) with a large-order prime and a base point $G \in CURVE(\mathbb{Z}_p)$ of order n defined over the prime field $\mathbb{Z}_p$, we can create a system that allows for the secure distribution of an ECC private key into key shares and its use without any participant being able to recreate the original private key from less than a threshold of shares.

For an unknown integer d_A, where $1 \leq d_A \leq (n-1)$, we know that it is extremely difficult to calculate d_A given $Q_A = d_A \times G$ [13].

Our fundamental technique is derived using the application of threshold cryptography. Using our technique, the ECDSA private key only exists potentially and need never be recreated on any system. Each of these multiple shares is distributed to multiple participants $[p_{(i)}]$ in a manner that is extensible and allows for the introduction of both group and individual party signature formats. Thus, the signing process differs from that deployed within bitcoin. In this process, a coordinating participant $p_{(c)}$ creates a transaction and a message signature that is distributed to the group. Each participant can vote on the use of its private key share by either computing a partial signature or passing. In effect, passing would be equivalent to a no vote. The coordinating participant $p_{(c)}$ will collate the responses and combine these to form a full signature if they have received the minimum threshold number of partial signatures.

The coordinating participant $p_{(c)}$ can either accept the no vote and do the calculation based on a null value from the other party or can seek to lobby the party and convince them to sign the message. The protocol can be implemented with a set coordinator, or any individual or group can form this role and propose a transaction to the threshold group to be signed. Our system extends the work of Ibrahim et al. [11] providing a completely distributed ECDSA private key generation algorithm. This paper also presents a distributed key re-sharing algorithm and a distributed ECDSA signing algorithm for use with bitcoin. The key re-sharing algorithm may be used to invalidate all private key shares that currently exist in favour of new ones or for the reallocation of private key shares to new participants. This protocol extends to the sharing of not only the ECDSA private key but to private key shares as well. The consequences of this mean that shares can be constructed and voted on as a group process.

Such a system removes all requirements for a trusted third-party to exist. Consequently, it is possible to create the new overlay and wallet for bitcoin that are entirely compatible with the existing protocol and yet remove any remaining single points of failure while also allowing greater extensibility. This system can also be extended to allow for the introduction of blind signatures.

As our system does not require a private key ever to be loaded into memory, we not only remove the need for a trusted third-party but further remove a broad range of common attacks. The protocol is extensible allowing the required number of shares and the distribution of shares to be decided by the use case, economic scenario and risk requirements.

The system we have implemented mitigates all side-channel attacks and thus any cache-timing attacks. This system takes the work of Gennaro et al. [10] and extends it from DSS such that it can be successfully used in any ECDSA-based application.

ECDSA: Bitcoin uses ECDSA based on the secp256k1 curve. ECDSA was first standardised by NIST in 2003 [17] varying the requirements for Diffie–Hellman-based key exchanges using elliptic curve cryptography (ECC). The creation of ECC was particularly important due to the reduction in key size and processing power when compared to other public/private key systems. No sub-exponential time algorithm has been discovered for ECDLP. ECDLP is known to be intractable and refers to the elliptic curve discrete logarithm problem [12].

The parameters used throughout the paper are documented in Table 1.

Security considerations: The system is bounded by the security of ECDSA which is a current limitation within bitcoin. At present, ECDSA remains secure if a private key can be securely deployed. Our system mitigates side-channel attacks and memory disclosure attacks up to the threshold value requiring that the threshold number of participants has been compromised before the rekeying event. Additionally, any uncompromised threshold majority will be able to identify compromised participants at less than the threshold value.

Halting problems: Service disruption is a form of attack that can be engaged in by a malicious adversary attempting to create a denial of service attack against the participants. This attack would require the participants to either receive invalid signatures that they would expend processing time analysing or through flooding network messages that would be subsequently dropped.

Table 1 Definitions

m	The message incl. bitcoin transaction
$e = H(m)$	The hash of the message
CURVE	The elliptic curve and field deployed (summarised as E)
G	The elliptic curve base point. This point is a generator of the elliptic curve with large prime order n
n	Integer order of G such that $n \times G = \emptyset$ This is defined as the number of rational points that satisfies the elliptic function and represents the order of the curve E
k	The threshold value for the key splitting algorithm. This value represents the number of keys needed to recover the key. The secret is safe for $(k - 1)$ shares or less and hence can be retrieved with k shares
d_A	A private key integer randomly selected in the interval $[1, (n - 1)]$
Q_A	The public key derived from the curve point $Q_A = d_A \times G$[a]
$\times$	Represents elliptic curve point multiplication by a scalar
j	The number of participants in the scheme

[a]Bitcoin uses secp256k1. This defines the parameters of the ECDSA curve used in bitcoin and may be referenced from the Standards for Efficient Cryptography (SEC) (Certicom Research, http://www.secg.org/sec2-v2.pdf)

The requirement to encrypt messages to the participants using either ECC or signcryption-based ECC mitigates this attack vector. Before an attacker can send invalid partially signed messages, they would need to have already compromised a participant, making this form of attack no longer necessary.

Randomness: Algorithm 2 provides a scenario where sufficient randomness is introduced even if $(n - 1)$ participants fail to choose random values. A possible addition to this protocol is the introduction of group oracles designed solely for the introduction of random values to the signing and rekeying process. In this optional scenario, each of the key slices can be generated using the same protocol. For instance, if we have an m of n primary slice requirement, each of the underlying key slices can also be generated and managed using an m' of n' threshold condition.

A participant using this system would be able to have the addition of an external Oracle that does nothing other than injecting randomness into the protocol. A user with m' key slices (where $m' < n - 1$) could choose to recreate and process their signature solution based on the key slices they hold or may introduce an external Oracle that is unnecessary other than for the introduction of randomness.

Each slice could be likewise split for robustness and security. The key slice could be distributed such that the user has a slice on an external device such as a mobile phone, smart card and a software program running on a computer such that the combination of sources would be required for them to create a partial signature.

It is important that a unique random ephemeral key D_k is produced or it would be possible to use the information to recreate the private key d_A.

Public signing: The primary purpose of transactional signing using this protocol is to enable the distributed signing of the bitcoin transaction. Any transaction that has not been published to the blockchain can be maintained privately by the participants. Therefore, if a coordinating participant $p_{(c)}$ on any occasion has not been able to achieve the required level of votes to sign a transaction successfully, it is not necessary to create a new bitcoin transaction. The ownership of any settled transaction remains secure if the key slices are themselves secure to the threshold value.

If the system is deployed well, the ability to compromise up to $(k - 1)$ participants leaves the system secure to attack below the threshold value. When coupled with a periodic rekeying protocol (Algorithm 2), our system can withstand side-channel attacks and memory disclosures.

Method and implementation: As the protocol encrypts the secret information required to be sent between participants using ECC based on a hierarchical derivation [22], it is both possible and advisable to collate all messages into a single packet sent to all users such that validation can be done against potentially compromised or hostile participants when necessary.

Signature generation is proposed by a coordinating participant $p_{(c)}$. By default, any key slice can act as the coordinating participant, and the requirements come down to the individual implementation of the protocol. The algorithms used are documented below, and a later section provides detail as to their deployment.

Algorithm 1 Key Generation
Domain Parameters (CURVE, Cardinality n, and Generator G)
   Input: NA
   Output: Public Key Q_A

$$\text{Private Key Shares } d_{A(1)}, d_{A(2)}, \ldots, d_{A(j)}$$

For our threshold of k slices from (j) participants, we have a constructed key segment $d_{A(i)}$ which is associated with participant (i), and $(j-1)$ participants nominated as participant (h) that are the other parties that participant (i) exchanges secrets with to sign a key (and hence a bitcoin transaction).

- In the scheme, j is the total number of participants where $k \leq j$ and hence $h = j-1$
- Hence, we have a (k, j)—threshold sharing scheme.

The method for Algorithm 1 follows:

(1) Each participant $p_{(i)}$ of (j) where $1 \leq i \leq j$ exchanges an ECC public key (or in this implementation, a bitcoin address) with all other participants. This address is the group identity address and does not need to be used for any other purpose. This can be formed of a 'Type 42' address of any form or level of the hierarchy (Patent 222). The exchange of symmetric keys for participants is completed using the method in patent 42.

Note: this is a derived address [22] and key based on a shared value between each of the participants from the process of 'Determining a common secret for two blockchain nodes for the secure exchange of information'.

(2) Each participant $p_{(i)}$ selects a polynomial $f_i(x)$ of degree $(k-1)$ with random coefficients in a manner that is secret from all other parties.

Such a function is subject to the participant's secret $a_0^{(i)}$ that is selected as the polynomial free term. This value is not shared. This value is calculated using a derived private key [22].

We define $f_i(h)$ to be the result of the function, $f_{(x)}$ that was selected by participant $p_{(i)}$ for the value at point $(x = h)$, and the base equation for participant $p_{(i)}$ is defined as the function: $f_{(x)} = \sum_{p=0}^{(k-1)} a_p x^p \bmod n$

In such an equation, a_0 is the secret for each participant $p_{(i)}$ and is not shared. Hence, each participant $p_{(i)}$ has a secretly kept function $f_i(x)$ that is expressed as the degree $(k-1)$ polynomial with a free term $a_0^{(i)}$ being defined as that participant's secret such that:

$$f_{i(x)} = \sum_{\gamma=0}^{(k-1)} a_\gamma x^\gamma \bmod n$$

(3) Each participant $p_{(i)}$ encrypts $f_i(h)$ to participant $P_{(h)}$ $\forall h = \{1, \ldots, (i-1), (i+1), \ldots, j\}$ using $P_{(h)}$'s public key.[1] [22] as noted above and exchanges the value for $P_{(h)}$ to decrypt.

Note that $n \times G = \emptyset$ for any basic point $G \in E(Z_p)$ of order n for the prime p.[2]

As such, for any set of integers B : $\{b_i \in \mathbb{Z}_n\}$ that can be represented as $(b, b_1, b_2, \ldots)$, if $bG = [b_1G + b_2G + \cdots] \bmod p$, then $b = [b_1 + b_2 + \cdots] \bmod n$. Further, if $bG = [b_1 b_2 \ldots]G \bmod p$, then $b = [b_1 b_2 \ldots] \bmod n$.

Given that $\mathbb{Z}_n$ is a field and we can validly carry out Lagrange interpolation modulo n over the values selected as ECC private keys, we have a condition which leads to the conclusion that Shamir's secret sharing scheme SSSS [5] can be implemented over $\mathbb{Z}_n$.

(4) Each participant $P_{(i)}$ broadcasts the values below to all participants.

(a) $a_\kappa^{(i)}G$ $\forall \kappa = \{0, \ldots, (k-1)\}$
(b) $f_i(h)G$ $\forall h = \{1, \ldots, j\}$

The value associated with the variable h in the equation above can either be the position of the participant $P_{(h)}$ such that if participant $P_{(h)}$ represents the third participant in a scheme, then $h = 3$ or equally may represent the value of the ECC public key used by the participant as an integer. Use cases and scenarios of applications exist for either implementation. In the latter implementation, the value $h = \{1, \ldots, j\}$ would be replaced by an array of values mapped to the individual participant's utilised public key.

(5) Each participant $P_{(h \neq i)}$ verifies the consistency of the received shares with those received from each other participant.

That is $\sum_{\kappa=0}^{(k-1)} h^\kappa a_\kappa^{(i)} G \overset{?}{=} f_i(h)$

And that $f_i(h)G$ is consistent with the participant's share.

(6) Each participant $P_{(h \neq i)}$ validates that the share owned by that participant ($P_{(h \neq i)}$) and which was received is consistent with the other received shares:

---

[1] A related research paper will detail how the process of 'Determining a common secret for two blockchain nodes for the secure exchange of information,' and the sharing of keys may be integrated.

[2] In the case of bitcoin, the values are:

Elliptic curve equation: $y^2 = x^3 + 7$;

Prime modulo: $2256 - 232 - 29 - 28 - 27 - 26 - 24 - 1 =$ FFFFFFFF FFFFFFFF FFFFFFFF FFFFFFFF FFFFFFFF FFFFFFFF FFFFFFFE FFFFFC2F;

Base point = 04 79BE667E F9DCBBAC 55A06295 CE870B07 029BFCDB 2DCE28D9 59F2815B 16F81798 483ADA77 26A3C465 5DA4FBFC 0E1108A8 FD17B448 A6855419 9C47D08F FB10D4B8 and

Order = FFFFFFFF FFFFFFFF FFFFFFFF FFFFFFFE BAAEDCE6 AF48A03B BFD25E8C D0364141.

$$a_0^{(i)}G = \sum_{h \in B} b_h f_i(h)G \quad \forall P_{(h \neq i)}$$

If it is not consistent, the participant rejects the protocol and starts again.

(7) Participant $p_{(i)}$ now either calculates their share $d_{A(i)}$ as

$$\text{SHARE}(p_{(i)}) = d_{A(i)} = \sum_{h=1}^{j} f_h(i) \bmod n$$

Where $\text{SHARE}(p_{(i)}) \in \mathbb{Z}_n$ and $d_{A(j)}$
and
Where $Q_A = Exp - Interpolate(f_1, \ldots, f_j) \quad \rhd [= G \times d_A]$
Return $(d_{A(i)}, Q_A)$

Participant $p_{(i)}$ now uses the share in calculating signatures. This role can be conducted by any participant or by a party $p_{(c)}$ that acts as a coordinator in the process of collecting a signature. The participant can vary and does not need to be the same party on each attempt to collect enough shares to sign a transaction. Hence, private key shares $d_{A(i)} \leftarrow \mathbb{Z}_n^*$ have been created without knowledge of the other participant's shares.

Algorithm 2 Updating the private key
Input: Participant P_i's share of private key d_A denoted as $d_{A(i)}$.
Output: Participant P_i's new private key share $d_{A(i)}$.
Algorithm 2 can be used to both update the private key as well as to add randomness into the protocol. In a follow-up paper related to this one, we will demonstrate how this algorithm can be extended to allow group additive and subtractive processes. In this manner, it will be possible to create distributed key sets where the private key is never created or issued and yet transactional signatures can be completed. More importantly, this system will be extensible through the addition and removal of members within a hierarchy of groups.

Using US patent number 15087315 [22] format keys, this process can lead to the recalculation of hierarchical sub-keys without the reconstruction or even calculated existence of the private keys. In this manner, we can construct hierarchies of bitcoin addresses and private key slices that when correctly deployed will remove any large-scale fraud or database theft as has occurred in the past.

(1) Each participant selects a random polynomial of degree $(k-1)$ subject to zero as it's free term. This is analogous to Algorithm 1 but that the participants must validate that the selected secret of all other participants is zero.

Note that $\emptyset G = nG = 0$ where 0 is a point at infinity on the elliptic curve.
Using this equality, all active participants validate the function:

$$a_0^{(i)}G = \emptyset \quad \forall i = \{1, \ldots, j\}$$

See Feldman [9] for an analogy.

Generate the zero-share: $z_i \leftarrow \mathbb{Z}_n^*$

(2) $d'_{A(i)} = d_{A(i)} + z_i$

(3) **Return**: $d'_{A(i)}$

The result of this algorithm is a new key share that is associated with the original private key. A variation of this algorithm makes the ability to both increase the randomness of the first algorithm or to engage in a re-sharing exercise that results in new key slices without the need to change the bitcoin address possible. In this way, our protocol allows a group to additively mask a private key share without altering the underlying private key. This process can be used to minimise any potential key leakage associated with the continued use and deployment of the individual key shares without changing the underlying bitcoin address and private key.

Algorithm 3 Signature Generation

Domain Parameters: CURVE, Cardinality n and Generator G

Input: Message to be signed $e = H(m)$

              Private Key Share $d_{A(i)} G \mathbb{Z}_n^*$

Output: Signature $(r, s) \in \mathbb{Z}_n^*$ for $e = H(m)$

(A) Distributed Key Generation

    (1) Generate the ephemeral key shares using Algorithm 1:

$$D_{k(i)} \leftarrow \mathbb{Z}_n^*$$

    (2) Generate mask shares using Algorithm 1:

$$\alpha_i \leftarrow \mathbb{Z}_n$$

    (3) Generate mask shares with Algorithm 2:

$$b_i, c_i \leftarrow \mathbb{Z}_n^2$$

(B) Signature Generation

    (4) $e = H(m)$ Validate the hash of the message m

    (5) Broadcast

$$\vartheta_i = D_{k(i)} \alpha_i + \beta_i \bmod n$$

    And

$$\omega_i = G \times \alpha_i$$

(6) $\quad \mu = Interpolate(\vartheta_i, \ldots, \vartheta_n) \bmod n$

$$\triangleright [= Dk\alpha \bmod n]$$

(7) $\quad \theta = Exp - Interpolate(\omega_1, \ldots, \omega_n)$

$$\triangleright [= G \times \alpha]$$

(8) Calculate (R_x, R_y) where $r_{x,y} = (R_x, R_y) = \theta \times \mu^{-1}$

$$\triangleright = [G \times D_k^{-1}]$$

(9) $\quad r = r_x = R_x \bmod n$
    If $r = 0$, start again (i.e. from the initial distribution)
(10) Broadcast $S_i = D_{k(i)}(e + D_{A(i)}r) + C_i \bmod n$
(11) $S = Interpolate(s_i, \ldots, s_n) \bmod n$
    If $s = 0$ redo Algorithm 3 from the start (A.1).
(12) Return (r, s)
(13) In bitcoin, reconstruct the transaction with the (r, s) pair to form a standard transaction.

4 The Model—Threshold ECDSA (T.ECDSA)

In our model, we allow for the system of n groups or individuals that we designate as participants. Each player can be an individual is a soul participant or a group or combination of the above. Participant $p_{(i)}$ may be mapped against an identity using a commonly derived public key calculation, or participant $p_{(i)}$ may be left as a pseudonymous entity with the public key of the participant used only for this protocol without being mapped back to the individual.

The system introduces a dedicated broadcast channel allowing for the recognition of other participants as a valid player and a member of the scheme while simultaneously allowing members within the group to remain unidentified. When a message is broadcast from participant $p_{(i)}$, the members within the group will recognise the message as coming from an authorised party without necessarily being able to identify the end-user or individual that is associated with the key. It is also possible to link the identity of the key to an individual, if such a system were warranted.

We see this process flow summarised below:
Step (1)

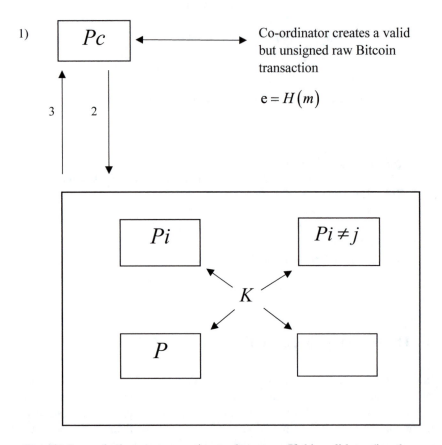

1) Pc Co-ordinator creates a valid
                                   but unsigned raw Bitcoin
                                   transaction

                                   $$e = H(m)$$

Step (2) *Pc* sends the raw transaction to the group. If this validates (i.e. the raw transaction matches the hash to be signed), the participant votes by signing it.

Step (3) If *Yes*, each participant returns the partially signed transaction.

Step (4) *Pc* (or any other participant) reconstructs the complete signature if a threshold of partially signed transactions is received.

Step (5) *Pc* broadcasts the transaction as a signed bitcoin transaction.

The calculation of a message signature can be initiated by an individual who does not change, or through a temporary broadcasting party. The role of the protocol coordinator can be conducted by any participant or by a party $p_{(c)}$ that acts as a coordinator in the process of collecting a signature. The individual scenario can be constructed ahead of time in a manner that allows for any member of the system as participant $p_{(i)}$ to step forth as the coordinating party $p_{(c)}$ in proposing a scenario that leads to a transaction signature if the required votes are obtained or may be based on the necessity to get authorisation through a set coordinator that does not change throughout the process. A subsequent paper will document this process in greater detail.

Key Generation: **We use a modified ECDSA key generation algorithm to make a signature scheme that is fully distributed. In this scheme, the private key is communally selected by the distributed group using a combination of hidden random secrets.** The threshold key derivation algorithm is given in Algorithm 1.

The algorithm is extensible, and each step of the algorithm can be executed by every participant synchronously without a dealer or in groups or individuals or dealers. This implementation is fully compatible with the current bitcoin protocol. Any signatories will appear to an outside observer or verifier as if they were signed in the standard manner. Consequently, there is no way to tell if a key has been generated in the standard format or using our enhanced protocol.

Signature Generation: The concept of threshold signature generation is described in [20]. Algorithm 3 is related to a procedure reported in [9] that was based on DH-based systems and has been modified to allow for ECDSA.

We've extended this process such that it is fully compatible with both bitcoin transaction processing and signing. This also extends to multisig transactions where it is possible to require distributed keys for each of the multiple signatures that are necessary.

Re-sharing the Private Key: Such a process can be extended to introduce an entirely distributed key re-sharing scheme. This re-distribution is completed when the current participants execute one round of Algorithm 2 adding the resulting zero-share to the participant's private key share. The new shares will be randomly distributed if one participant has introduced a random value.

Such a process allows us to additively mask the private key share while not altering the **actual private key**.

Threshold ECDSA Signature Derivation: The threshold ECDSA signature creation system derived using the ideas related to the threshold DSS signature generation protocol found in [9] which followed the scheme developed in [20].

Verification: Our system allows for the off-line signing and verification of messages before any value being transferred to a known bitcoin address. Each of the parties can calculate and validate an address independently using the processes noted in Algorithm 1. Hence, all participants are aware that their share is valid before any exercise that requires funding a bitcoin address. For this process, although verification schemes are possible, they are unnecessary. Any threshold participant who chooses to send an invalid signature slice is in effect voting for the negative. That is, a vote to not sign the message and hence not complete the transaction in bitcoin is achieved from inaction. The impact is as if they did not sign a message at all.

Algorithm 2 provides a method where participants can have their share consistency verified. If a threshold of non-malicious participants has been maintained, it is possible to exclude any known malicious participants on rekeying. Hence, key slices can be updated while not allocating fresh slices to known malicious participants, allowing for the refreshing of the key in a manner that also allows for reallocations of slices.

In environments where trust is particularly scarce and malicious adversaries are to be expected as the norm, we can further enhance the robustness of our verification process increasing the ability to defend against an $j/2$ passive and an $j/3$ active adversary [4, 19] when completing secure multi-party computations.

We can enhance the robustness of the system using the additional process:

1. Let D_a be the secret shared among the j participants on a polynomial $A(x)$ of degree $(k-1)$.
2. Separately, participants $p_{(i)}$ have a share $D_{a(i)}$ of D_a and $D_{a(i)}G \forall (i \in 0, \ldots, j)$ which are made available to the group.
3. All participants next share a secret b using Algorithm 2 such that each participant $p_{(i)}$ has a new hidden share $D_{b(i)}$ of D_b on a polynomial of degree $(k-1)$.
   Note: $D_{b(i)} = \sum_{h=1}^{j} D_{b(i)}^{(h)}$ where $D_{b(i)}^{(h)}$ is the sub-share submitted to participant $p_{(i)}$ from participants $p_{(h \neq i)}$.
4. The participants use Algorithm 2 so that each participant $p_{(i)}$ has a new hidden share $Z_{(i)}$ on a polynomial of degree $(2k-1)$ of which the free term equals zero.
5. Each participant $p_{(i)}$ publishes $D_{a(i)} D_{b(i)}^{(h)} G \forall (h \in 0, \ldots, j)$ and $D_{a(i)} D_{b(i)} G$ to the group.
6. Each participant $p_{(h \neq i)}$ can verify the validity of $D_{a(i)} D_{b(i)}^{(h)} G$ as they have $D_{b(i)}^{(h)}$ and $D_{a(i)} G$.
7. Also, participant $p_{(i)}$ can further verify that $D_{a(i)} D_{b(i)} G = \sum_{h=1}^{j} D_{a(i)} D_{b(i)}^{(h)}$.

Any participants can determine if other participants are acting maliciously with such a system.

5 Distributed Key Generation

It is possible to complete the implementation of both distributed autonomous corporations (DACs) and distributed autonomous social organisations (DASOs) in a secure manner through this scheme. We have shown that any k members can represent such a group through an identification scheme (including through digital certificates signed and published by a certification authority) and that any k members can construct a digital signature on behalf of the organisation. This system extends to the signing of bitcoin transactions that verify without any distinguishing feature and provide for the transfer of value. These authentication schemes are proven secure.

Method and implementation: As the protocol encrypts the secret information required to be sent between participants using ECC based on a patent 42 derivation, it is both possible and advisable to collate all messages into a single packet sent to all users such that validation can be done against potentially compromised or hostile participants when necessary.

Signature generation is proposed by a coordinating participant $p_{(c)}$. By default, any key slice can act as the coordinating participant, and the requirements come down to the individual implementation of the protocol. On the creation of a valid raw transaction by $p_{(c)}$, the transaction and the message hash of the transaction are broadcast to all participants $p_{(i \neq c)}$ using an encrypted channel.

A. **Generate ephemeral key shares $D_{k(i)}$**

The participants generate the ephemeral key $\mathbf{D}_k$, uniformly distributed in $\mathbb{Z}_n^*$, with a polynomial of degree $(k - 1)$, using Algorithm 1, which creates shares

$$(D_{k(1)}, \ldots, D_{k(j)}) \overset{((k-1),j)}{\leftrightarrow} D_k \bmod n.$$

Shares of D_k are maintained in **secret** being held individually by each participant.

B. **Generate mask shares α_i**

Each participant generates a random value α_i, uniformly distributed in $\mathbb{Z}_n^*$ with a polynomial of degree $(k - 1)$, using Algorithm 1 to create shares

$$(\alpha_1, \ldots, \alpha_j) \overset{((k-1),j)}{\leftrightarrow} \alpha \bmod n.$$ These are used to multiplicatively mask $\mathbf{D}_{k(i)}$.

The shares of α_i are **secret** and are maintained by the corresponding participant.

C. **Generate mask shares β_i, c_i**

Execute Algorithm 2 twice using polynomials of degrees $2(k - 1)$.

Denote the shares created in these protocols as $(\beta_1, \ldots, \beta_j) \overset{(2(k-1),j)}{\leftrightarrow} \beta \bmod n$ and $(c_1, \ldots, c_j) \overset{(2(k-1),j)}{\leftrightarrow} c \bmod n$. These are used as additive masks. The polynomial must be of degree $2(k - 1)$ because the numbers being masked involve the products of two polynomials of degree $(k - 1)$. This doubles the required number of shares needed to recover the secret.

The shares of b and c are to be kept **secret** by the participants.

D. **Compute** digest of message m: $e = H(m)$

This value is checked against the received hash of the transaction obtained from $p_{(c)}$.

E. **Broadcast** $v_i = D_{k(i)}\alpha_i + \beta_i \bmod n$ and $\omega_i = G \times \alpha_i$

Participant P_i broadcasts $v_i = D_{k(i)}\alpha_i + \beta_i \bmod n$ and $\omega_i = G \times \alpha_i$. If no response is received from P_i, the value used is set to *null*.

Note: $(v_1, \ldots, v_j) \overset{(2(k-1),j)}{\leftrightarrow} D_k\alpha \bmod n$

F. **Compute** $\mu = $ **Interpolate** $(v_1, \ldots, v_j) \bmod n$

Interpolate() [2]:

Where $\{v_1, \ldots, v_n\}(j \geq (2k - 1))$ forms a set, such that at most of $(k - 1)$ are *null,* and all the residual values reside on a $(k - 1)$-degree polynomial $F(\cdot)$, then $\mu = F(0)$.

The polynomial can be computed using ordinary polynomial interpolation. The function 'Interpolate()' is the Berlekamp–Welch Interpolation [2] and is defined[3] as an error-correcting algorithm for BCH and Reed–Solomon codes.

G.   **Compute** $\theta = \mathbf{Exp} - \mathbf{Interpolate}(\omega_1, \ldots, \omega_j)$

Exp-Interpolate() [10]:

If $\{\omega_1, \ldots, \omega_j\}(j \geq (2k - 1))$ is a set where at most $(k - 1)$ values are *null*, the remaining values are of the form $G \times \alpha_i$ and each α_i *exists on* some $(k - 1)$-degree polynomial $H(\cdot)$, then $\theta = G \times H(0)$.

The value can be computed by $\theta = \sum_{i \in V} \omega_i \times \lambda_i = \sum_{i \in V} (G \times H(i)) \times \lambda_i \, \beta$ for which V is a (k)-subset of the correct ω_i *values*, and further, λ_i represents the resultant Lagrange interpolation coefficients. The polynomial can be computed by using the Berlekamp–Welch decoder.

H.   **Compute** $(R_x, R_y) = \theta \times \mu^{-1}$

I.   **Assign** $r = R_x \bmod q$ If $r = 0$, go to step A.

Each participant $p_{(i)}$ computes their slice of r_i in step J. The coordinator $p_{(c)}$ can use these values to reconstruct s if they have received a threshold number of responses.

J.   **Broadcast** $s_i = D_{k(i)}(e + D_{A(i)}r) + c_i \bmod n$

If a response is not solicited/received from P_i, the values used are set to *null*.

Note: $(s_1, \ldots, s_j) \overset{(2(k-1),j)}{\longleftrightarrow} D_k(m + D_A r) \bmod n$ $(s_1, \ldots, s_n) \longleftrightarrow k(m + D_A r) \bmod n$.

K.   **Compute** $s = \mathbf{Interpolate} \, (s_1, \ldots, s_n) \bmod n$

If $s = 0$, go to step I.

See above for the meaning of the function *Interpolate()*.

Each participant $p_{(i)}$ computes their slice of s_i in step J. The coordinator $p_{(c)}$ can use these values to reconstruct s if they have received a threshold number of s_i responses.

L.   Return (r, s)

M.   Replace the signature section of the raw transaction and broadcast this to the network.

Dealer distribution of slices: The system derived above can be made much more flexible through the introduction of group shares. In this manner, the allocation of shares can be split between a dealer, multiple dealers, a group containing no dealers or any possible combination of the above in any level of hierarchical depth.

By replacing the value d_A and its corresponding key slice $d_{A(i)}$ with a value derived using the same algorithm, we can create hierarchies of votes. For example, we can create a scheme that simultaneously integrates shares that are derived from:

---

[3]http://mathworld.wolfram.com/LagrangeInterpolatingPolynomial.html. See also [21]. https://jeremykun.com/2015/09/07/welch-berlekamp/.

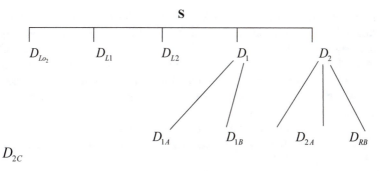

Fig. 1 Secrets can be distributed in reconstructive hierarchies

(1) Dealer-based distributions
(2) Multiple Dealers
(3) No Dealer

Hence, the scheme is extensible and can be made to incorporate any business structure or organisational system. The allocation of slices is also extensible. Deploying an uneven allocation process allows us to add weighting on shares. In the scheme displayed in Fig. 1, we can create a hypothetical organisation with five top-level members. This, however, does not require setting the value of $n = 5$ equally weighted shares. In our hypothetical organisation, we could set the voting structure for the top-level schema as follows:

• Threshold(0)	61 shares
• D_{L0_2}	15 shares
• D_{L1}	15 shares
• D_{L2}	15 shares
• D_1	45 shares
• D_2	10 shares

Here, we have set $n= 100$. As noted, this is an arbitrary value that can reflect any organisational structure. The organisation in Fig. 1 allows for a veto scenario (D_1) and through the introduction of multi-layered allocation that allows for any voting structure that can be imagined. What is often missed in a multilevel hierarchical structure is that although we have allocated slices of the secret, these do not need to be evenly distributed, and further, the ownership of subgroups does not need to mirror that of other levels. In Fig. 1, we have a seemingly powerful block controlling 45% of the total number of shares in 75% of the threshold. If we then look at the lower-level allocation of the shares, the scenario becomes far more complex. We can create cross-ownership with individuals holding voting shares in multiple levels and positions on the table.

The distributions in Table 2 are defined as (shares held, threshold, allocation {**n**}).

From Table 2, we see that participants P1 and P2 each hold sway over the votes but that they coalition with participant P4 provides either P1 or P2 with a sufficient voting block as long as P1 or P2 does not veto the vote. As there are no limits to the implementation and structure of the voting format in our system, we can use this to create any organisational hierarchy that can be imagined as well as ensuring secure backup and recovery methodology.

The result is that we can have veto powers and voting rights that are assigned to higher-level signing shares. In our example, ownership of the shares in S could be held at D_{L0_2}, D_{1A} and D_{2A}.

Secure Multi-party Computation: Secure multi-party function computation with n participants, $p_{(1)}, \ldots, p_{(i)}, \ldots, p_{(n)}$ is a problem based on the need to assess a function $\mathbb{F}(x_1, \ldots, x_i, \ldots, x_n)$, involving $x_{(i)}$, a secret value provided by $p_{(i)}$ that is required to be maintained in confidence such that no participant $p_{(j \neq i)}$ or external party gains any knowledge of $x_{(i)}$. Hence, the objective is to preserve the confidentiality of each participant's values while being able to guarantee the exactness of the calculation.

In this scenario, the trusted third-party T collects all the values $x_{(i:1\ldots n)}$ from the various participants $p_{(i:1\ldots n)}$ and returns the calculation. This design works only in an idealised world where we can implicitly trust T. Where there is any possibility that T could either be malicious, rogue or compromised, the use of a trusted third-party becomes less viable. This scenario mirrors existing elections where the participants are the voters, and the trusted third-party is played by government.

It has been proven [1] that any value that can be computed in a secure manner using a trusted third-party may also be calculated without a trusted party while maintaining the security of the individual secrets $x_{(i)}$. The protocols presented in this paper are secure against private computation and provide secure computation even where a non-threshold group of compromise participants can collaborate.

Simple multiplication: Where we have two secret values, x and y that a distributed among n participants $p_{(i:1\ldots n)}$, it is possible to compute the product xy while simultaneously maintaining the secrecy of both input variables x and y as well as ensuring that the individual secrets $x_{(i:1\ldots n)}$ and $y_{(i:1\ldots n)}$ are maintained by participant $p_{(i)}$ retaining the confidentiality.

Table 2 Hypothetical organisation structure

Participant	Level 0	Level 1	S. votes (max)	S. votes (min)
P1	D_{L0_2} (15, 61, 100)	D_{1A} (5, 6, 10) D_{2A} (3, 8, 10)	70	15
P2	D_{L1} (15, 61, 100)	D_{1B} (5, 6, 10) D_{2B} (3, 8, 10)	70	15
P3	D_{L2} (15, 61, 100)		15	0
P4		D_{2C} (6, 8, 10)	10	0

In this scheme, x and y are each shared between a threshold group of participants using a polynomial of degree $(k-1)$. Each participant $p_{(i)}$ can multiply their share of $x_{(i:1...n)}$ on a polynomial of degree $(k-1)$ of x and $y_{(i:1...n)}$ in a polynomial of degree $(k-1)$ on y.

Introducing Algorithm 2, returns the participant $p_{(i)}$ share of $z_{(i)}$, a polynomial of degree $(2k-1)$. With such a value, each participant $p_{(i)}$ calculates the value $x_{(i)}y_{(i)} + z_{(i)}$. The return value for $x_{(i)}y_{(i)} + z_{(i)}$ represents a valid share of the calculation for $x.y$ on a polynomial of degree $(2k-1)$. Any participant or a coordinator acting for the threshold number of shares can use the return value held by each participant to calculate the true value of $x.y$ without obtaining any knowledge of the individual shares.

Simple addition: Where we have two secret values, x and y, that are distributed among n participants $p_{(i:1...n)}$, it is possible to compute the product $x + y$ while simultaneously maintaining the secrecy of both input variables x and y as well as ensuring that the individual secrets $x_{(i:1...n)}$ and $y_{(i:1...n)}$ that are maintained by participant $p_{(i)}$ retain the confidentiality.

As per the process for simple multiplication, each participant $p_{(i)}$ calculates the value $x_{(i)} + y_{(i)} + z_{(i)}$. The calculation of $z_{(i)}$ is not necessary, but adds a further level of randomness and confidentiality to the process.

The return value for $x_{(i)} + y_{(i)} + z_{(i)}$ represents a valid share of the calculation for $x + y$ on a polynomial of degree $(2k-1)$. Any participant or a coordinator acting for the threshold number of shares can use the return value held by each participant to calculate the true value of $x + y$ without obtaining any knowledge of the individual shares. If the participants are less hostile, this can be simplified as an $x_{(i)} + y_{(i)}$ addition without the additional step.

Inverse or reciprocal: For a distributed secret value, $x \bmod n$ which is distributed confidentially between j participants as $x_{(i:1...j)}$, it is possible to generate shares of the polynomial associated with the value for $x^{-1} \bmod n$ while not revealing any information that could disclose the values $x_{(i)}, x$ or x^{-1} [10]. Again, each participant $p_{(i)}$ maintains a share of the value x represented by $x_{(i)}$ over a polynomial of degree $(k-1)$.

Using Algorithm 1, each participant creates a share $x_{(i)}$ of an unknown secret $x.y$ on a polynomial of degree $(k-1)$. Each participant then runs Algorithm 2 to calculate $(k-1)$ of a zero secret on a polynomial of degree $(2k-1)$. Each participant $(2k-1)$ performs the calculation to compute the value $x_{(i)}y_{(i)} + z_{(i)}$.

Using the **Interpolate()** routine presented above, each participant can calculate the value of $\mu = x_{(i)}y_{(i)} + z_{(i)}$ returning the value μ from the collected values of μ_i. Each participant can then calculate the value of $\mu^{-1} \bmod n$. Such values are sufficient such that any participant $p_{(i)}$ can compute the associated share of x_i^{-1} using $\zeta_i = \gamma_i \mu^{-1}$ one a polynomial of degree $(2k-1)$. The Berlekamp–Welch decoding scheme [2] provides one of several methods that can be used to complete this process.

Assignment: **The ability to sign a transaction in a verifiable and provable manner provides the opportunity to prove ownership privately and even relinquish or exchange ownership of the bitcoin private key and associated bitcoin address without publicly moving anything on the blockchain. In this manner, a bitcoin**

address can be funded, and the contents of that address may be transferred or sold without leaving a public record. As this process is a threshold system, the assignment of key slices can be achieved securely without further settlement recorded on the blockchain. In such a way, we can separate the ownership of a note already settled on the blockchain from the process of transacting that note.

CLTV.: A bitcoin message or, in more common parlance, transaction can be created with the inclusion of a CLTV [5] entry. With such an addition, the transaction can be made recoverable even in the catastrophic loss of all key slices or if multiple slices from an entity are deemed untrustworthy or lost in a manner that does not allow for the secure reconstruction of a signature with a minimum threshold.

This is further possible where an entity is using a third-party service and desires to ensure that that service cannot hold or deny access to the keys. In constructing a bitcoin transaction with a time-based fail-safe, the user knows that a malicious third-party or a compromised exchange site or bank cannot extort them for access to their keys. As a worst-case scenario, the compromise to a catastrophic level would lead to the time-based reversal of a transaction to a predefined address based on a CLTV condition. This predefined address can be created using the protocols noted within this paper. As such, it is possible to construct a series of transactions and keys that cannot be readily compromised.

6 Security Considerations

Benger et al. [3] offered one example of ECDSA private key recovery using a flash and reload methodology. This occurrence is but one example of attacks against system RAM and cache. Such methods leave the use of procedures such as that of Shamir's SSS [20] wanting as they reconstruct the private key. Also, in any scenario where a private key is reconstructed at any point in time, a requirement for trust is introduced. It is necessary in such a scenario to rely on the systems and processes of the entity holding the private key.

Even if the trusted party is not malicious, there is a necessity to rely on their processes. As we have seen from many recent compromises, such reliance on reconstructing the private key leaves avenues of attack.

As both a drop-in replacement for the existing ECDSA implementations as well as being completely transparent and compatible with the current bitcoin protocol, no hard fork or soft fork is required for its implementation, and the implementation is indistinguishable from any current transaction. Our application can treat individuals as separate participants allowing for the group signing of keys with a recovery function. As an example, a two of two scheme can be implemented using four key slices where the online wallet provider or exchange maintains two key slices, and the end-user maintains two slices. The exchange and the user would each have a two of two process over their key slices which would then be used in conjunction with each other for the secure signing of a message when required.

7 Conclusions

The scheme forms the foundation of what bitcoin sought to achieve with the introduction of a group signature process. The addition of a fault-tolerable signing system with the coupling of a distributed key-creation system removes all centralisation and trust requirements. Many systems will evolve with the need for trust. The removal of the need to trust the underlying infrastructure powering such systems enables them to differentiate themselves based on their real merits.

Moreover, the introduction of an implicitly decentralised system allows for the creation of more robust and resilient protocols. The compatibility between ECDSA [12] and Shamir's SSS [20] has allowed us to introduce a system that extends bitcoin with a new verifiable secret sharing scheme. This system is far more efficient than anything derived by Feldman [9] or Pedersen [18] while losing nothing in security.

In this paper, we have presented a system that extends the functionality of bitcoin without the requirement for a change in the base protocol. Using our system

1. a trusted third-party is no longer required for the selection or distribution of a key secret,
2. a distributed banking exchange system can be created that does not rely on third-party trust,
3. each member or group of members may independently verify that the share of the secret key that is held corresponds to the bitcoin address and public key advertised,
4. a protocol exists to refresh the private key slices to mitigate the effects of eavesdropping and related attacks and
5. no trusted third-party is required for the group signing of transactions and messages.

As our system prevents sensitive data from ever appearing in memory, we have completely solved many extant security risks.

This paper will be extended to allow for the automated group allocation of shares to patent number 15087315 format hierarchical wallets [22]. This previous group calculation will be extended from simple signing into a hierarchical group structure allowing for the automated creation of change addresses and hierarchical sub-keys that are associated with an initial shared secret group.

Additionally, it will be possible to incorporate key reconstruction within a bitcoin message and transaction. This system would allow the creation of a P2SH address associated directly with a standard single address transaction. The merging of these forms of transactions will open many opportunities for smart contracts.

The full ramifications of this paper will be extended in a series of daughter papers. In these, we will introduce a fully distributed ecosystem that can coexist within the bitcoin blockchain as a fully distributed allocation system based on blind signatures that extends the work of Chaum [6] in creating a fully distributed model of what was sought with Digi-cash.

A further paper will introduce a series of multi-party computations allowing bitcoin to act as a black-box computational algorithm. The use of zero-knowledge proofs for programming when coupled with our group distributed share system allows for a flexible design of computational systems and smart contracts that can be stored in the blockchain while also being run without knowledge of other parties.

The final edition will be the extension into a dynamic threshold group management structure for e-wallets and calculational entities. This will allow for the secure addition and removal of parties involved in the sharing and distribution of keys in a manner that provides external group allocations without the need to enact settlement on the bitcoin blockchain.

Appendix

To compute $\frac{\beta = Exp - Interpolate(w_i, \ldots, w_n)}{Exp - Interpolate() \quad \rightarrow *}$

If $\{w_i, \ldots, w_n\}(n \geq 2t + 1)$ is a set of values, such that at most, t are null and the remaining are of the form $G \times a_i$ where the a_i's lie on some $(k-1)$-degree Polynomial $H(\cdot)$, then $\beta = G \times H(\emptyset)$. This is computed using:

$$\beta = \Sigma_{i \in v} w_i \times \lambda_i$$
$$= \Sigma_{i \in v} (G \times H(i)) \times \lambda_i$$

where v is a k-subset of the correct w_i's and λ_i's are the corresponding Lagrange interpolation coefficients. For more background information, see the following: [7, 14, 16, 19, 22].

References

1. Bar-Ilan, J.B.: Non-cryptographic fault-tolerant computing in a constant number of rounds. In: Proceedings of 8th PODC, pp. 201–209 (1989)
2. Berlekamp, E.R.: Algebraic Coding Theory. McGraw-Hill, New York (1968)
3. Benger, N., van de Pol, J., Smart, N.P., Yarom, Y.: "Ooh Aah… Just a Little Bit": a small amount of side channel can go a long way. In: Batina, L., Robshaw, M. (eds.) Cryptographic Hardware and Embedded Systems I CHES 2014. LNCS, vol. 8731, pp. 75–92. Springer, Heidelberg (2014)
4. Ben-Or, M., Goldwasser, S., Wigderson, A.: Completeness theorems for noncryptographic fault-tolerant distributed computation. In: Proceedings of the Twentieth Annual ACM Symposium on Theory of Computing, STOC'88, pp. 1–10. ACM, New York (1988)
5. BIP 65. Github Homepage. https://github.com/bitcoin/bips/blob/master/bip-0065.mediawiki. Accessed 26 Jan 2019
6. Chaum, D.: Blind signatures for untraceable payments (PDF). Adv. Cryptol. Proc. Crypto **82**(3), 199–203 (1983)
7. Dawson, E., Donovan, D.: The breadth of Shamir's secret-sharing scheme. Comput. Secur. **13**, 69–78 (1994)
8. Desmedt, Y.: Society and group oriented cryptography: a new concept. In: Pomerance, C. (ed.) A Conference on the Theory and Applications of Cryptographic Techniques on Advances in Cryptology, CRYPTO'87, pp. 120–127. Springer, London (1987)

9. Feldman, P.: A practical scheme for non-interactive verifiable secret sharing. In: Proceedings of the 28th IEEE Annual Symposium on Foundations of Computer Science, pp. 427–437. Computer Society Press of the IEEE, Washington, DC (1987)

10. Gennaro, R., Jarecki, S., Krawczyk, H., Rabin, T.: Robust threshold DSS signatures. In: Maurer, U. (ed.) Proceedings of the 15th Annual International Conference on Theory and Application of Cryptographic Techniques, EUROCRYPT'96, pp. 354–371. Springer, Berlin, Heidelberg (1996)

11. Ibrahim, M., Ali, I., Ibrahim, I., El-sawi, A.: A robust threshold elliptic curve digital signature providing a new verifiable secret sharing scheme. In: 2003 46th Midwest Symposium on Circuits and Systems, vol. 1, pp. 276–280. IEEE, Cairo (2003)

12. Johnson, D., Menezes, A., Vanstone, S.: The elliptic curve digital signature algorithm (ECDSA). Int. J. Inf. Secur. **1**(1), 36–63 (2001)

13. Kapoor, V., Abraham, V.S., Singh, R.: Elliptic Curve Cryptography. ACM Ubiquity **9**, 1–8 (2008)

14. Knuth, D.E.: The Art of Computer Programming, II: Seminumerical Algorithms, 3rd edn., p. 505. Addison-Wesley, Reading (1997)

15. Koblitz, N.: An elliptic curve implementation of the finite field digital signature algorithm. In: Advances in Cryptology—Crypto'98. Lecture Notes in Computer Science, vol. 1462, pp. 327–337. Springer, Berlin (1998)

16. Liu, C.L.: Introduction to Combinatorial Mathematics. McGraw-Hill, New York (1968)

17. National Institute of Standards and Technology: Digital Signature Standard (DSS). FIPS PUB 186-4 (2003). CSRC Homepage. https://csrc.nist.gov/publications/detail/fips/186/4/final. Accessed 26 Jan 2019

18. Pedersen, T.: Non-interactive and information-theoretic secure verifiable secret sharing. In: Feigenbaum, J. (ed.) Advances in Cryptology—CRYPTO'91. LNCS, vol. 576, pp. 129–140. Springer, Berlin, Heidelberg (1992)

19. Rabin, T., Ben-Or, M.: Verifiable secret sharing and multiparty protocols with honest majority. In: STOC'89 Proceedings of the Twenty-First ACM Symposium on Theory of Computing, pp. 73–85. ACM, Seattle (1989)

20. Shamir, A.: How to share a secret. Commun. ACM **22**(11), 612–613 (1979)

21. Whittaker, E.T., Robinson, G.: Lagrange's formula of interpolation. In: The Calculus of Observations: A Treatise on Numerical Mathematics, 4th ed., pp. 28–30. Dover, New York (1967) (Sect. 17)

22. Wright, C., Savanah, S.: Determining a Common Secret for the Secure Exchange of Information and Hierarchical, Deterministic Cryptographic Keys. International Patent Application Number: HRP20181373 (T1). WIPO. 2019/01/11. Espacenet Homepage. https://worldwide. espacenet.com/publicationDetails/biblio?DB=EPODOC&II=0&ND=3&adjacent=true& locale=en_EP&FT=D&date=20190111&CC=HR&NR=P20181373T1&KC=T1. Accessed 26 Jan 2019

Comparative Analysis of ML Classifiers for Network Intrusion Detection

Ahmed M. Mahfouz, Deepak Venugopal and Sajjan G. Shiva

Abstract With the rapid growth in network-based applications, new risks arise, and different security mechanisms need additional attention to improve speed and accuracy. Although many new security tools have been developed, the fast growth of malicious activities continues to be a severe issue, and the ever-evolving attacks create serious threats to network security. Network administrators rely heavily on intrusion detection systems to detect such network intrusive activities. Machine learning methods are one of the predominant approaches to intrusion detection, where we learn models from data to differentiate between abnormal and normal traffic. Though machine learning approaches are used frequently, a deep analysis of machine learning algorithms in the context of intrusion detection is somewhat lacking. In this work, we present a comprehensive analysis of some existing machine learning classifiers regarding identifying intrusions in network traffic. Specifically, we analyze classifiers along various dimensions, namely feature selection, sensitivity to hyperparameter selection, and class imbalance problems that are inherent to intrusion detection. We evaluate several classifiers using the NSL-KDD dataset and summarize their effectiveness using a detailed experimental evaluation.

Keywords IDS · Machine learning · Classification algorithms · NSL-KDD dataset · Network intrusion detection · Data mining · Feature selection · WEKA · Hyperparameters · Hyperparameter optimization

A. M. Mahfouz (✉) · D. Venugopal · S. G. Shiva
The University of Memphis, Memphis, TN 38152, USA
e-mail: amahfouz@memphis.edu

D. Venugopal
e-mail: dvngopal@memphis.edu

S. G. Shiva
e-mail: sshiva@memphis.edu

© Springer Nature Singapore Pte Ltd. 2020
X.-S. Yang et al. (eds.), *Fourth International Congress on Information and Communication Technology*, Advances in Intelligent Systems and Computing 1027, https://doi.org/10.1007/978-981-32-9343-4_16

193

1 Introduction

Because of the massive volume of data on the network, the content of the network becomes vulnerable to a variety of attacks, and different intrusions are increasing day after day. Detecting intrusions is an essential step to stopping the intruders from breaking into or misusing network data. To defend against numerous network intrusions and malicious activities, many methods have been developed. Network intrusion detection is considered one of the most promising methods to protect the network from different dynamic intrusion behaviors. Intrusion detection system differentiates intrusive behaviors from normal network activities by classifying data into various categories [1].

Several machine learning (ML) methods have been proposed to develop effective and intelligent intrusion detection [2]. However, there have been very few systematic studies that evaluate ML approaches for intrusion detection. Specifically, ML classifiers are typically used as a "black-box," where reported results may be obtained as a result of overfitting the model for a specific dataset [3]. Thus, the results are not very generalizable and hard-to-replicate. Our main contribution in this paper is to provide a detailed experimental evaluation of a group of supervised ML methods in the context of intrusion detection. Specifically, we evaluate several well-known ML methods for intrusion detection along the following dimensions: feature selection, sensitivity to hyperparameter tuning, and effect of the class imbalance. All these three dimensions are critical to effective use of ML methods in intrusion detection. Specifically, feature selection chooses the optimal subset of features to avoid building a complex classifier that may overfit the data. A large hyperparameter sensitivity indicates that tuning the detection system optimally may be difficult for other datasets, and algorithms that handle the class imbalance problem more effectively are more viable in practice for intrusion detection.

The rest of the paper is organized as follows; we provide the background in Sect. 2 where we talk about intrusion detection systems and machine learning and introduce one of the most popular network traffic datasets. In Sect. 2 also, we discuss the feature selection concept as well as the concept of hyperparameter optimization. The performance evaluation metrics are discussed in Sect. 3. Related work is produced in Sect. 4, and the experimental results are reported in Sect. 5. Finally, in Sect. 6, we present the conclusion and mention the future work.

2 Background

2.1 Intrusion Detection System

An intrusion is a malicious activity that aims to compromise the confidentiality, the integrity, or the availability of any of the network components as an attempt to disrupt the security policy of the network [1].

Based on the analysis strategies and the detection methods, IDSs are categorized into misuse intrusion detection (MID) systems and anomaly intrusion detection (AID) systems. Based on the data source, IDSs are categorized into network-based (NIDS) and host-based (HIDS) detection systems [4, 5].

2.2 Machine Learning Classifiers

Several ML methods have been proposed to monitor and analyze network traffic for different anomalies. Most of these methods (classifiers) identify the anomaly by looking for variations from a basic normal traffic model. Usually, these models are trained with a set of attack-free traffic data that is collected over a long period. Any ML anomaly detection method is one of three broad categories that are supervised, unsupervised, or semi-supervised learning methods. In this paper, we will focus on the supervised learning classifiers.

Supervised learning is the type of models that take both of input variable (X) and output variable (Y) to provide a learning basis to support future judgments by learning the mapping function $Y = f(X)$. Supervised learning uses a training data which is a set of examples with paired input records and their desired outputs. In this learning, the correct answer is known in advance, and the learner algorithm iteratively makes predictions on the training data and stops only when an acceptable level of performance is achieved. Thus, this method is appropriate when there is a specific target value. Supervised learning problem can be defined as either a classification problem or a regression problem. The output variable of the classification problem is a category, like "white" or "black" and "disease" or "no disease." On the other hand, the output variable of the regression problem is a real value, such as "the number of dollars" or "the height" [6]. The most famous supervised learning algorithms are decision trees (DT), support vector machines (SVM), artificial neural network (ANN), K-nearest neighbors (KNN), logistic regression (LR), random forests (RF), Naive Bayes (NB), etc.

2.3 Datasets

To support the assessment of different intrusion detection methods, researchers have introduced several network traffic datasets. These datasets are either public, private, or network simulation dataset. Most of these datasets were generated using several tools that helped in capturing the traffic, launching different types of attacks, and monitoring traffic patterns. In this paper, we use NSL-KDD dataset which is one of the most popular benchmark datasets in the domain of intrusion detection.

NSL-KDD. The NSL-KDD dataset [7] is a refined offline version of the well-known KDDcup99 dataset. Many researchers have carried out different types of analysis on the NSL-KDD and have employed different methods and tools to develop effective IDSs [8]. The NSL-KDD dataset has 41 attributes plus one class attribute. A full description of these attributes can be found in [9]. We present a statistical summary of the NSL-KDD dataset in Sect. 5.1.

2.4 Feature Selection

Feature selection is the process of selecting a subset of the original features so that the feature space is optimally reduced to the evaluation criterion [10]. A feature selection method selects a subset of relevant features. The relevance definition varies from technique to another. Based on its notion of relevance, a feature selection technique mathematically formulates a criterion for evaluating a set of features generated by a scheme that searches over the feature space. Kohavi et al. [11] define two degrees of relevance, strong and weak. A feature s is strongly relevant if removal of s deteriorates the performance of a classifier. A feature s is called weakly relevant if it is not strongly relevant and removal of a subset of features containing s deteriorates the performance of the classifier. A feature is irrelevant if it is neither strongly nor weakly relevant.

2.5 Hyperparameters

Parameters versus hyperparameters. Model parameters are the set of configuration variables that are internal to the model and can be learned from the historical training data. The value of those parameters is estimated from the input data. Model hyperparameters are the set of configuration variables that are external to the model. They are the properties that govern the entire training process and cannot be directly trained from the input data. The model parameters specify how input data is transformed into the desired output, while the hyperparameters define the structure of the model.

 Hyperparameter optimization. Hyperparameter optimization (also known as hyperparameter tuning) is the process of finding the most optimal hyperparameters for the learning algorithm in ML. A set of different measures for a single ML model can be used to generalize different data patterns. This set is known as the hyperparameters set which should be optimized such that the ML model can solve the assigned problem as optimally as possible. The optimization process locates the hyperparameters tuple and then produces a model that minimizes the predefined loss function on the given data. The objective function takes the hyperparameters tuple and returns the associated loss. The generalization performance is often estimated using cross-validation [12]. Hyperparameter optimization techniques mostly use any one of optimization algorithms: grid search, random search, or Bayesian optimization.

3 Performance Evaluation

Various performance measurements have been proposed in the literature. Following are the most popularly used parameters in evaluating an ML model performance that can be used in ML-based IDS:

3.1 Confusion Matrix

The efficiency of an ML model is usually determined by metrics called sensitivity and specificity measure. The sensitivity is referred to as the true positive rate (TPR), while specificity is known as a true negative rate (TNR). However, there is often a trade-off between these metrics in "real-world" applications.

$$\text{Sensitivity} = \frac{(TP)}{(TP + FN)} \tag{1}$$

$$\text{Specificity} = \frac{(TN)}{(TN + FP)} \tag{2}$$

$$\text{Accuracy} = \frac{(TP + TN)}{(TP + TN + FP + FN)} \tag{3}$$

3.2 Precision, Recall, and F-Measure

Precision and recall are two recognized evaluation metrics in the information retrieval area. Precision refers to the portion of the relevant instances among the retrieved instances. Recall refers to the portion of relevant retrieved instances from the total number of the relevant instances. F-measure is the precision and recall harmonic mean.

$$\text{Precision} = \frac{TP}{(TP + FP)} \tag{4}$$

$$\text{Recall} = \frac{(TP)}{(TP + FN)} \tag{5}$$

$$\text{F-measure} = \frac{(2 * \text{Precision} * \text{Recall})}{(\text{Precision} + \text{Recall})} \tag{6}$$

3.3   Receiver Operating Characteristic (ROC)

The use of ROC curves is a well-known evaluation measure that visualizes the relation between true positive (TP) and false positive (FP) rates of IDSs. It is also used to compare two or more ML classifiers regarding accuracy effectively.

3.4   K-Fold Cross-Validation

One of the most famous statistical methods in evaluating and comparing ML models is K-fold cross-validation. It works by first separating the dataset into K equally sized folds (instances). K-1 folds are used to train the model, and the last one is left out for prepared model testing. The procedure is then reiterated so that every fold gets the chance to act as the test dataset. Finally, the capability of the model on the problem is estimated by averaging the performance measures across all folds. The K-folds number is decided based on the size of the dataset, but the most used numbers are 3, 5, 7, and 10. The goal is to choose a number that makes a good balance between the size and representation of data in your train and test sets.

4   Related Work

MeeraGandhi [13] used DARPA–Lincoln dataset to evaluate and compare the performance of four supervised ML classifiers regarding detecting the four categories; DoS, R2L, Probe, and U2R attacks. Their results show that the J48 classifier outperforms the other three classifiers IBK, MLP, and NB in prediction accuracy. In [14], Nguyen et al. performed an empirical study to evaluate a comprehensive set of ML classifiers on the KDD'99 dataset to detect attacks from the four attack classes. Abdeljalil et al. [15] tested the performance of three ML classifiers, namely J48, NN, and SVM using the KDD'99 dataset and found that the J48 algorithm outperformed the other two algorithms.

Dhanabal et al. [9] analyzed and used the NSL-KDD dataset to measure the effectiveness of ML classifiers in detecting the anomalies in the network traffic patterns. In their experiment, 20% of the NSL-KDD dataset has been used to compare the accuracy of three classifiers. Their results show that with CFS being used for dimensionality reduction, J48 outperforms SVM and NB regarding accuracy. In [17], Belavagi et al. tried to check the performance of four supervised ML classifiers, namely SVM, RF, LR, and NB for intrusion detection over the NSL-KDD dataset. From the results, it was found that the RF classifier has 99% accuracy.

Unlike the above studies, our paper concentrates on evaluating and comparing the performance of a group of well-known, supervised ML classifiers over the full

NSL-KDD dataset for intrusion detection along the following dimensions: feature selection, sensitivity to hyperparameter tuning and class imbalance.

5 Experimental Results

In this section, we present the experimental setup and the results of comparing six ML classifiers regarding classification accuracy, TPR, FPR, precision, recall, F-measure, and ROC area. We selected the six classifiers from various classifier families and applied them to NSL-KDD dataset. The selected classifiers are Naïve Bayes, logistic, multilayer perceptron (NN), SMO (SVM), IBK (KNN), and J48 (DT).

5.1 Statistical Summary of NSL-KDD

Each record in the NSL-KDD dataset unfolds different features of the traffic with 41 attributes plus an assigned label classifying each record as either normal or attack. The features of the dataset are three types: nominal, numeric, and binary. The nominal features are 2, 3, and 4, while the binary features are 7, 12, 14, 15, 21, and 22, and the rest of the features are a numeric type. Authors in [9] listed the details of those attributes that are the attributes names, description, and sample data.

Attack types in the dataset can be grouped into four main classes, namely DoS, U2R, Probe, and R2L [16]. Table 1 maps different attack types with its attack class,

Table 1 NSL-KDD attack types and classes

Attack class	Attack type	Sample relevant feature	Example
DoS	Apache2, Back, Pod, Process table, Worm, Neptune, Smurf, Land, Udpstorm, Teardrop	Percentage of packets with errors—source bytes	Syn flooding
Probe	Satan, Ipsweep, Nmap, Portsweep, Mscan, Saint	Source bytes—duration of the connection	Port scanning
R2L	Httptunnel, Snmpgetattack, Snmpguess, Guess_Password, Imap, Warezclient, Ftp_write, Phf, Multihop, Warezmaster, Spy, Xsnoop, Xlock, Sendmail	Number of shell prompts invoked—the number of file creations	Buffer overflow
U2R	Buffer_overflow, Xterm, SQL attack, Perl, Loadmodule, Loadmodule, Ps, Rootkit	Service requested—connection duration—num of failed login attempts	Password guessing

Table 2 No of samples for normal and attack classes

Class	Training set	Occurrences percentage (%)	Testing set	Occurrences percentage (%)
Normal	67,343	53.46	9711	43.08
DoS	45,927	36.46	7460	33.08
Probe	11,656	9.25	2421	10.74
R2L	995	0.79	2885	12.22
U2R	52	0.04	67	0.89
Total	125,973	100.0	22,544	100.0

while Table 2 shows the number of occurrences for normal and different attack classes.

Table 2 shows that the number of attack records associated with the R2L and U2R attack classes in the dataset is very low compared to the normal and other attack classes, which leads to the imbalanced problem. Classification process for any imbalanced dataset is always a challenging issue for researchers. Most standard ML and data mining methods consider balanced datasets. When the methods are used with an imbalanced dataset, they produce biased results toward the samples from the majority classes. The classification accuracy for the majority classes is much higher than for the minority classes [17].

5.2　Experimental Setup

Our experimental setup went through three phases. In the first phase, we compared the performance of the classifiers with their default settings and without any preprocessing for the dataset. We trained the classifiers on the training dataset provided by NSL-KDD using stratified cross-validation of 10-folds and used the trained models with the testing dataset. The testing datasets were also provided by NSL-KDD to compare the performance. The results of this phase are summarized in Tables 3 and

Table 3 Classifiers trained models accuracy metrics of Phase 1

Classifier	Accuracy (%)	TPR	FPR	Precision	Recall	F-measure	ROC area
NB	80.20	0.841	0.087	0.890	0.841	0.852	0.968
Logistic	91.55	0.955	0.039	0.952	0.955	0.952	0.983
MLP	94.98	0.969	0.026	0.973	0.969	0.970	0.992
SMO	91.81	0.957	0.027	0.972	0.957	0.973	0.977
IBK	94.62	0.986	0.002	0.996	0.996	0.996	0.988
J48	94.74	0.987	0.002	0.997	0.997	0.997	0.987

4 where Table 3 summarizes performance metrics for the trained models and Table 4 summarizes the same performance metrics for the trained models on the test dataset.

In the second phase, the NSL-KDD dataset was preprocessed to reduce its dimension by selecting the most relevant features. We applied the InfoGainAttributeEval algorithm with ranker which ranked the attributes by their evaluation and resulted in selecting 14 out of 41 features suggested by NSL-KDD. The selected features are 3, 4, 5, 6, 9, 12, 14, 25, 26, 29, 30, 37, 38, and 39. We also used CVParameterSelection to perform the hyperparameter optimization for each classifier. Finally, we compared the performance as we did in the first phase. The results of this phase are summarized in Tables 5 and 6.

Table 4 Classifiers trained models accuracy metrics on the test dataset of Phase 1

Classifier	Accuracy (%)	TPR	FPR	Precision	Recall	F-measure	ROC area
NB	76.12	0.916	0.337	0.673	0.916	0.776	0.837
Logistic	75.60	0.928	0.380	0.649	0.928	0.763	0.653
MLP	77.60	0.929	0.393	0.642	0.929	0.759	0.886
SMO	75.39	0.926	0.393	0.641	0.926	0.758	0.625
IBK	79.35	0.927	0.353	0.665	0.927	0.775	0.802
J48	81.69	0.972	0.318	0.698	0.972	0.813	0.818

Table 5 Classifiers trained models accuracy metrics of Phase 2

Classifier	Accuracy (%)	TPR	FPR	Precision	Recall	F-measure	ROC area
NB	90.41	0.898	0.083	0.947	0.898	0.922	0.957
Logistic	95.48	0.965	0.064	0.958	0.965	0.961	0.975
MLP	96.50	0.973	0.051	0.965	0.973	0.969	0.973
SMO	95.73	0.966	0.060	0.960	0.966	0.963	0.953
IBK	97.83	0.979	0.022	0.979	0.979	0.979	0.978
J48	97.89	0.979	0.022	0.979	0.979	0.979	0.979

Table 6 Classifiers trained models accuracy metrics on the test dataset of Phase 2

Classifier	Accuracy (%)	TPR	FPR	Precision	Recall	F-measure	ROC area
NB	78.15	0.782	0.083	0.821	0.782	0.794	0.889
Logistic	81.51	0.815	0.142	0.851	0.815	0.832	0.889
MLP	78.15	0.782	0.173	0.818	0.782	0.799	0.889
SMO	79.83	0.798	0.161	0.832	0.798	0.814	0.856
IBK	84.35	0.824	0.134	0.860	0.824	0.841	0.838
J48	82.67	0.807	0.157	0.837	0.807	0.821	0.883

In the third phase, we worked on mitigating the dataset imbalance problem by undersampling the dominant classes and oversampling the minority classes. For undersampling, we used WEKA's resample filter which takes a random subsample. By setting the bias toward the uniform class to 1, we ensured that the output subsample was balanced. For oversampling, we used WEKA's SMOTE filter. SMOTE stands for synthetic minority oversampling technique. It works by generating synthetic instances based on the existing instances in the minority class to balance the data. The synthetic instances are generated by taking random points along a line between an existing minority instance and its nearest neighbors.

All experiments have been carried out using WEKA [18], a data mining tool running on a PC with Intel(R) CORE(TM) i5-6600K CPU @ 3.50 GHz, 3.50 GHz, 8 GB RAM installed and running a 64-bit Windows 10 OS, x64-based processor.

5.3 *Experimental Results*

For the evaluation purpose of each of the classifiers, we considered stratified cross-validation more important. The evaluation is performed using the training dataset, stratified cross-validation of 10-fold, and the testing dataset provided by NSL-KDD.

Figures 1, 2, 3, 4, and 5 summarize the performance of the tested classifiers according to accuracy and ROC area in the first two experimental phases. Tables 3, 4, 6, and 7 present a comprehensive comparison of the classifiers regarding classification accuracy, precision, recall, TPR, FPR, F-measure, and ROC Area. Table 7 shows the accuracy of each classifier in classifying different types of attacks.

Fig. 1 Training versus testing accuracy of Phase 1

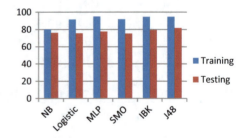

Fig. 2 Training versus testing accuracy of Phase 2

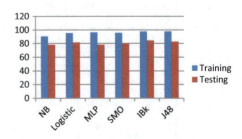

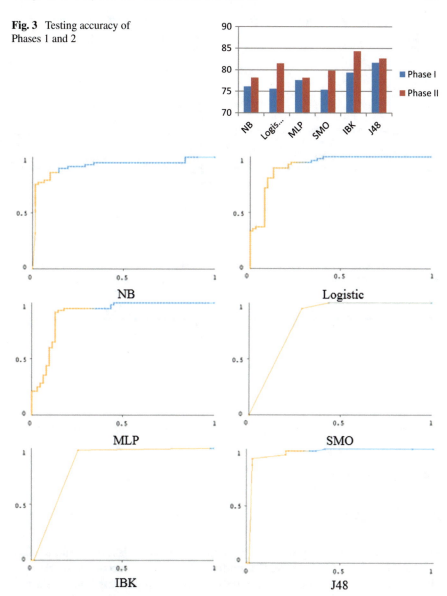

Fig. 3 Testing accuracy of Phases 1 and 2

Fig. 4 ROC curves for Phase 1

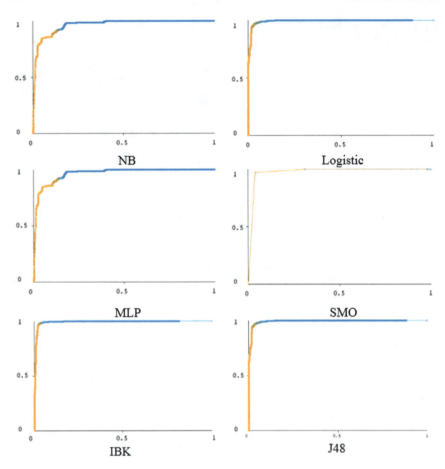

Fig. 5 ROC curves for Phase 2

5.4 Discussion

Our experimental results show that J48 outperforms other classifiers with the best accuracy in the first phase, while IBK performs better in the second phase. Figure 3 shows that the best performance improvement when applying the feature selection methods is for SMO, logistic, and IBK classifiers. Moreover, the results shown in Table 7 indicate that all the classifiers give good accuracy for the dominant classes, while it is not the case for the R2L and U2R classes. It also shows that the imbalance mitigation method improves limitations in detecting R2L and U2R attacks.

Table 7 Classifiers accuracy detection for different classes of attacks

Classifier	Class	Phase I (%)	Phase 2 (%)	Phase 3 (%)
NB	Normal	76.1	86.0	89.9
	DoS	75.2	83.8	91.8
	Probe	76.1	81.8	83.9
	R2L	10.1	26.7	39.0
	U2R	30.3	30.8	32.1
Logistic	Normal	75.6	84.4	96.4
	DoS	74.9	90.7	96.7
	Probe	75.1	69.6	88.7
	R2L	00.0	00.0	26.2
	U2R	22.3	26.7	53.2
MLP	Normal	77.6	82.4	97.5
	DoS	80.5	86.3	97.4
	Probe	68.9	63.2	93.9
	R2L	00.0	00.0	60.6
	U2R	08.9	09.7	30.2
SMO	Normal	75.3	83.3	96.7
	DoS	74.7	91.9	97.5
	Probe	55.4	60.0	87.6
	R2L	00.0	00.0	02.7
	U2R	00.0	00.0	04.9
IBK	Normal	79.3	86.8	99.4
	DoS	80.5	90.7	99.5
	Probe	71.8	76.2	99.0
	R2L	00.0	00.0	53.2
	U2R	00.0	00.0	41.5
J48	Normal	81.6	84.8	99.5
	DoS	80.1	89.2	99.2
	Probe	67.9	63.2	91.6
	R2L	18.9	18.2	55.1
	U2R	00.0	00.0	39.3

6 Conclusion and Future Work

Our analysis results for the performance of the six different classifiers on the NSL-KDD dataset show that J48 and IBK are the best two classifiers in terms of accuracy detection, but IBK is much better when applying feature selection techniques. For future work, we propose to carry out an exploration on how to employ optimization techniques to develop an intrusion detection model with a better accuracy rate.

References

1. Roy, D.B., Chaki, R.: State of the art analysis of network traffic anomaly detection. In: Applications and Innovations in Mobile Computing (AIMoC), IEEE, pp. 186–192 (2014)
2. Buczak, Anna L., Guven, Erhan: A survey of data mining and machine learning methods for cyber security intrusion detection. IEEE Commun. Surv. Tutorials **18**(2), 1153–1176 (2016)
3. Papernot, N., McDaniel, P., Goodfellow, I.: Transferability in Machine Learning: From Phenomena to Black-Box Attacks Using Adversarial Samples. arXiv preprint arXiv:1605.07277 (2016)
4. Alkasassbeh, M.: An Empirical Evaluation for the Intrusion Detection Features Based on Machine Learning and Feature Selection Methods. arXiv preprint arXiv:1712.09623 (2017)
5. Potluri, S., Diedrich, C.: High performance intrusion detection and prevention systems: a survey. In: ECCWS2016-Proceedings of the 15th European Conference on Cyber Warfare and Security. Academic Conferences and Publishing Limited (2016)
6. Fabris, F., De Magalhães, J.P., Freitas, A.A.: A review of supervised machine learning applied to ageing research. Biogerontology **18**(2), 171–188 (2017)
7. NSL-KDD dataset [online] available: http://www.unb.ca/cic/datasets/nsl.html. Accessed on 21 Oct 2018
8. Ingre, B., Yadav, A.: Performance analysis of NSL-KDD dataset using ANN. In: 2015 International Conference on Signal Processing and Communication Engineering Systems (SPACES), IEEE (2015)
9. Dhanabal, L., Shantharajah, S.P.: A study on NSL-KDD dataset for intrusion detection system based on classification algorithms. Int. J. Adv. Res. Comput. Commun. Eng. **4**(6), 446–452 (2015)
10. Karimi, Z., Kashani, M.M.R., Harounabadi, A.: Feature ranking in intrusion detection dataset using combination of filtering methods. Int. J. Comput. Appl. **78**(4) (2013)
11. Kohavi, R., John, G.H.: Wrappers for feature subset selection. Artif. Intell. **97**(1-2), 273–324 (1997)
12. Claesen, M., De Moor, B.: Hyperparameter Search in Machine Learning (2015). arXiv:1502.02127
13. MeeraGandhi, G.: Machine learning approach for attack prediction and classification using supervised learning algorithms. Int. J. Comput. Sci. Commun. **1**(2) (2010)
14. Nguyen, H.A., Choi, D.: Application of data mining to network intrusion detection: classifier selection model. In: Asia-Pacific Network Operations and Management Symposium. Springer, Heidelberg (2008)
15. Jalil, K.A., Kamarudin, M.H., Masrek, M.N.: Comparison of machine learning algorithms performance in detecting network intrusion. In: 2010 International Conference on Networking and Information Technology (ICNIT), IEEE, 2010
16. Revathi, S., Malathi, A.: A detailed analysis on NSL-KDD dataset using various machine learning techniques for intrusion detection. Int. J. Eng. Res. Technol. ESRSA Publications (2013)

17. He, H., Garcia, E.A.: Learning from imbalanced data. IEEE Trans. Knowl. Data Eng. **21**(9), 1263–1284 (2009)
18. Frank, E., Hall, M.A., Witten, I.H.: The WEKA Workbench. Online Appendix for "Data Mining: Practical Machine Learning Tools and Techniques". Fourth Edition, Morgan Kaufmann, (2016)

Achieving Wellness by Monitoring the Gait Pattern with Behavioral Intervention for Lifestyle Diseases

Neha Sathe⊙ and Anil Hiwale⊙

Abstract Obesity is becoming one of the prevalent lifestyle diseases across the globe; need to be deal with behavioral intervention through self-management and motivation in line with rigorous physical exercise. When it is a matter to handle obesity to regain the health, parameters like: self-motivation, self-control, guided treatment, counseling, monitored exercise, and medical assistance are essential in consideration list. Paper proposes the model encompassing three-dimensional care including nutritional intake, counseling, and gait monitoring during exercise. Considering the effect of obesity on biomechanics of foot, gait pattern analysis of obese person provides greater information regarding variations in spatio-temporal parameters. Ever-increasing contribution of behavioral intervention will maintain the line of action in the perfect direction. A selection of accelerometer, gyroscope, and electromyography sensors is appropriate for the cause to derive basic hardware. MSP430 processor and ZigBee module are used for processing information and establishing communication. Within close proximity and placement of nodes at a different level, nodes are able to achieve 90–94% packet delivery ratio in actual environment compare to the 100% packet delivery in simulation environment. Result suggests that with the adaption of accurate classification process, system could be useful for controlled exercise monitoring or for daily activity monitoring, which is working at low-power level with affordable wearable technology in achieving wellness.

Keywords Wearable device · Gait analysis · Behavioral intervention · Obesity

N. Sathe (✉)
MIT World Peace University, Pune, India
e-mail: neha.sathe@mitwpu.edu.in

A. Hiwale
MIT College of Engineering, Pune, India
e-mail: anil.hiwale@mitcoe.edu.in

© Springer Nature Singapore Pte Ltd. 2020
X.-S. Yang et al. (eds.), *Fourth International Congress on Information and Communication Technology*, Advances in Intelligent Systems and Computing 1027, https://doi.org/10.1007/978-981-32-9343-4_17

1 Introduction

Considering the fast-changing world along with the development of technology; the lifestyle adapted by majority of the young community is moving toward diverse health-related issues. While engineering technology and medical science are going hand in hand, problems to be handled in these areas are opening various challenges in healthcare. On the same note, collaborative study of behavioral health, lifestyle diseases, nutritional control, and appropriate exercise is the need of the time. Behavioral healthcare sector is having broad and diversified angles toward study of it. The paper considers only lifestyle factors out of it, such as habitats and behavior, while cognition and environmental factors are not considered. Behavioral intervention for lifestyle diseases is of major concern at the present. Unawareness or less attention toward preventive healthcare is the biggest hurdle. In most of the cases, it was difficult to understand the linking between lifestyle disease, behavioral aspect, and health-related problems. The biggest challenge the world is facing today is obesity observed in various age groups, which could be figured out as the alarming state of affairs raised due to the changed lifestyle and behavioral impact. According to the World Health Organization (WHO), "Overweight and obesity may impair health." The foremost identified causes are an intake of fatty food; increased physical inactivity with different forms of work, variation in transportation modes, and increasing urbanization. The WHO survey highlights that the worldwide prevalence of obesity nearly tripled between 1975 and 2016, see Fig. 1 [1].

The most preferred and suitable exercise is to have a power walk every day, which helps in maintaining health. This most commonly adapted exercise becomes beneficial to balance health when accompanied with perfect nutritional intake. Although,

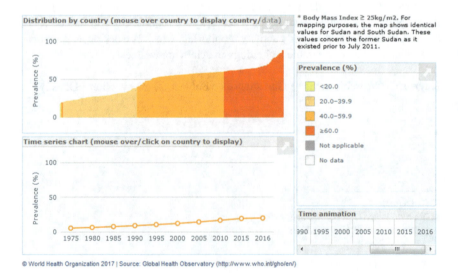

Fig. 1 Gradual increase in obesity in India

Fig. 2 Pervasive model component

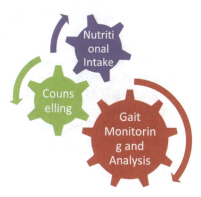

when it is a matter to regain the health from obesity, then the crucial point is to have a perfectly monitored walk. Perfect walk is concerned with monitoring crucial spatio-temporal parameters of gait pattern. Monitoring the spatio-temporal parameter is of great help to assure that the initiative taken to handle obesity must not lead to any other health issue. Use of gait pattern monitoring, in case of recovery process of patients undergone critical medical conditions is common. However, using this aspect of analysis in handling obesity is seamlessly going to be a new approach. Figure 2 shows the proposed pervasive model toward achieving the wellness in behavioral intervention for lifestyle disease. It encompasses three factors: gait monitoring and analysis, nutritional intake and counseling sessions as shown in Fig. 2, each factor is interlinked to achieve the three-dimensional care.

1. **Gait Monitoring and Analysis**: An obese and a non-obese person shows variation in the walking pattern, sometimes visually distinguishable. While to understand the effect of obesity on excessive loading on the foot needed to be measured through biomechanical measurement. At the primary level comparative analysis is needed to assure with basic parameters like normal heel to toe pattern, position of arch during walk, coordination of upper and lower limb, swing, stance, cadence, velocity, step time and length, in/out angle, etc.

2. **Counseling**: Includes describing the need of adopting patient-centric approach for assessing lifestyle-related risk factors and necessary plan of action to be adopted. The best way to handle these issues is appropriate counseling. It may include brief or intensive levels of counseling, depending on the case to be handled. For example, to handle obesity, ways of counseling might include briefing about related critical illness, relief from addiction, uplifting the thought process toward positivity, and convincing for routine exercise associated with nutritional intake.

3. **Nutritional intake**: One of the strongest reasons resulting in obesity is inapt food intake at irregular intervals. Food with high-fat content and lower-nutritional values leads to cumulative effect toward obesity. Managing the calorie intake in consultation with nutritionist is the first expected step. Considering the physical health condition and prior medical history capacity to undergo nutritional training

will get monitored in gradual manner in such process. Abrupt decision, without consulting any nutritionist regarding adaption of any diet plan might lead to unexpected issues instead of health gain. A nutritionist upgrades the nutritional plan in consultation with the physio instructor whenever needed [17].

Considering the necessity of involving various factors toward maintenance of health, to handle lifestyle diseases is becoming essential. The paper suggests a collaborative solution to overcome obesity through gait analysis in behavioral intervention for lifestyle disease. The approach suggested here is to use the body-worn sensors for monitoring the gait pattern of the person.

2 Previous Findings

1. Miller-Rosales et al. suggest the four-component model for behavioral intervention of the patient. Components include increasing patient's activity level, improving patient's engagement skill, screening for behavioral risk, and care planning. They are able to represent evidence-based, patient-centered for patients not responding to traditional disease management [2].

2. Jegede et al. perform the comparative analysis of obese person on weight reduction plan with normal weight person on specific spatio-temporal gait parameters. They recommend the adaption suggested model for improving spatio-temporal gait parameters in obesity-related issues [3].

3. Koushyar et al. investigated the effect of obesity and age on the flexion strength at the hip, knee, and ankle. In conclusion, the absolute strength in obese participants is higher when compared with healthy weight participants. Relative strength in obese participants is lower than healthy participants [4].

4. Rosso et al. perform test and re-test for transverse plane measurement using H-gait, a wearable system having magnetic and inertial sensors. Based on static and anatomical calibration the roto-translation matrix between body and sensors are established. Within a finding they specified that the knee angle is higher in overweight, the area of knee and ankle joint center trajectory is greater in overweight in the middle of the stance and swing [5].

5. Milner et al. Address the issue of knee osteoarthritis and risk of obese adults toward it. They focus is to determine the relation of velocity and step length in obese and healthy weight person. In conclusion, they stated that the decreasing step length can reduce knee joint load during daily walking [6].

6. Kathirgamanathan et al. they have highlighted the fact that the obesity can cause damage to soft tissue and lead to foot discomfort and the development of foot pathology. They had created a three-dimensional anatomical model of a foot; nonlinear finite element analysis was performed. In the conclusion, they stated that the greater magnitude of contact pressure is experienced by obese foot, it may lead to formation of foot pathology and can cause discomfort to the person [7].

7. Boateng et al. paper suggests wrist-worn device that monitors daily activity using metabolic equivalence. The evaluation is performed on the basis of leave-one-subject-out (LOSO) cross-validation. Along with observations and readings, they conclude that activity detection model is able to provide accuracy up to 94% [8].

8. Andreu-Perez et al. it highlights the contribution of low-power electronics, nanofabrication, and invention in biocompatible material for foreign body acceptance in case of implants. Selection of arm or feet-fitted sensors and its use in obesity management and clinical gait analysis is described with the use of sensors like accelerometer, gyroscope, bend and pressure sensor, air pressure, activity level monitoring, and electromyography. Database storage, availability of clinical information, data processing through evolution in algorithm emphasis of machine learning in training and classification is also covered [9].

9. Misgeld et al. author shows the result for the model developed for spasticity detection using body-worn sensors. With the Integrated Posture and Activity Network by Medit Aachen body sensor network, electromyography sensor was developed and deployed for monitoring locomotion. Considering limitations selection of hardware and software is done. MSP430 is selected for its on chip solution and various low-power application modes and during implementation of algorithm few pre-assumptions are made to minimize calculation without compromising on quality of sensitive information. Implementation is able to distinguish five groups of hemiparetic gait as: minimal gait deviation, equinus foot and flexed knee in stance; equines foot and flexed knee and hip in stance; equinus foot and hyperextension of the knee; and hyperextension of the knee with normal ankle kinematics [10].

10. Chinchole et al. paper suggest real-time body mass index tracking system using sensor, data analytics and cloud. Concept targets the root cause of obesity as unhealthy lifestyle, irregular exercise, and unhealthy intake. System executes weight and body mass index algorithm to convert weight into precise reading of BMI [11].

11. Hegde et al. specified the use of footwear-based sensors for gait monitoring is put forward within the paper. System is able to classify various body positions as lie down, stand, sit, walk, etc., with precision of 100% [12].

12. Hegde et al. paper provides the concept of providing validation to daily activity monitoring with the use of insole-based sensor monitoring, verified through wrist-worn sensors and able to provide accuracy of 81% [13].

Numerous thoughts being put forwarded by various authors regarding monitoring of gait pattern with diverse techniques for the range of reasons. More or less, literature review included highlighting the varied angles of gait analysis, starting from need of the gait analysis, sensors, location of sensors, hardware part, and software development [14, 15]. The summery about the selection of sensor for the specific activity to be monitored in provided in Table 1 [13].

Table 1 Comparative study for types of sensors used for gait monitoring [13]

S. no	Name of author	Pattern considered	Sensors selected
1	L. Bao and S. S. Intille	Walk, run, stand, stairs	Bi-axial accelerometer
2	J.-Y. Yang et al.	Walk, run, stand	Tri-axial accelerometer
3	A. G. Bonomi et al.	Walk, run, stand	Tri-axial accelerometer
4	A. Moncada-Torres et al.	Walk, run, stand, stairs	Tri-axial accelerometer, gyroscope, and barometric sensor
5	F. Attal et al.	Walk, run, stand, stairs	Tri-axial accelerometer, gyroscope, and magnetometer
6	W. Tang and E. S. Sazonov	Walk, run, stand, stairs	Tri-axial accelerometer, gyroscope, and pressure sensor
7	M. Altini et al.	Walking, standing	Tri-axial accelerometer and heart rate sensor
8	S. Pirttikangas et al.	Walk, run, stand, stairs	Tri-axial accelerometer and heart rate sensor
9	L. Atallah et al.	Walk, run, stand, stairs	Tri-axial accelerometer and visual sensor

3 Typical System Model

The paper suggests the three-dimensional model for achieving wellness to overcome obesity. As shown in Fig. 3, it is comprised of various peers working in parallel to regain the health. Nutritional information for deciding intake, it is frequent

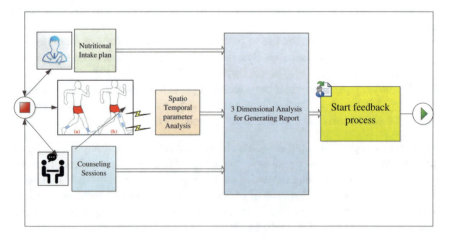

Fig. 3 Proposed three-dimensional architecture of wellness

up-gradation, inclusion of counseling sessions for self-management and monitored workout.

Body-worn tiny model consisting of accelerometer, gyroscope, and electromyography sensors associated with MSP430 processor, and GPS tracker with the provision of wireless connectivity. Lot of options is available in the process of monitoring, but the specified set is adequate for the initial cause. For on board computation, MSP430 processor is selected having good processing speed, simple connectivity, and communication option. The basic cause of selection is its low-power operating modes, which is a key requirement in wireless sensing technology. The MSP development boards are going to have all the required facility onboard with simple packing structure. Attaching or detaching any unit would be an easy job with MSP. Establishing wireless connectivity will be possible with Bluetooth, ZigBee or Wi-Fi connection. Depending on the area to be covered the selection of the connecting device is going to change. Adaption of latest possible option could not be the economical solution. So, considering the fact that, a well-defined coverage area does not exist for wireless media because propagation characteristics are dynamic and uncertain. Small changes in the placement position or direction may result in drastic difference in the signal strength.

4 Experimental Result

Basic experimental target is to establish communication among various sensing nodes and data collection module, normally called as a sink node. At the initial level, testing of establishing communication within closed environment is done with the help of ZigBee module attached along with the MSP430 processor working at 2.4 GHz ISM band with inbuilt 64 Byte buffer for receiver [16]. Each node is associated with processing element and communication device; actual sensors are not in place. Basic testing for the hardcoded data is done with the selected hardware and result analysis is done with the Dock Light Simulator. The primary intention of the testing is to validate the system for continuous reception of data with minimal loss of information. Assumed scenario is handling information in small chunks, with maximum of 3-byte size and varying reporting rate range: 1–4 packets/sec. System is tested with 2 and 4 transmitting nodes located at different positions considering the possibility to handle different parameters from different sensing elements. Results are shown in Figs. 4 and 5 show the relative relation among the positions. Results may differ while performing testing with actual sensors in comparison with hardcoded data. While experimenting, it is necessary to consider the various options suggested within different deployment modules discussed in literature survey, the location of body-worn sensors contributes countless within findings. Two different locations are experimented for communication establishment at the level of ankle and in-between ankle and knee position level.

Fig. 4 Two nodes as a function of position

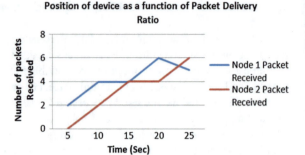

Fig. 5 Four nodes as a function of position

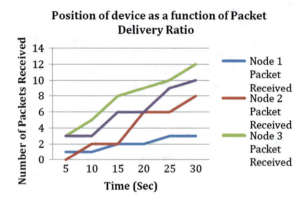

4.1 Experimental Setup 1

It uses two nodes, situated at different positions as one at ankle level and second at knee level. With the hardcoded data system is able to achieve 94% packet delivery ratio. Contribution of two nodes is shown in Fig. 4.

4.2 Experimental Setup 2

It uses four nodes situated at different positions as two at ankle level and two at knee level. With the hardcoded data system can achieve 90% packet delivery ratio. Contribution of four nodes is shown in Fig. 5.

5 Conclusion and Future Work

Based on experimental results, system can provide result within a range of 90–94% with limited information transferred. Selected numbers of sensors are three; with this count of information generator, it is expected to maintain ratio of packet delivery around 90%. Emphasizing on the fact that the position of placement of sensor on the human body shows different result for same setup. For each person, the selection of appropriate position of placement of sensor may vary. The proposed solution is going to be affordable with low-power consumption and ease to wear. In future work, the major contributory part is software development in conjunction with best classifier selection; system needs to inculcate the role and finding of counseling and nutritional intake plan in completion of analysis. Depending on the history and current condition of each person, the role of three caring factors is differing from each other. Albeit, it will help to overcome obesity with the suggested pervasive three-dimensional model.

References

1. Author, F.: Article title. Journal 2(5), 99–110 (2016). https://www.who.int/topics/obesity/en/
2. Miller-Rosales, C., et al.: CREATE Wellness: a multi-component behavioral intervention for patients not responding to traditional cardiovascular disease management. Contemp. Clin. Trials Commun. 8, 140–146 (2017)
3. Jegede, J.A., Adegoke, B.O.A., Olagbegi, O.M.: Effects of a twelve-week weight reduction exercise programme on selected spatiotemporal gait parameters of obese individuals. J. Obes. (2017)
4. Koushyar, H., et al.: Relative strength at the hip, knee, and ankle is lower among younger and older females who are obese. J. Geriatr. Phys. Ther. (2001) 40(3), 143 (2017)
5. Rosso, V. et al.: Gait measurements in the transverse plane using a wearable system: an experimental study of test-retest reliability. In: Instrumentation and Measurement Technology Conference (I2MTC), 2017 IEEE International
6. Milner, C.E. et al.: Walking velocity and step length adjustments affect knee joint contact forces in healthy weight and obese adults. J. Orthop. Res. (2018)
7. Kathirgamanathan, B., Silva, P., Fernandez. J.: Does obesity affect the biomechanics of the foot? A preliminary computational and experimental study. In: Engineering Research Conference (MERCon), 2017 Moratuwa. IEEE 2017
8. Boateng, G. et al.: GeriActive: wearable app for monitoring and encouraging physical activity among older adults. BSN 2018
9. Andreu-Perez, J., et al.: From wearable sensors to smart implants—toward pervasive and personalized healthcare. IEEE Trans. Biomed. Eng. 62(12), 2750–2762 (2015)
10. Misgeld, B.J.E., et al.: Body-sensor-network-based spasticity detection. IEEE J. Biomed. Health Inform. 20(3), 748–755 (2016)
11. Chinchole, S., Samir, P.: Cloud and sensors based obesity monitoring system. In: 2017 International Conference on Intelligent Sustainable Systems (ICISS). IEEE 2017
12. Hegde, N. et al.: One size fits all electronics for insole-based activity monitoring. In: Engineering in Medicine and Biology Society (EMBC), 2017 39th Annual International Conference of the IEEE. IEEE 2017
13. Hegde, N., et al.: Automatic recognition of activities of daily living utilizing insole-based and wrist-worn wearable sensors. IEEE J. Biomed. Health Inform. 22(4), 979–988 (2018)

14. Sazonov, E., et al.: Posture and activity recognition and energy expenditure prediction in a wearable platform. IEEE J. Biomed. Health Informat. **19**(4), 1339–1346 (2015)
15. Debes, C., et al.: Monitoring activities of daily living in smart homes: understanding human behavior. IEEE Sig. Process. Mag. **33**(2), 81–94 (2016)
16. Sathe, N., Anil H., Rashmi P.: Pre-habilitation and wellness through gait analysis using body worn sensors. In: Proceedings of the 2018 International Conference on Communication Engineering and Technology. ACM 2018
17. Looney, S.M., Raynor. H.A.: Behavioral lifestyle intervention in the treatment of obesity. Health Serv. Insights **6**, HSI-S10474 (2013)

A Temporal-Causal Modelling Approach to Analyse the Dynamics of Burnout and the Effects of Sleep

Hendrik von Kentzinsky, Stefan Wijtsma and Jan Treur◉

Abstract In this paper, a temporal-causal network model is introduced for a burnout in relation to sleep. The network model approach shows the impact of different lifestyle, personal and job factors on the development of a burnout. This model, for instance, can be used to schedule night shifts in order to preserve the needed recovery of exhaustive, irregular sleeping patterns or to investigate the effects of certain in lifestyles induced triggers on burnout.

Keywords Network model · Burnout · Sleep

1 Introduction

Most workers experience stress sometimes, and when this continues over longer periods of time, people can experience physical consequences of this. Examples include sickness and the feeling of being burn out or emotional exhaustion. In 2017, 15.9% of the Dutch population older than 15 reported burnout-related symptoms [4]. In light of social change and transformations in work situations, interest in this topic has grown. A lot can and is being done to prevent or alleviate burnout syndrome, especially by changing lifestyle or habits. We build on top of existing research by Dujmić et al. [6] by extending their model through adding sleep-related factors. This paper focuses on the optimization of sleep habits as this was found as a protector from burnouts [15].

In contrast to the traditional paradigm for assessing psychological diseases, where these were seen as a latent construct that could be measured by their symptoms, this paper uses a newer paradigm. This newer paradigm defines psychological diseases

H. von Kentzinsky (✉) · S. Wijtsma · J. Treur
Behavioural Informatics Group, Vrije Universiteit Amsterdam, Amsterdam, The Netherlands
e-mail: hvkentzinsky@gmail.com

S. Wijtsma
e-mail: stefanwijtsma@gmail.com

J. Treur
e-mail: j.treur@vu.nl

© Springer Nature Singapore Pte Ltd. 2020
X.-S. Yang et al. (eds.), *Fourth International Congress on Information and Communication Technology*, Advances in Intelligent Systems and Computing 1027, https://doi.org/10.1007/978-981-32-9343-4_18

219

as the relation or dynamic interplay between symptoms [3]. This fits well to the Network-Oriented Modelling approach described in [23, 24], which is based on temporal-causal network modelling in which networks with dynamic and cyclic causal relations are addressed. This approach can be considered as a branch in the causal modelling area which has a long tradition in AI, e.g. see [11, 12, 18]. This paper aims to help understand the causes, elements, consequences, risk factors and protective factors of burnout syndrome gathered from the literature through network-oriented modelling. The dynamic interaction of these states forms (cyclic) patterns resembling real patterns.

In this study, a non-burnout scenario will be simulated, as well as a burnout scenario and a recovery from burnout scenario. Furthermore, we will show a cyclic pattern with a periodical sleeping pattern. Here, we will make use of a trigger as well (having kids after some point in time) that could potentially disrupt this cyclic pattern.

First, some background literature will be discussed, based on which the network model is constructed, which happens in Sect. 3. Simulations for some example scenarios are shown in Sect. 4, followed by verification of the model by mathematical analysis (Sect. 5) and validation by parameter tuning in Sect. 6. In Sect. 7, the results will be discussed.

2 Background Literature

The classical definition of a burnout, as given by Maslach and Jackson [13], is nowadays still relevant: emotional exhaustion, depersonalization and personal accomplishment are since then still described as the core elements of the burnout syndrome [10]. Maslach created the first measurement instrument for burnouts called the Maslach Burnout Inventory (MBI). She describes various job characteristics that form risk factors for developing a burnout. The most important risk factor is *experienced overload*, a term describing subjective stress. Long exposure to job-related stress can lead to cynicism and emotional exhaustion, in their turn increasing the chance of a burnout. Other risk factors include *job sacrifice*, describing how much someone is willing to sacrifice for achieving his or her ideals, and *role ambiguity*, pointing to the unclarity of what and how much is expected from a worker.

Furthermore, there are job-related factors that either suppress the effects of these risk factors or decrease the core elements of burnout syndrome that are called protective factors. An example of a protective factor is *social contact with co-workers*, which is found to decrease depersonalization and cynicism [5]. Also, *work experience* is an important protective factor, since the more experience someone has, the more sense of personal accomplishment one experiences [13]. Experience also lowers the experience of role ambiguity [5]. This study builds on earlier work [6], where similar job-related factors were used, and also some personal characteristics including *Neuroticism, Openness* and *Hardiness*. This study aims to extend this research by adding sleep-related factors that can either suppress or enforce burnout causes. Neuroticism is a risk factor associated with higher amounts of stress and experienced overload [9]. Openness, on the other hand, is shown to be negatively correlated with depersonalization and emotional exhaustion [7], and positively correlated with physical exercise [22], and therefore, more of a protective factor. The last personal characteristic we included is hardiness, which is found to protect from the effects of stress, but also is linked to the core elements of burnout [2, 8].

By analysing a questionnaire concerning stress at work, health, sleep and lifestyle factors of 676 employees, [21] identifies four factors as significant burnout predictors: work demands, thoughts of work during leisure time, sleep quality and getting too little sleep (less than six hours). The latter was identified as the main risk factor for clinical burnout. Söderström et al. [21] concludes that getting insufficient sleep and having difficulties detaching from thoughts of work during leisure time are stronger predictors of burnout syndrome than stressful work demands, highlighting the importance of recovery from stress—and not that much stress itself—in the process of developing a burnout. As described by Miró et al. [15], sleep quality, next to job-related factors, is a significant predictor of a burnout. Their research showed a higher impact on emotional exhaustion than job demands in our study described as 'feeling of charged work'. Also, sleep quality interacts with other job-related factors. Furthermore, Rosen et al. [20] showed that not only sleep quality but also sleep quantity has a serious impact on the development of burnout. Analogically, whether a person has uniform working hours plays an important role in merely sleep quantity. Akerstedt and Wright [1] confirmed this and more importantly showed the impact of habitual sleep efficiency, a term to describe the percentage of time in bed that a person is asleep. Pagnin et al. [17] showed that daytime dysfunction is an important predictor for emotional exhaustion and cynicism. A meta-analysis by Pilcher and Huffcutt [19] showed that lack of sleep significantly impairs human functioning during daytime activities. Obviously, earlier named sleep factors have an influence on daytime dysfunction. To complete the sleep-related factors, sleeping pill intake and caffeine intake were added to the model. Sleeping pill intake has a positive effect on daytime dysfunction [25] and may vary depending on the sleep quality and sleep quantity. McGeary et al. [14] also showed a relation between caffeine intake and burnout, while caffeine intake can also reduce daytime dysfunction. These factors can result in a cyclic pattern when affecting other factors.

3 The Designed Network Model

Based on the literature discussed in the previous section, in this section, we describe the temporal-causal network model we used. This model is based on the Network-Oriented Modelling approach described in [23, 24] and was implemented using the software environment described in [16]. Table 1 provides a concise overview of a temporal-causal network model in general. The differential equations in the last row in Table 2 can be used for simulation and mathematical analysis.

Table 1 Representations of a temporal-causal network; adopted from [24]

Concepts	Notation	Explanation
States and connections	$X, Y, X \rightarrow Y$	Describes the nodes and links of a network structure (e.g. in graphical or matrix format)
Connection weight	$\omega_{X,Y}$	*Connection weight* $\omega_{X,Y} \in [-1, 1]$ represents the strength of the impact of state X on state Y through connection $X \rightarrow Y$
Aggregating multiple impacts	$\mathbf{c}_Y(..)$	For each state Y, a *combination function* $\mathbf{c}_Y(..)$ is chosen to combine the causal impacts of other states on state Y
Timing of the causal effect	η_Y	For each state Y, a *speed factor* $\eta_Y \geq 0$ is used to represent how fast a state is changing upon causal impact

Concepts	Numerical representation	Explanation
State values over time t	$Y(t)$	At each time point t, each state Y has a real number value in [0, 1]
Single causal impact	$\mathbf{impact}_{X,Y}(t) = \omega_{X,Y} X(t)$	At t, state X with connection to state Y has an impact on Y, using weight $\omega_{X,Y}$
Aggregating multiple impacts	$\mathbf{aggimpact}_Y(t) = \mathbf{c}_Y(\mathbf{impact}_{X1,Y}(t),...,$ $\mathbf{impact}_{Xk,Y}(t)) = \mathbf{c}_Y(\omega_{X1,Y}X_1(t), ...,$ $\omega_{Xk,Y}X_k(t))$	The aggregated impact of multiple states X_i on Y at t is determined using combination function $\mathbf{c}_Y(..)$
Timing of the causal effect	$Y(t + \Delta t) = Y(t) + \eta_Y$ $[\mathbf{aggimpact}_Y(t) - Y(t)] \Delta t =$ $Y(t) + \eta_Y [\mathbf{c}_Y(\omega_{X1,Y}X_1(t),$ $..., \omega_{Xk,Y}X_k(t)) - Y(t)] \Delta t$	The impact on Y is exerted over time gradually, using speed factor η_Y

They can also be written in differential equation format:

$$Y(t + \Delta t) = Y(t) + \eta_Y$$
$$[\mathbf{c}_Y(\omega_{X_1,Y}X_1(t), \ldots, \omega_{X_k,Y}X_k(t)) - Y(t)]\Delta t \qquad (1)$$

$$\mathbf{d}Y(t)/\mathbf{d}t = \eta_Y[\mathbf{c}_Y(\omega_{X_1,Y}X_1(t), \ldots, \omega_{X_k,Y}X_k(t)) - Y(t)]$$

Based on the literature—and in particular on the work of [6]—we could identify 29 state variables relevant for a burnout, as shown in Table 2. The table includes the classification of the variables into the risk and protective factors, the elements of the syndrome and the consequent state and also shows the combination functions used and their parameters. Most of the combination functions were chosen as advanced logistic functions:

$$\mathbf{id}(V) = V \qquad (2)$$

Table 2 States used in the model with their combination functions and their parameters

State	Abbr.	Description	Type	Function	σ	τ	Speed
X_1	FW	Feeling charged work	Risk factor	Identity			0.1
X_2	EX	Work experience	Protective factor	Advanced logistic			0.1
X_3	HA	Hardiness	Protective factor	Identity			0.1
X_4	JS	Job sacrifice	Consequent	Identity			0.1
X_5	RA	Role ambiguity	Risk factor	Advanced logistic	10	0.5	0.1
X_6	NE	Neuroticism	Risk factor	Identity			0.1
X_7	OP	Openness	Protective factor	Identity			0.1
X_8	EO	Experienced overload	Risk factor	Advanced logistic	10	0.5	0.1
X_9	EE	Emotional exhaustion	Burnout element	Advanced logistic	10	0.5	0.1
X_{10}	PH	Physical health problems	Consequent	Advanced logistic	10	0.5	0.1
X_{11}	JR	Job resignation	Consequent	Identity			0.1
X_{12}	CY	Cynicism/ depersonalization	Burnout element	Advanced logistic	10	0.5	0.1
X_{13}	SC	Social contact with co-workers	Protective factor	Identity			0.1
X_{14}	JD	Job detachment	Consequent	Advanced logistic	10	0.5	0.1
X_{15}	JP	Job performance	Consequent	Advanced logistic	10	0.5	0.1
X_{16}	JA	Job attendance	Consequent	Advanced logistic	10	0.5	0.1
X_{17}	DU	Drug and alcohol use	Protective factor	Advanced logistic	10	0.5	0.1
X_{18}	PA	Personal accomplishment	Burnout element	Advanced logistic	10	0.5	0.1
X_{19}	PE	Physical exercise	Protective factor	Advanced logistic	10	0.5	0.1
X_{20}	SQ	Sleep quality	Protective factor	Advanced logistic	10	0.5	0.1
X_{21}	FJ	Feeling risk of losing job	Risk factor	Advanced logistic	10	0.5	0.1
X_{22}	SY	Sleep quantity	Protective factor	Advanced logistic	10	0.5	0.1
X_{23}	HS	Habitual sleep efficiency	Protective factor	Identity			0.1

(continued)

Table 2 (continued)

State	Abbr.	Description	Type	Function	σ	τ	Speed
X_{24}	DD	Daytime dysfunction	Risk factor	Advanced logistic	10	0.5	0.1
X_{25}	CI	Caffeine intake	Risk factor	Advanced logistic	10	0.5	0.1
X_{26}	PI	Sleeping pill intake	Risk factor	Advanced logistic	20	0.75	0.1
X_{29}	HK	Having kids	Risk factor	Identity			0
X_{30}	SD	Sleep disturbance	Risk factor	Advanced logistic	18	0.2	0.04
X_{31}	UW	Uniform working hours	Protective factor	Advanced logistic	50	0.8	1

$$\textbf{alogistic}_{\sigma,\tau}(V_1, \ldots, V_k) = \left[\frac{1}{1 + e^{-\sigma}(V_1 + \cdots + V_{k-\tau})} - \frac{1}{(1 + e^{\sigma\tau})} \right](1 + e^{-\sigma\tau})$$

The literature suggests many causal relations between the various states in our model, either positive or negative. We captured these relations in the graphical conceptual representation of the model shown in Fig. 1. The strength of these causal relations is represented by assigned connection weight values, see Table 3.

The green lines represent a high positive effect and the yellow line a slight positive effect. The red, orange and cyan lines show a negative effect, ranging from a high effect (red) to a slight effect (cyan). Note that X_{27} and X_{28} are external factors which will be explained below.

Examples of the difference and differential equations in the model are following:

Examples of a states using the identity (3) and the logistic (4) combination function

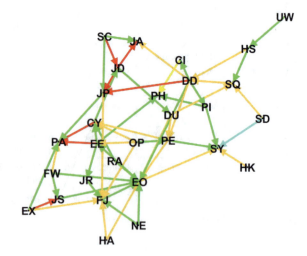

Fig. 1 Graphical conceptual representation of the temporal-causal network model

Table 3 Connection weights

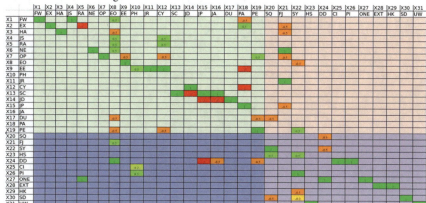

$$JS(t + \Delta t) = JS(t) + \eta_{JS}[\omega_{FW,JS}FW(t) - JS(t)]\Delta t \tag{3}$$

$$\frac{dJS(t)}{dt} = \eta_{JS}[\omega_{FW,JS}FW(t) - JS(t)]$$

$$JD(t + \Delta t) = JD(t)$$
$$+ \eta_{JD}[\mathbf{alogistic}_{\sigma,\tau}(\omega_{CY,JD}CY(t), \omega_{SC,JD}SC(t)) - JD(t)]\,\Delta t$$
$$\frac{dJD(t)}{dt} = \eta_{JD}[\mathbf{alogistic}_{\sigma,\tau}(\omega_{CY,JD}CY(t), \omega_{SC,JD}SC(t)) - JD(t)] \tag{4}$$

4 Simulations for Three Example Scenarios

To validate our model against typical burnout patterns described in the literature, we created three example scenarios and simulated them using the dedicated software environment described in [16]. First, Scenario 1 in which no burnout occurs, and next Scenario 2 in which burnout does occur. For Scenario 1, we expected a stable situation, where a person has high enough protective factors and low enough risk factors, such that the burnout elements diminish and all states will move into equilibrium. For the burnout Scenario 2, we expected that high-risk factors would overwhelm low protective factors, such that the burnout elements and its consequences would increase over time. Suitable initial values were assigned as shown in Table 4. For instance, a situation where a burnout occurs is probably accompanied with a low amount of sleep and a high daytime dysfunction in contrast to a no burnout scenario. In addition, we added Scenario 3 of 'recovery', namely where a person would experience some burnout elements, but then recovers and state values of burnout

Table 4 Initial values for the three different scenarios

State	X_1	X_2	X_3	X_4	X_5	X_6	X_7	X_8	X_9	X_{10}	X_{11}	X_{12}	X_{13}	X_{14}	X_{15}	X_{16}	X_{17}	X_{18}	X_{19}	X_{20}	X_{21}
Abbr.	FW	EX	HA	JS	RA	NE	OP	EO	EE	PH	JR	CY	SC	JD	JP	JA	DU	PA	PE	SQ	FJ
Scen. 1	0.1	0.8	0.9	0.2	0.15	0.23	0.75	0.1	0.1	0.2	0	0.1	0.9	0.1	0.85	0.99	0.1	0.85	0.8	0.85	0.2
Scen. 2	0.85	0.2	0.2	0.9	0.7	1	0.2	0.3	0.3	0	0	0.2	0.2	0.1	0.75	0.95	0.3	0.8	0.7	0.2	0.8
Scen. 3	0.1	0.8	0.9	0.9	0.95	0.23	0.75	0.99	0.98	0.99	0.98	0.99	0.9	0.95	0	0	0.99	0	0.99	0	0.99

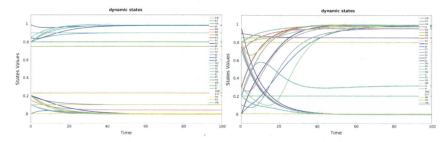

Fig. 2 Simulation of Scenario 1 (left, no burnout) and 2 (right, burnout)

elements would slowly fade away. To model this scenario, we created a mix of high- and low-risk factors—as also shown in Table 4.

In Scenario 1, there is not one sudden change in state values to be observed, as the protective factors continue to increase (if not constant), while the opposite holds true for the risk factors. Thus, the burnout elements drop even further until an equilibrium situation is reached. Scenario 2 displays a situation where a person gradually develops a burnout. Low protective factors like work experience, openness and social contact with co-workers are not enough to diminish the high-risk factors like daytime dysfunction, neuroticism and with feeling charged work. Thus, experienced overload increases and consequently with it the burnout elements emotional exhaustion and cynicism, while the third burnout element personal accomplishment is decreasing to zero. Consequently, job resignation and physical health problems both increase. Sleep quantity increases first, as the high daytime dysfunction is leading to an increase in sleeping pill intake, until daytime dysfunction and high experienced overload peak, causing the sleep quantity to settle on a lower level.

Scenario 3 shows a person who is starting to experience a burnout, but then it fades away due to high enough protective factors. Burnout elements and its consequences, like job detachment, are all high in the beginning, while sleep quality is low. In addition, protective factors like working experience and hardiness are high, while some risk factors like sleep deprivation and with feeling charged work are low. The combination of low-risk factors and high protective factors is enough for the burnout symptoms to lower over time, until a stable no burnout scenario is reached (Figs. 2 and 3).

5 Verification of the Model by Mathematical Analysis

For verification, we analysed the stationary points and equilibria occurring in the model. A *stationary point* of a state Y at time t occurs when $dY(t)/dt = 0$. The network model is in *equilibrium* at t when all states have a stationary point at t. As described in Sect. 3 Eq. (1) in a temporal-causal network model, the differential equation for all states is $dY(t)/dt = \eta_Y [\textbf{aggimpact}_Y(t) - Y(t)]$. As all speed factors η_Y in the model

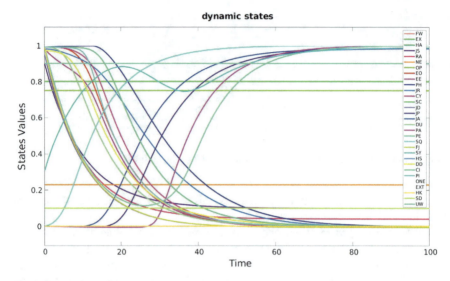

Fig. 3 Simulation of Scenario 3: recovery from burnout

are nonzero, all stationary points must follow the criterion **aggimpact**$_Y(t) = Y(t)$, also formulated as: in a temporal-causal network model, there is a stationary point for state Y at t if and only if $\eta_Y = 0$ or $\mathbf{c}_Y(\omega_{X1,Y}X_1(t), \ldots, \omega_{Xk,Y}X_k(t)) = Y(t)$. Using this criterion, first we looked at the burnout Scenario 1 from above and identified a stationary point for emotional exhaustion at $t = 3.3$. The aggregated impact, other states have on emotional exhaustion, is modelled with an advanced logistic function with steepness $\sigma = 10$ and threshold $\tau = 0.5$. Openness and experienced overload are the two states affecting emotional exhaustion with connection weights of -0.5 and 1, respectively. The values for emotional exhaustion at time $t = 3.3$ for the stationary point and at time $t = 100$ for the equilibrium are 0.25301 and 0.98177, respectively. In addition, the values for experienced overload are 0.49759 resp. 0.99997, while openness has a constant value of 0.2. We used the values for openness and experienced overload to compute the aggregated impact they have on emotional exhaustion, as shown in Table 6. We did this also with the stationary points and equilibria occurring for sleep quantity and drug use. The state values and aggregated impacts show little differences, which add to confidence that the model does what is expected (Table 5).

Table 5 Outcomes of the verification

State	EE	EE	SY	SY	DU	DU
Time point t	3.3	100	10.2	100	14.1	100
$X(t)$	0.25301	0.98177	0.57697	0.31436	0.09493	0.98863
aggimpact(t)	0.25927	0.98189	0.57364	0.31491	0.0997	0.98915
aggimpact(t) – $X(t)$	0.00626	0.00012	– 0.00333	0.00055	0.00477	0.00052

6 Validation of the Model by Parameter Tuning

For further validation of the model, we used simulated annealing based on acquired data for parameter tuning. For this, we evaluated the prevailing literature regarding the relation of daytime dysfunction and quality and quantity of sleep (e.g. [1, 15, 17, 20, 21]), as well as the use of caffeine and sleeping pills to overcome daytime dysfunction (e.g. [14, 15]). Through these scientifically based relations, we were able to create expected patterns of how those factors would interact. Based on that, we could create data points meeting those expected patterns, as shown in Table 6.

Over time, we expected a sequential pattern, where high daytime dysfunction would increase sleeping pill intake, leading to a subsequent rise in sleeping quantity. A rise in sleeping quantity in turn starts to have an effect on sleep quality and daytime dysfunction. With a lowered daytime dysfunction, the sleeping pill intake decreases, until sleep quantity decreases again and the causal chain begins all over again. We started to create this pattern for a person with high daytime dysfunction. For a high daytime dysfunction (0.92), we could expect an equally low sleep quality (0.07). For a sleep-deprived person, we expected a higher resistance using sleeping pills in comparison with caffeine intake thus the higher value for caffeine than for sleeping pill intake. The combination of high daytime dysfunction (0.92) and relatively high sleeping pill intake (0.74) should result in a medium sleep quantity (0.44). With a decreasing daytime dysfunction and recovering sleep quality, we expected the sleeping pill intake to decrease faster than the caffeine intake (e.g. time 10–20). Likewise, as daytime dysfunction increases again, sleeping pills should show a higher increase used over time, as they are more addictive than caffeine (e.g. time 40– 50). With the generated data obtained from interpolating our expected values, we could compare these to our model. Indeed, parameter tuning gave a higher steepness as well as a higher threshold for sleeping pill intake. The root mean square error (RMS) for the five selected factors is 0.028758, which is relatively low. Perhaps, this may be because our estimation of empirical values was partly biased by what we knew already about typical model patterns. With our tuned model, we validated our model against expected patterns based on personality and lifestyle factors. At first, our aim is to model a high demanding job situation, including a low value

Table 6 Expected pattern of sleeping factors

Time	Sleep quality	Sleep quantity	Daytime dysfunction	Caffeine intake	Sleeping pill intake
0	0.07	0.44	0.92	0.88	0.74
10	0.41	0.75	0.73	0.93	0.79
20	0.55	0.5	0.43	0.67	0.31
30	0.24	0.19	0.67	0.63	0.14
40	0.09	0.13	0.86	0.83	0.48
50	0.09	0.49	0.92	0.93	0.78

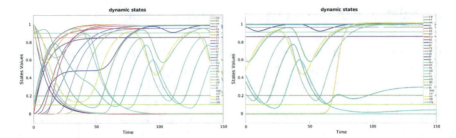

Fig. 4 Burnout Scenario 1 with periodic sleeping pattern (left) and having kids after some point in time lower the sleep quantity, thus breaking the cyclic pattern (right)

of personal accomplishment and high values of cynicism and emotional exhaustion. Low protective factors and high-risk factors lead to a scenario where a burnout occurs. In addition, through the high experienced overload and low physical exercise, the sleeping quantity decreases. This turns into a rise in daytime dysfunction, which is causing a higher intake of sleeping pills. In turn, the sleep quantity rises, until daytime dysfunction lowers and consequently the intake of sleeping pills. Thus, it causes daytime dysfunction to rise again and the pattern repeats itself. Simultaneously, personal health problems fluctuate with the sleeping pill intake (Fig. 4).

After the model settles down to its equilibria, we introduce an external state variable with impact on having kids at around $t = 210$ $(150 + 60)$, which in turn can negatively affect the amount of sleep. We implemented this via introducing the external factor EXT, which has an impact on having kids HK. This external factor EXT was modelled using a connection to itself and with an advanced logistic sum combination function with $\sigma = 18$, $\tau = 0.2$ and speed 0.04; the state HK has $\sigma = 50$, $\tau = 0.8$ and speed 1. Thus, the external factor EXT builds up very slowly and has very little impact on having kids HK, until the threshold is reached (at birth) and the state value of having kids HK changes quite abruptly. In this scenario, it is enough that even an increased sleeping pill intake can no longer match the negative effects on sleeping quantity, which results in increase of daytime dysfunction. This in turn causes sleep quantity and quality to fall to low equilibrium values.

7   Discussion

In this study, we introduced a temporal-causal network model for a burnout in relation to sleep. The Network-Oriented Modelling approach described in [23, 24] was used, and for implementation the dedicated software environment is described in [16]. The modelling approach showed the impact of different lifestyle, personal and job factors on the development of a burnout. This model can be the basis for different types of simulations. For instance, an agent-based model can be used to schedule night shifts in order to preserve the needed recovery. Also, our model can be used to simulate the effects of certain changes in lifestyle on the development of burnout, especially when even more states are added to resemble the real-world complexity.

For validity, it was critical to implement both risk and protective factors. To analyse computationally their roles in development and recovery, we designed different scenarios, such as the one of a burnout and a non-burnout, and a scenario where a burnout occurs and a periodic pattern between sleep-related factors arises. It showed that a small increase in sleep disturbance in an already demanding situation can cause a non-recoverable downwards movement. Unfortunately, there was no numerical empirical data available to test these patterns in more detail. For example, it is unclear to extract from the literature how strong and fast certain variables relate. Consequently, this gives room for improvement on our model regarding how the relations of the variables vary in type, speed and weight.

References

1. Åkerstedt, T., Wright, K.P.: Sleep loss and fatigue in shift work and shift work disorder. Sleep Med. Clin. **4**(2), 257–271 (2009)
2. Alarcon, G., Eschleman, K.J., Bowling, N.A.: Relationships between personality variables and burnout: a meta-analysis. Work Stress **23**(3), 244–263 (2009)
3. Borsboom, D., Cramer, A.O.: Network analysis: an integrative approach to the structure of psychopathology. Annu. Rev. Clin. Psychol. **9**, 91–121 (2013)
4. CBS, TNO: Psychosociale arbeidsbelasting (PSA) werknemers. Retrieved from http://statline.cbs.nl/StatWeb/publication/?VW=T&DM=SLNL&PA=83049NED&LA=NL (2017)
5. DePaepe, J., French, R., Lavay, B.: Burnout symptoms experienced among special physical educators: a descriptive longitudinal study. Adap. Phys. Act. Q. **2**(3), 189–196 (1985)
6. Dujmić, Z., Machielse, E., Treur, J.: A temporal-causal modeling approach to the dynamics of a burnout and the role of physical exercise. In: Samsonovich, A.V. (ed.) Biologically Inspired Cognitive Architectures 2018: Proceedings of the 9th International Conference on Biologically Inspired Cognitive Architectures, BICA'18, vol. 1, Advances in Intelligent Systems and Computing, vol. 848, pp. 88–100. Springer (2019)
7. Emilia, I., Gómez-Urquiza, J.L., Cañadas, G.R., Albendín-García, L., Ortega-Campos, E., Cañadas-De la Fuente, G.A.: Burnout and its relationship with personality factors in oncology nurses. Eur. J. Oncol. Nurs. **30**, 91–96 (2017)
8. Eschleman, K.J., Bowling, N.A., Alarcon, G.M.: A meta-analytic examination of hardiness. Int. J. Stress Manage. **17**(4), 277–307 (2010)
9. Huang, L., Zhou, D., Yao, Y., Lan, Y.: Relationship of personality with job burnout and psychological stress risk in clinicians. Chin. J. Ind. Hygiene Occup. Dis. **33**(2), 84–87 (2015)
10. Kabadayi, A.: Investigating the burn-out levels of turkish preschool teachers. Proc.-Soc. Behav. Sci. **197**, 156–160 (2015)
11. Kuipers, B.J.: Commonsense reasoning about causality: deriving behavior from structure. Artif. Intell. **24**, 169–203 (1984)
12. Kuipers, B.J., Kassirer, J.P.: How to discover a knowledge representation for causal reasoning by studying an expert physician. In: Proceedings of the Eighth International Joint Conference on Artificial Intelligence, IJCAI'83. William Kaufman, Los Altos, CA (1983)
13. Maslach, C., Jackson, S.E.: The measurement of experienced burnout. J. Organ. Behav. **2**(2), 99–113 (1981)
14. McGeary, C.A., Garcia, H.A., McGeary, D.D., Finley, E.P., Peterson, A.L.: Burnout and coping: veterans health administration posttraumatic stress disorder mental health providers. Psychol. Trauma: Theory Res. Pract. Policy **6**(4), 390 (2014)

15. Miró, E., Solanes, A., Martínez, P., Sanchez, A.I., Rodríguez, J.M.: Relationship between burnout, job strain, and sleep characteristics. Psicothema **19**(3), 388–394 (2007)
16. Mohammadi Ziabari, S.S., Treur, J.: A modeling environment for dynamic and adaptive network models implemented in matlab. In: Proceedings of the Fourth International Congress on Information and Communication Technology, ICICT'19. Advances in Intelligent Systems and Computing. Springer (2019)
17. Pagnin, D., de Queiroz, V., Carvalho, Y.T.M.S., Dutra, A.S.S., Amaral, M.B., Queiroz, T.T.: The relation between burnout and sleep disorders in medical students. Acad. Psychiatry **38**(4), 438–444 (2014)
18. Pearl, J.: Causality. Cambridge University Press (2000)
19. Pilcher, J.J., Huffcutt, A.I.: Effects of sleep deprivation on performance: a meta-analysis. Sleep **19**(4), 318–326 (1996)
20. Rosen, I.M., Gimotty, P.A., Shea, J.A., Bellini, L.M.: Evolution of sleep quantity, sleep deprivation, mood disturbances, empathy, and burnout among interns. Acad. Med. **81**(1), 82–85 (2006)
21. Söderström, M., Jeding, M., Akerstedt, T., Ekstedt, M., Perski, A.: Insufficient sleep predicts clinical burnout. J. Occup. Health Psychol. **17**(2), 175–183 (2012)
22. Sutin, A.R., Stephan, Y., Luchetti, M., Artese, A., Oshio, A., Terracciano, A.: The five-factor model of personality and physical inactivity: a meta-analysis of 16 samples. J. Res. Pers. **63**, 22–28 (2016)
23. Treur, J.: Network-Oriented Modeling: Addressing Complexity of Cognitive. Affective and Social Interactions. Springer, Cham (2016)
24. Treur, J.: The ins and outs of network-oriented modeling: from biological networks and mental networks to social networks and beyond. Trans. Comput. Collect. Intell. **32**, 120–139 (2019). Based on Keynote Lecture at the 10th International Conference on Computational Collective Intelligence, ICCCI'18 (2019)
25. Whitney, C.W., Enright, P.L., Newman, A.B., Bonekat, W., Foley, D., Quan, S.F.: Correlates of daytime sleepiness in 4578 elderly persons: the Cardiovascular Health Study (1998)

Modeling Cultural Segregation of the Queer Community Through an Adaptive Social Network Model

Pieke Heijmans, Jip van Stijn and Jan Treur

Abstract In this study, the forming of social communities and segregation is examined through a case study on the involvement in the queer community. This is examined using a temporal-causal network model. In this study, several scenarios are proposed to model this segregation and a small questionnaire is set up to collect empirical data to validate the model. Mathematical verification provides insight into the model's expected behavior.

Keywords Queer community · Temporal-causal network · Social hardship · Social contagion · Homophily principle · Social network · Cultural segregation

1 Introduction

In a developed world that contains increasingly pluralistic and diverse societies, the establishment of subcultures seems inevitable. In the West, subcultures are associated with an increased identification with in-group individuals, leading to a caring environment. However, subcultures are at risk of a high degree of segregation and misunderstanding of and by out-group individuals, possibly leading to discrimination and aggression. In view of stimulating and maintaining peaceful and democratic processes in these societies, it can be useful to investigate this behavior. This can be performed through network-oriented modeling, which describes the behaviors and opinions of people in relation to the connections among them. In graphical representations of social networks, individuals and their states are depicted by nodes which are connected by uni- or bidirectional links.

P. Heijmans (✉) · J. van Stijn · J. Treur
Behavioural Informatics Group, Vrije Universiteit Amsterdam, Amsterdam,
The Netherlands
e-mail: piekeheijmans@gmail.com

J. van Stijn
e-mail: jipvanstijn@gmail.com

J. Treur
e-mail: j.treur@vu.nl

© Springer Nature Singapore Pte Ltd. 2020
X.-S. Yang et al. (eds.), *Fourth International Congress on Information and Communication Technology*, Advances in Intelligent Systems and Computing 1027, https://doi.org/10.1007/978-981-32-9343-4_19

233

In this paper, firstly some background about cultural segregation of the queer community is discussed. Next, the network model is explained including the homophily and social contagion principle. In addition, some different scenarios for the model are set up and different simulations with the model are discussed. It will be shown how mathematical verification clarifies how different parameters influence the outcomes of the model. The current paper investigates subculture identification and cultural segregation of the queer community. Being born differently from the heteronormative society that they grow up in, queer people have to deal with a number of factors that contribute to a permanent level of distress. Meyer [9] frames this psychological distress as minority stress. Due to a continuous internalized homophobia, stigma, which relates to society's expectations, and experiencing actual discrimination or violence, minority stress is established. She found a strong connection between experiencing minority stress and dealing with psychological distress. She adds that hiding and concealing, expectation of rejection, and ameliorative coping processes contribute to this psychological minority stress as well [10].

To deal with psychological distress, a social support system can help. For example, successful coming-out stories of other queer people can help reduce the anxiety of getting negative reactions from friends and family. Wright and Perry claim that support systems are necessary as they influence the development of young people's self-concept and self-esteem [10]. Queer community and queer community spaces can function as this social support system. Beemyn [2] examines the historical role of queer spaces and states that queer people needed their own space, not only to escape from governmental pressures such as police harassment, but also to not deal with constant territorial struggle, a place where they could escape the dominant cultural order. These processes lead to cultural segregation, as consequently queers will distantiate themselves from the dominant cultural order by collectively grouping together in their own communities. Another strong reason to get involved in such a community is collectivism, in the sense that people want to benefit their group. The more you are involved with and identify with this subgroup, the greater your sense of collectivism [1], and thus, the stronger the effect of cultural segregation will be.

From these theories, we expect queer people with a greater experience of hardship and psychological distress to get involved more in the queer community as the community serves as a support system, resulting in a stronger cultural segregation. In contrast, queer people that experience no to little hardship are not inclined to look for a social support system; however, they may be involved with the community for other reasons, for example, relating to their peers. Finally, straight people that experience hardship may look for community support, but not necessarily for the queer community as they do not particularly identify with this group and do not have a strong sense of collectivism. Resulting from these conclusions, it can be expected that a certain group of queer people will get involved with their community, finding comfort, support, and finding equals. In contrast, straight people will not share these needs and will not identify strongly with the queer community. The result is cultural segregation, in which queer people's identification with their community is opposed to straight people's identification.

These processes were analyzed computationally by designing an adaptive network model based on the homophily principle that describes bonding between persons that consider each other similar in some respect(s); see, for example, [8]. This principle works in combination with the principle of social contagion [3] in a circular mutual causal relationship, also called co-evolution [5, 18]. In Sect. 2, the adaptive temporal-causal network model based on these two principles is introduced. Section 3 illustrates the model by example simulations. In Sect. 4, it is shown that the simulation outcomes are in accordance with what is predicted by a mathematical analysis of the model. Section 5 describes validation of the model by comparing simulation outcomes to empirical data and applies parameter tuning. Finally, Sect. 6 is a conclusion.

2 The Adaptive Temporal-Causal Network Model

Thus, the following difference and differential equation for state Y are obtained:A network-oriented modeling approach based on temporal-causal networks [13, 14, 17] was used to analyze the type of processes described in Sect. 1. This approach can be considered as a branch in the causal modeling area which has a long tradition in AI; e.g., see [6, 7, 11]. It distinguishes itself by a dynamic perspective on causal relations, according to which causal relations exert causal effects over time, and these causal relations themselves can also change over time. The type of network models that form the basis is called a *temporal-causal network model*. These network models can be used to translate informally described theories from a variety of human-directed disciplines into adaptive and dynamical numerical models. It takes into account states and their causal effects on other states. The strengths of causal relations from a state X to a state Y are indicated by differences in connection weights $\omega_{X,Y}$. These connection weights can be combined with activation levels $Y(t)$ of states Y and used as input for combination functions $c_Y(...)$ to determine the aggregated impacts on the states. The precise dynamics of the network are also defined using speed factors η_Y of states Y. The network becomes adaptive when connection weights are dynamic as well. The conceptual representation basically is a graph of states and their causal relations; a graphical overview of the network is represented as depicted in Fig. 2. The numerical representation is a translation of this conceptual representation in the way described in Table 1.

$$Y(t + \Delta t) = Y(t) + \eta_Y\big[c_Y\big(\omega_{X_1,Y}X_1(t), \ldots, \omega_{X_k,Y}X_k(t)\big) - Y(t)\big]\Delta t$$
$$dY(t)/dt = \eta_Y\big[c_Y\big(\omega_{X_1,Y}X_1(t), \ldots, \omega_{X_k,Y}X_k(t)\big) - Y(t)\big] \tag{1}$$

The adaptive social network model used here is based on two fundamental principles. Firstly, the notion of social contagion is used to explain the causal influence of one state on another through the connection between the two [3]. This principle accounts for the change of state values over time, and its numerical representation is (where

Table 1 From conceptual representation to numerical representation of a temporal-causal network model; adopted from [17]

Concept	Representation	Explanation
State values over time t	$Y(t)$	At each time point t, each state Y in the model has a real number value in $[0, 1]$
Single causal impact	$\mathbf{impact}_{X,Y}(t) = \omega_{X,Y}\, X(t)$	At t, state X with connection to state Y has an impact on Y, using connection weight $\omega_{X,Y}$
Aggregating multiple impacts	$\mathbf{aggimpact}_Y(t)\ _Y(\mathbf{impact}_{X1,Y}(t),...,$ $\mathbf{impact}_{Xk,Y}(t)) = \mathbf{c}_Y(\omega_{X1,Y}X_1(t),$ $..., _{Xk,Y}X_k(t))$	The aggregated causal impact of multiple states X_i on Y at t is determined using combination function $\mathbf{c}_Y(..)$
Timing of the causal effect	$Y(t + \Delta t) = Y(t) + \eta_Y$ $[\mathbf{aggimpact}_Y(t) - Y(t)]\,\Delta t = Y(t)$ $+\,\eta_Y\,[\mathbf{c}_Y(\omega_{X1,Y}X_1(t), ...,$ $\omega_{Xk,Y}X_k(t)) - Y(t)]\,\Delta t$	The causal impact on Y is exerted over time gradually, using speed factor η_Y; here, the X_i are all states with connections to state Y

X_{Ai} and X_B are the states of persons A_i and B):

$$dX_B/dt = \eta_B\big[c_B\big(\omega_{A_1,B}X_{A_1}, \ldots, \omega_{A_k,B}X_{A_k}\big) - X_B\big]$$
$$X_B(t + \Delta t) = X_B(t) + \eta_B[c_B\big(\omega_{A_1,B}X_{A_1}(t), \ldots, \omega_{A_k,B}X_{A_k}(t)\big) - X_B(t)]\Delta t \quad (2)$$

One option for the combination functions for modeling the aggregated impact of multiple states on another is the scaled sum function:

$$\mathbf{ssum}_\lambda(V_1, \ldots V_k) = (V_1 + \cdots + V_k)/\lambda \quad (3)$$

Usually, a normalized scaled sum is used: The value of λ is the sum of all incoming weights $\omega_{X_i,Y}$. In cases that these connection weights change over time an adaptive version of the scaled sum can be used:

$$\mathbf{adapssum}_\lambda(V_1, \ldots V_k) = (V_1 + \cdots + V_k)/\lambda(t) \quad (4)$$

where $\lambda(t)$ is the sum of all incoming weights $\omega_{X_i,Y}(t)$ at t. This version was used to model social contagion in the work reported here. Another function used for social contagion in this research is the advanced logistic sum function:

$$\mathbf{alogistic}_{\sigma, \tau_{\log}}(V_1, \ldots V_k) = \left[\frac{1}{1 + e^{-\sigma(V_1 + \cdots + V_k - \tau_{\log})}} - \frac{1}{(1 + e^{\sigma\tau_{\log}})}\right]$$
$$(1 + e^{-\sigma\tau_{\log}}) \quad (5)$$

Fig. 1 A conceptual representation of the homophily principle

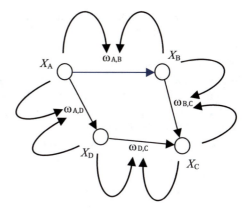

where σ is the steepness factor and $\tau_{\log}$ the logistic threshold.

Secondly, the principle of homophily describes the change of connection weights between the states X_A and X_B of two persons A and B; e.g., [8]. According to this principle, when the values of two nodes are similar, the connection between them becomes stronger (represented by a higher connection weight). Conversely, the lower the similarity between the (values of) the two nodes, the smaller their connection weight. In Fig. 1, the homophily principle is depicted by the striped arrows. Numerically, this principle can be represented as follows:

$$\omega_{A,B}(t + \Delta t) = \omega_{A,B}(t) + \eta_{A,B}[c_{A,B}(X_A(t), X_B(t), \omega_{A,B}(t)) - \omega_{A,B}(t)] \Delta t$$
$$d\omega_{A,B}/dt = \eta_{A,B}[c_{A,B}(X_A, X_B, \omega_{A,B}) - \omega_{A,B}] \tag{6}$$

in which X_A and X_B represent the states of person A and person B.

In the current paper, the combination function $c_{A,B}(V_1, V_2, W)$ used for the homophily principle is the following:

$$\mathbf{slhom}_{\tau_{hom} \cdot \alpha}(V_1, V_2, W) = W + \alpha\, W(1 - W)(\tau_{hom} - |V_1 - V_2|) \tag{7}$$

3 Simulations of Example Scenarios

In this section, the adaptive network model is described for three scenarios. In the first example Scenario 1 for this model, a social network of 10 nodes is used, consisting of three communities of three or four nodes. Each community contains one node that is the so-called bridge node, which has connections to the two bridge nodes of the other communities. Within the three communities, there is maximal connectedness: All nodes are connected to the other nodes within the community. As the connections represent social interactions, they are all assumed to be bidirectional. All connections are presumed to be relatively strong, so all connection weights were set initially to 0.8. The state values represent the individual's identification with the queer culture, with 0 being minimal identification and 1 being maximal identification. The initial values of identification with the queer culture are assumed to be spread evenly on the spectrum of 0.1–1, as depicted in Fig. 2. For social contagion, in this scenario the adaptive normalized scaled sum function was used, with a dynamic scaling factor of the sum of all the connection weights per state at that time. Table 2 shows the values used for the parameters. In this scenario, two simulations were carried out using two different state speed factors. The simulation of this model shows a classic example of the interplay of social contagion and homophily; sometimes also called co-evolution [5, 18]. In Scenario 1.1, a speed factor of 0.2 was used for all states, and the three communities all converged to their own equilibrium value. The connection weights of the within-community connections converged to 1, while the weights of the bridge connections converged to 0. This result is illustrated in Fig. 3.

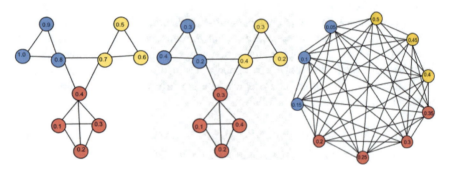

Fig. 2 Left hand: conceptual representation of the network in Scenario 1. Middle: for Scenario 2. Right hand: for scenario 3. Each node represents an individual, with the initial state value illustrated in the node. Each line represents a bidirectional connection. The three communities are labeled with different colors

Table 2 Parameters and their values used in the simulation of Scenarios 1 and 2

Parameters Scenario 1	Values	Parameters Scenarios 2 and 3	Values
State speed factor η_Y	0.2/0.8	State speed factor η_Y	0.2
ω speed factor η_ω	0.5	ω speed factor η_ω	0.5
slhom threshold factor τ_{hom}	0.1	slhom threshold factor τ_{hom}	0.08
slhom amplification factor α	8.0	slhom amplification factor α	8.0
		Alogistic steepness factor σ	2.5
		Alogistic threshold factor τ_{log}	0.18

Except for the initial connection weights mentioned above, all parameters are equal for all states and connections. (ω = connection weight, slhom = simple logistic homophily function, alogistic = advanced logistic function as defined above)

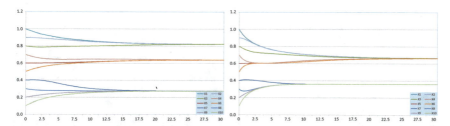

Fig. 3 A graphical representation of the state values in the simulation of Scenarios 1.1 resp. 1.2 for state speed factors 0.2 resp. 0.8. The y-axis represents the state values, while the x-axis represents time. Every community converges to its own equilibrium value. Social contagion is quicker in Scenario 1.2 than in Scenario 1.1. This leads to the fact that the first two communities converge toward each other. The third community still converges to its own equilibrium

The final example Scenario 3 concerns a fully connected network in which each node is connected to every other node. Figure 5 shows a conceptual representation of this network. Again, the advanced logistic function is used to model social contagion between the nodes, and the simple homophily function alters the connection weights over time. This scenario is somewhat more life-like than the previous examples, as it is reasonable to assume that, in a group of 10 people, every person knows all others to some degree. All initial connection weights are set to 0.8. In Scenario 1.2, when the state speed factor of 0.8 was used, two of the communities first converge within themselves, and then converged to a shared equilibrium state value. The third community converged to its own equilibrium value. The bridge connection between states 3 and 4 now converged to 1 instead of 0. This shows that the effect of the social contagion function is now quicker than in simulation 1.1 and influences the homophily of the connection weights.

Note that in case that no homophily principle is applied but only the social contagion, according to Theorems 3 and 4 in [15] for the so-called strongly connected network using normalized scaled sum combination functions (which are strictly monotonically increasing and scalar-free [15]) all states will converge to the same value, and this value lies between the minimal and maximal initial state value. The emergence of communities is a result of the homophily, and the faster the contagion in comparison with the homophily principle, the lower the number of communities that emerge, as shown here in Fig. 3.

In the example Scenario 2, the number of nodes and their connectedness is equal to the first scenario. However, the initial values range from 0.1 to 0.4 and this time the advanced logistic function is used for social contagion. The results of this scenario shows that, when using the advanced logistic function to model social contagion, the equilibrium values can end up higher or lower than any initial values of the states, in contrast to what holds for normalized scaled sum functions; see [15]. Most of the states converge to an equilibrium of 1 or 0.982, whereas all the initial values lay between 0.1 and 0.4. This may represent a real-life process in which a sentiment is strengthened and amplified beyond its original level, because it is shared with others.

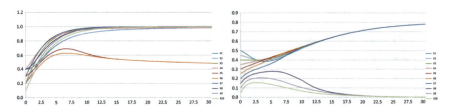

Fig. 4 A graphical representation of the state values in the simulation of Scenarios 2 and 3. The
y-axis represents the state values, while the x-axis represents time. The initial values of all states
range between 0.1 and 0.4, while most of the values converge to 1 or 0.982. However, state 4 and
5 converge to 0.415

Yet, the equilibrium values of state 5 and 6 (initial values 0.2 and 0.3) converge to
a different, much lower equilibrium of 0.415. The connection weights of all initial
connections converge to 1, except for those of state X_4 with X_5 and X_6. This shows
that: (1) when using the advanced logistic function in combination with homophily,
the initial value of a state does not necessarily determine in which equilibrium it ends
up; and (2) that this combination function can strongly influence the connectivity in
a network, possibly leading to the change or dissolvement of communities over time.
Thus, this simulation demonstrates the phenomenon of segregation within a network
(and in a specific community within the network) regardless of a similarity in initial
values of the states. Figure 4 shows these results in a graphical representation.

Again, the simulation resulted in the separating behavior into two groups. The
seven states with the highest initial values all converge to a value of 0.8, spiraling
past their original values. The remaining three states converge to a value of 0.036,
well below their initial values. The connection weights within the groups all converge to 1, while the intergroup connections end up with a weight value of 0. This
simulation not only shows separating behavior of one group into two communities,
as with the previous scenario. It also hints at a notion of extremism, in which the two
groups increasingly push each other off. This may be comparable to the process of
'othering', in which the shaping of an identity depends on the supposition with other
people's behavior or convictions. This combined with group behavior can then lead
to polarization, which can be observed in many social and political situations.

4   Model Verification by Mathematical Analysis

In order to verify the model, first a mathematical analysis of stationary points was
performed, in particular for the third scenario. A *stationary point* of a state Y at time
t occurs when $dY(t)/dt = 0$. A *stationary point* of a connection weight ω at time t
occurs when $d\omega(t)/dt = 0$. The network model is in an *equilibrium* at t when all states
and all connection weights have a stationary point at t. As described in Sect. 2, in
a temporal-causal network model the differential equation for all states is: $dY(t)/dt
= \eta_Y [\textbf{aggimpact}_Y(t) - Y(t)]$. As all speed factors in the model are nonzero, all

stationary points must follow the criterion: **aggimpact**$_Y(t) = Y(t)$ also formulated as: In a temporal-causal network model, there is a stationary point for state Y at t if and only if $\eta_Y = 0$ or $c_Y(\omega_{X_1,Y} X_1(t), \ldots, \omega_{X_k,Y} X_k(t)) = Y(t)$.

From the modeled data in Scenario 3, stationary points were gathered from several states, and the aggregated impact at that time was calculated per state using the advanced logistic function described earlier. If the state values and the calculated aggregated impact are equal, the stationary point equation above is fulfilled. This mathematically verifies the model. The results are presented in Table 3. As appears in the table, the deviations between the observed state values and the calculated aggregated impact on that state at that time are very low. This indicates that the model does what is expected; it calculates the expected state values with high precision.

When the stationary point Eq. (9) mentioned above applies to all network states and connection weights at a single time, the model is in equilibrium. In the third scenario, the model appears to be in equilibrium at $t = 300$. The state values at this time were read, and the aggregated impact at that moment was calculated per state. The results are presented in Table 4. As is visible in the table, the deviation between the state values and the aggregated impact of all states at $t = 300$ is very low, indicating again that the model calculates the state values in a proper way with high precision.

Additionally, the dynamic connection weights were analyzed. Recall the following combination function for the homophily principle (7):

$$\textbf{slhom}_{\tau,\alpha}(V_1, V_2, W) = W + \alpha\, W(1 - W)(\tau_{\text{hom}} - |V_1 - V_2|) \qquad (8)$$

The stationary point criterion of $\textbf{slhom}_{\tau_{\text{hom}},\alpha}(V_1, V_2, W) = 0$ provides the equation:

$$W(1 - W)(\tau_{\text{hom}} - |V_1 - V_2|) = W \qquad (9)$$

meaning that

$$|V_1 - V_2| = \tau_{\text{hom}} \text{ or } \omega_{A,B} = 0 \text{ or } \omega_{A,B} = 1 \qquad (10)$$

Table 3 An overview of stationary point values and the aggregated impact values in the simulation of Scenario 3

State	X_1	X_2	X_3	X_8	X_9	X_{10}
Time point	3.95	3.55	2.10	4.65	3.10	2.55
State value	0.387	0.397	0.397	0.261	0.197	0.145
Aggregated impact	0.388	0.396	0.397	0.262	0.197	0.145
Deviation	0.001	0.001	0	0.001	0	0

States 4–7 did not have a temporary stationary point at the considered time interval. The bottom row shows the deviation between these values

Table 4 State values in the simulation of Scenario 3 at $t = 300$

State	X_1	X_2	X_3	X_4	X_5	X_6	X_7	X_8	X_9	X_{10}
State value	0.802	0.802	0.802	0.802	0.802	0.802	0.802	0.040	0.036	0.038
Agg. impact	0.803	0.802	0.803	0.803	0.802	0.801	0.802	0.038	0.037	0.037
Deviation	0.001	0	0.001	0.001	0	0.001	0	0.002	0.001	0.001

However, the solution $|V_1 - V_2| = \tau_{hom}$ turns out non-attracting, eliminating this solution as a possible stationary point in the simulation. This corresponds to the connection weight values at $t = 300$ in the simulation of Scenario 3: All connections eventually are either 1 or 0. These connections define the two clusters that appeared, with full connectivity within the communities, and no connection to nodes outside the community.

In a wider context, such an analysis of limit behavior for some classes of homophily combination functions has been presented in [16]. The above analysis fits in that more general approach. In addition to the 0 or 1 values as limit for connection weights, one of the results is that independent of the size of the network there can be at most $1 + 1/\tau_{hom}$ groups; see [16], Theorem 1a). Indeed, the actual number of the groups in the simulations is less than that predicted maximal number.

5 Validation Using Empirical Data[1]

To validate the proposed model, we acquired a data set for which we set up a questionnaire that participants had to fill out online. The questionnaire consisted of some general introductory questions like age, gender, education, and most importantly sexual orientation. Secondly, the participant had to indicate to what extent they agreed to statements (using a Likert scale, from 1 till 5, 1 indicating 'strongly disagree' and 5 indicating 'strongly agree'). The first 10 questions related to how involved they are in the queer community now. The second 10 questions related to how involved they were in the queer community 5 years ago. A final 10 questions related to how much social hardship the participant experienced in their youth, based both on a general level and in relation to their sexuality.

To explore the suitability of our data, we used SPSS to perform a statistical analysis regarding the following hypotheses: (1) that queer people would generally score higher on involvement in the queer scene than straight people, (2) that queers would be involved more in the community now than 5 years ago, and finally (3) that hardship would make up for explaining this difference between involvement in the queer community now versus 5 years ago.

By exploring the differences in scores on involvement in the queer scene between queer people and straight people, two of our hypotheses were confirmed: Queer people generally score higher on involvement in the queer scene than straight people, and queers are involved more in the community now than 5 years ago. Another analysis was needed to check the assumption that differences in scores were dependent on the hardship that people experienced by verifying a (possible) a correlation of the scores with the scores on the hardship questions. However, Pearson's correlation r of 0,056 indicated no correlation for the variables. This refutes the hypothesis that the increase in identification with the queer community is mediated by the degree of social hardship experienced when young.

---

[1] A more detailed account of this section can be requested from the third author.

Next, the empirical data were reshaped in a format compatible with the format of the simulation outcomes so that parameter tuning would provide the best possible solution for the model and the empirical data to fit together. Scenario 3 was chosen for the parameter tuning as it would be the best possible fit for real-life scenarios: as a case where each individual knows each other individual resembles a real-world scenario most. The connection weights for the model were set under the assumption that some segregation was already in place: Straight people were set to a connection weight of 0.8 to other straight people, knowing mostly other straight people and a connection weight of 0.3 to queers. For queers, it was the other way around, setting connection weights to other queers at 0.8 and to straights at 0.3. This leaves some parameters of the model to still be tuned: the speed factor η_X for each state X, the steepness σ of the alogistic combination function, the threshold $\tau_{\log}$ of the alogistic combination function, the threshold $\tau_{\text{hom},X,Y}$ of the simple linear homophily, and the amplification factor $\alpha_{X,Y}$ of the simple linear homophily.

The way we estimated these parameters was through exhaustive search. The exhaustive search method is a problem-solving technique in which all possible candidate solutions for parameter values are investigated on how well they make the model fit to the data. Because testing each parameter with a grain size of 0.05 would lead to combinatorial explosion, the way to execute the exhaustive search was by iterative refinement, starting with larger grain sizes, for example, 0.5 or 0.1 (depending on the parameter) and narrowing down the grain sizes until the model fits the empirical data best. Speed factors, steepness, and thresholds were investigated with grain size 0.1, and then in a second phase tuned with a grain size of 0.05; however, amplification was investigated with a grain size of 1 and then further tuned with a grain size of 0.5. The values found are $\sigma = 0.53$, $\tau_{\log} = 0.2$, $\tau_{\text{hom}} = 0.1$, $\alpha = 3.5$, and $\eta_{X_1} = 0.1$, $\eta_{X_2} = 0.1$, $\eta_{X_3} = 0.1$, $\eta_{X_4} = 0.35$, $\eta_{X_5} = 0.3$, $\eta_{X_6} = 0.3$, $\eta_{X_7} = 0.3$, $\eta_{X_8} = 0.3$, $\eta_{X_9} = 0.3$, $\eta_{X_{10}} = 0.3$. Simulation outcomes for the tuned parameters are depicted in Fig. 5. An overview of the remaining errors is shown in Table 5.

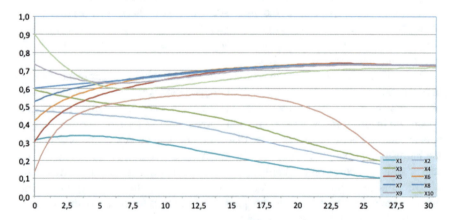

Fig. 5 Simulation modeling of the empirical data. The X-axis represents time, and the Y-axis represents the identification with the community

Table 5 Calculated error: average absolute deviation and root mean square

State	X_1	X_2	X_3	X_4	X_5	X_6	X_7	X_8	X_9	X_{10}	Total	
Empirical equilibrium	0.357	0.597	0.583	0.601	0.694	0.718	0.778	0.856	0.745	0.88		
Model equilibrium	0	0	0	0	0.747	0.747	0.747	0.747	0.747	0.747		
Absolute deviation	0.357	0.597	0.583	0.601	0.053	0.029	0.031	0.109	0.002	0.133	0.2495	Average deviation
Square of deviation	0.127	0.356	0.340	0.361	0.003	0.001	0.001	0.012	0.000	0.018	1.219	SSR
									Root mean square		0.3492	

Naturally, some variances exist between the proposed model and the empirical data. Using the root mean square to calculate the differences between the values of the model equilibria and equilibria of the empirical data (the final state of involvement in the queer community), this difference should be kept to a minimum in tuning the parameters. The achieved results of the calculated average (absolute) deviation (0.2495) and root mean square (0.3492) are depicted in Table 5, last column.

6 Conclusion

This paper investigated the interplay or co-evolution of the social contagion principle and the homophily principle in their application in an adaptive temporal-causal network model. Both principles were modeled and applied in three model scenarios. State values represented personal convictions, while the connections between states represented real-life social interaction, leading to influencing behavior.

The first scenario used an adaptive normalized scaled sum function to model social contagion. It showed the segregation behavior of a group of ten people into either two or three communities, depending on parameters such as speed factors, threshold factors, and steepness factors. The second scenario used the advanced logistic sum function and showed a separation into two communities that were not entirely defined by the initial grouping of connections. Furthermore, it demonstrated the spiraling of values beyond the range of initial values: a pattern that is not achievable with linear functions such as scaled sum functions. This pattern is sometimes called 'emotion amplification' [3], where emotions or opinions are amplified through sharing them.

The simulation of the third scenario also used the advanced logistic sum function, with the addition of connections to all other nodes in the network. This scenario is slightly more life-like. This scenario showed the segregation into two groups, and the amplifying behavior as discussed in Scenario 2. Furthermore, the third scenario showed a pattern of polarization, in which the values of the two groups increasingly move apart. In social terms, this can be compared to group identification, in which the more the other team disagrees with you, the stronger your opinion gets. It also shows signs of a process called 'othering', in which an individual's or group's identity strongly depends on what they are *not*. These phenomena can be observed in many social situations, with the political system of the USA being a classic example. This process may well contribute to segregating behavior, including political 'echo chamber' that is online social media [12], or the taking shape of subgroups or communities.

The final section of this paper described our attempt to gather empirical data regarding the shaping of the queer community, and the segregating behavior that it relates to. Before using the data in the network model, a statistical analysis showed that there was a significantly higher identification with the queer community of sexual queers than heterosexual participants. It also revealed that queer people had a higher number of queer friends and showed a higher increase in identification with the queer community than heterosexuals. The hypothesis stating that the level of identification increase is mediated by social hardship experienced in youth was rejected. The empirical data were transformed to fit the model (based on Scenario 3), and an exhaustive search tuning method was performed in order to tune the parameters to best fit the empirical data. The lowest average linear deviation was found to be 0.386. This is relatively high, which can partly be attributed to the fact that the empirical data used an average of scores, leading to values that lie relatively close to one another but are still significantly different. In the model, however, every state represented the average of a group of 5 or 6 people. Moreover, the combination of the advanced logistic sum function and the simple homophily function has a polarizing tendency (as described in Scenario 2), which was not clearly visible in the empirical data.

The current research may be improved by using more than 10 nodes, preferably 50 or more. Not only would this overcome the limitation of having to group participants together and losing their unique trends, but it also might be expected that the interplay between 50 nodes is vastly different from that of 10 nodes, in the way that a classroom with fifty children acts different than one with ten.

Additionally, the empirical data regarding the correlation between youth social hardship and connectedness to the queer community did not show a correlation (or even a trend). This may be due to the survey used, in which only 10 questions were focused on the general social hardship, while a focus on sexuality-related hardship would have been more useful. Secondly, the survey answers are strongly constrained by the snowball effect that was used for gathering the data: Many people who filled in the questionnaire knew others, which results in bias when attempting to study the formation of communities.

Finally, the study would be highly improved if the individual connections between people would be investigated over time. Now, questions about the number of queer people in the individual's network were used to approximate the average effect, while the combination of using the real individual interactions in a network of 50 nodes would give deeper insight into the workings of human group behavior.

Overall this research might be considered as a step in the direction of understanding more about segregation and polarizing behavior. Especially, the adaptiveness, using homophily, is indispensable for the future of creating temporal-causal models of human interactions and their consequences. The work reported here contributes by exploring the use of variants of such network models in the specific real-world context addressed.

References

1. Batson, C.D., Ahmad, N., Tsang, J.-A.: Four motives for community involvement. J. Soc. Issues 5(8.3), 429–445 (2002)
2. Beemyn, B.: Creating a place for ourselves: lesbian, gay, and bisexual community histories. Routledge (2013)
3. Bosse, T., Duell, R., Memon, Z.A., Treur, J., van der Wal, C.N.: Agent-based modeling of emotion contagion in groups. Cogn. Comput. 7, 111–136 (2015)

4. Harcourt, J.: Sexual identity distress, social support, and the health of gay, lesbian, and bisexual youth. In: Current issues in lesbian, gay, bisexual, and transgender health, pp. 97–126. Routledge (2013)
5. Holme, P., Newman, M.E.J.: Nonequilibrium phase transition in the coevolution of networks and opinions Phys. Rev. E **74**(5), 056108 (2006)
6. Kuipers, B.J.: Commonsense reasoning about causality: deriving behavior from structure. Artif. Intell. **24**, 169–203 (1984)
7. Kuipers, B.J., Kassirer, J.P.: How to discover a knowledge representation for causal reasoning by studying an expert physician. In: Proceedings Eighth International Joint Conference on Artificial Intelligence, IJCAI'83, Karlsruhe. William Kaufman, Los Altos, CA (1983)
8. McPherson, M., Smith-Lovin, L., Cook, J.M.: Birds of a feather: homophily in social networks. Ann. Rev. Soc. **27**(1), 415–444 (2001)
9. Meyer, I.H.: Minority stress and mental health in gay men. J. Health Soc. Behav. **36**, 38–56 (1995)
10. Meyer, I.H.: Prejudice, social stress, and mental health in lesbian, gay, and bisexual populations: conceptual issues and research evidence. Psychol. Sex. Orient. Gend. Diversity **1**(S), 3–26 (2013)
11. Pearl, J.: Causality. Cambridge University Press (2000)
12. Sunstein, C.R.: # Republic: Divided Democracy in the Age of Social Media. Princeton University Press (2018)
13. Treur, J.: Dynamic modeling based on a temporal–causal network modeling approach. Biol. Inspired Cogn. Archit. **16**, 131–168 (2016)
14. Treur, J.: Network-Oriented Modeling: Addressing Complexity of Cognitive, Affective and Social Interactions. Springer, Cham (2016)
15. Treur, J.: Relating emerging network behaviour to network structure. In: Proceedings of the 7th International Conference on Complex Networks and their Applications, Complex Networks'18, vol. 1. Studies in Computational Intelligence, vol. 812, pp. 619–634. Springer (2018a)
16. Treur, J.: Relating an adaptive social network's structure to its emerging behaviour based on homophily. In: Proceedings of the 7th International Conference on Complex Networks and their Applications, Complex Networks'18, vol. 2. Studies in Computational Intelligence, vol. 813, pp. 341–356. Springer (2018b)
17. Treur, J.: The ins and outs of network-oriented modeling: from biological networks and mental networks to social networks and beyond. Trans. Comput. Collect. Intell. **32**, 120–139. Based on Keynote Lecture at the 10th International Conference on Computational Collective Intelligence, ICCCI'18 (2019)
18. Vazquez, F.: Opinion dynamics on coevolving networks. In: Mukherjee, A., Choudhury, M., Peruani, F., Ganguly, N., Mitra, B.: (eds.) Dynamics on and of Complex Networks, vol. 2, Modeling and Simulation in Science, Engineering and Technology, pp. 89–107. Springer, New York (2013)

Derivation of a Conceptual Framework to Assess and Mitigate Identified Customer Cybersecurity Risks by Utilizing the Public Cloud

David Bird[ID]

Abstract The number of end points connecting to the cloud can increase distributed attack vectors due to vulnerable devices connecting from the front end. The risk is also enhanced due to the technological abstractions associated with public cloud computing models at the back end. On the one hand, cloud service providers make sets of defined service criteria and supporting documentation, publicly available to assist customers with their public cloud deployments. However, on the other hand, a cacophony of security incidents over the past five years involving vulnerable cloud customer instantiations reveals that cloud security risks may not be completely comprehended. Essentially, the fundamental principle of cloud computing is the 'shared security responsibility' model. It is argued in this paper that from a cloud customer perspective, there is either too much reliance upon legacy risk assessment methods and/or standards orientated compliance-mapping approaches when trying to apply due diligence for cybersecurity. This can be amplified by different cloud service providers using terms like 'core services' and 'managed services' rather than traditional terms such as Infrastructure-as-a-Service and Platform-as-a-Service. This extended paper describes the myriad of techniques used to derive a conceptual framework through post-graduate research. Based around a defense-in-depth model, the proposed conceptual framework is a proof of concept to enable customers to focus on the contextualized risks when using the public cloud. A method of reducing the risks using mitigation categories is also proposed. Consequently, a method of calculating residual risk against the identified risks levels is theoretically defined and dependent upon the rigor of counter-measure selection.

Keywords Cloud risk · Cloud security · Cybersecurity · Conceptual risk framework · Risk mitigation model

D. Bird (✉)
British Computer Society, Swindon, UK
e-mail: david.bird@bcs.org

© Springer Nature Singapore Pte Ltd. 2020
X.-S. Yang et al. (eds.), *Fourth International Congress on Information and Communication Technology*, Advances in Intelligent Systems and Computing 1027, https://doi.org/10.1007/978-981-32-9343-4_20

1   Introduction

This extended paper is a continuation of two previously published papers. Traditionally, customers have had a desire for a holistic data center security understanding—ranging from physical, policy, personnel, and technical perspectives [1]. Cloud computing is a step change compared to normal enterprise solutions where customarily the responsibility lies either with customer infrastructure support teams or a contracted and outsourced service provider utilizing dedicated infrastructure; in such cases, customers usually articulate, as part of their contract with the supplier, specific requirements and criteria that can be audited by the customer. Inherently, the data center and configuration of the underlying hardware or software infrastructure for cloud computing are under the control of the third-party cloud service provider (CSP). In the cloud domain, CSPs are audited by trusted third-parties (TTP) such as Ernest and Young (EY) [2] rather than by the customers themselves. Therefore, it is up to the customer, as part of due diligence routines, to check TTP certifications prior to contracting with a CSP. However, cloud computing works as a 'shared security responsibility,' where a combination of CSP responsibilities vary per service model and customers are required to allay some elements of risk associated with their leased cloud instances.

The high-profile attack on Code Spaces hosted on Amazon Web Services (AWS) provides a warning about the consequences of not applying robust technical authentication and authorization mechanisms; in this case, a weakness ultimately enabled hackers to gain a foothold and cause unrecoverable damage to Cloud Spaces' business [3]. Additionally, an old representational state transfer (REST) application programming interface (API) key, probably compromised from a Github account, enabled One More Cloud to be attacked [4]. An attack against Deloitte compromised the company's email server through the administrator account, which enabled email details to be exposed; in this case, multi-factor authentication (MFA) was not utilized in the Microsoft Azure cloud [5]. Essentially, these examples are embryonic of *laissez-faire* security implementations and vulnerable dispositions within the cloud amplified by weaknesses or software misconfigurations [6].

The new European Union (EU) Network and Information Systems Regulation 2018/151 pertain to CSPs like AWS and Microsoft Azure who are designated digital service providers and, therefore, are required to provide a baseline level of security that is appropriate to the risks [7]. However, White [8] has stated that the risks to the CSP also reside with the customers that it hosts. The use of a checklist mentality can indirectly encourage customer implementation gaps because of the disparate configuration decisions that are necessary for operating in the cloud [9]. Perhaps this is why customers are now publicly exposing AWS snapshots, and this is occurring even though AWS warns of the potential risks related to snapshot sharing. Certificates, security credentials, API keys, and proprietary source code can all be inadvertently compromised through snapshot exposures [10]. Such eventualities demand customer organizations to specify adequate backup and business continuity policies.

Although compliance-orientated approaches are well established, a study by Westervelt [11] in 2013 found that some customers had been negligent with their cloud configurations; and could effectively expose their data held on public cloud storage services to the wider Internet. A lack of understanding of this fundamental security precept and a supplier dependent mindset for security implementation can be corroborated through a more recent study; this study by McAfee has declared that one in every 20 AWS Simple Storage Service (S3) buckets have been left exposed by customers to the public Internet [12]. This has caused AWS to implement S3 encryption by default [13] to mitigate such misconfiguration errors. These incidents are examples of mostly organizations in the United States (US) making common mistakes or lapses of judgement. However, similar mistakes could be made by EU and United Kingdom (UK) customer DevOps teams attracted by the flexibility, scalability, and power afforded to them through public cloud computing.

2 Problem Defined

Traditional compliance-centric information assurance approaches have their place to provide some confidence that CSPs take information security seriously and can be trusted. One of the criticisms that can be levied is that there is an over-reliance on standard-orientated approaches, which tend to obscure the relevance of the more technical control-sets themselves, and consequently, organizations could unknowingly become non-compliant with legislation and regulations [14]. A RedLock study of cloud compromises indicated that root-causes of security incidents comprised: (a) inadvertent data and access key/credential exposures, (b) user account compromises, (c) ability of suspicious Internet Protocol addresses (IPs) to probe hosted databases, (d) unpatched host vulnerabilities, and (e) databases lacking encryption mechanisms [14].

Not only that existing risk assessment methods do not necessarily capture the complexity of the cloud computing domain and the abstractions from both CSP and customer perspectives [15]. A reliance on risk matrices to assess enterprise or operational risk has recently been criticized as a failed construct of business and the use of alternatives for technical risk establishment has been called for [16]. This is further exacerbated by existing risk assessment methods being deemed in effective for assessing technical risk in rapidly evolving technological capabilities [17].

Standards provide an assessment benchmark for external auditors of certifying bodies. However, auditing approaches used by Booz Allen did not thwart the newsworthy security violation made by this US defense company [18] from its use of AWS. This is a particularly worrisome situation when the provisions of General European Data Protection Regulation 2016/679 are now enshrined in the new Data Protection Act 2018 from a UK perspective. Within this directive, CSPs have obligated responsibilities and customers are also accountable to protect personally identifiable information in the cloud [19].

Therefore, there is a gap that could be filled by qualifiedly contextualizing security controls applicable to the public cloud by considering a cloud-native approach; that is ascertaining the potential security risks to virtual instances under customer-delegated control. Therefore, logic presupposes that cybersecurity approaches need to be re-evaluated [20] to apply proportionate and continually relevant counter-measures in the cloud. In line with this vision, the proposed conceptual framework (CF) was designed to bridge the deficiencies laid bare by legacy risk and compliance approaches, and to generate a better understanding of the security lines of demarcation and controls under the customers' area of responsibility.

3 Preliminary Analysis

The underlying aim of this research is to discern a way of understanding cloud risks where security control measures or dependency controls may be weak or vulnerable; thereby, enabling customers to be more risk aware and allow them to modify/revise their risk mitigation decisions in a more informed and proportionate manner.

3.1 Attack Tree Analyses

Following an initial literature review, the problem statement was represented as a rich picture of perceived attack vectors and issues. Scholarly interpretations of risks were researched to formulate a baseline of criteria needed for data collection during secondary research activities. Attack tree analyses (ATA) were derived that considered (a) attack methods, (b) attack vectors, (c) vulnerabilities that could be exploited, and (d) proposed mitigations. The outcomes created three threat profiles consisting of CSP insider threat, customer insider threat, and external threats. Proposed or perceived control sets to mitigate the threats were represented as an initial systems map; this approach then evolved 12 control-set groupings (CSG) for both customers and CSPs.

3.2 Control-Set Group Analysis

Secondary research was used to collate qualitative data from reputable industry, scholarly, and learned sources. A two-case study approach was used to provide confirmatory cases of presumed replications of the same phenomenon. The data was used to establish a representative sample of CSP controls represented in the form of instrumental systems maps (ISM). Data was sourced from documentation originated by the CSPs themselves, academia and industry subject matter experts and third-parties, with a vested interest in the cloud. Subsequently, the CSGs were used as a means to collect and correlate relevant data about the selected exemplars: AWS and Microsoft Azure. Data collection was focused on their Infrastructure-as-a-Service (IaaS) and Platform-as-a-Service (PaaS) offerings; this is because the Software-as-a-Service stack is almost entirely under the control of the CSP with very limited flexibility apportioned to the consumer compared to the previous two models. Identifiable controls were categorized by CSG for the ISM of each case study candidate.

3.3 Systems Thinking

The cloud security paradigm presents a number of intangible controls from a customer perspective such as configuration or software abstractions within the cloud stack. Primary research comprised systems thinking to dissect the complexities of abstracted cloud architectures. Based on the assimilated data, a formal systems model (FSM) was created to define the influencers and disruptors of the cloud system of systems. However, associations between the control-sets themselves also had to be expressed to better understand control-measure relevance [21].

Therefore, the inter-dependencies between CSGs at the control-set level required a novel way to discern them. So, a set of architectural principles were used based on system-agnostic criteria defined by the Information Security Forum; the criterion used was security-by-design, simplicity, defense-in-depth, least privilege, default deny, fail secure, and do not trust external systems [22]. These were cross-referenced against the applicable CSGs and used to establish separate representations of the salient contexts called security contexts (SC); the resultant SCs consisted of: oversight, build, privilege, access control, and trust verification [23].

Security factors (SF) were derived from the two-case study ISMs using non-vendor specific language so they would be CSP agnostic, and this is because the CF outcomes needed to be cross-platform compatible by nature. Soft systems methodology was then used to build upon root definitions and stakeholder perspectives for each cloud SC. Transformations and environment definitions were transposed from the FSM outcomes. Subsequently, relevant systems were created for each of the SCs by using multi-cause diagramming techniques. This approach thus enabled the analysis to simplistically slice through cloud abstractions to diffuse relevant CSG linkages and express the interrelating relationships between the contexts.

4 Conceptual Framework Formation

4.1 Red Teaming

Red team techniques—analysis of competing hypotheses (ACH) and string of pearls analysis (SoPA) modeling [23, 24]—were then used against real-world cyberattack or compromise scenarios to discern likelihood and impact criteria. Therefore, in order to calculate the probability of attack or compromise, hypotheses (H) of known cases were defined and represented in Fig. 1. The ACH analysis expanded upon the examples previously identified in the ATA by conducting further analysis of various real-world scenarios covering the past five years. Hence, the likelihood of occurrence was established for each threat (T) using the eight hypotheses [23].

The SoPA technique was then used to contextualize orders of effects enabling impact variables to be formulated as shown in Fig. 2; the SoPA technique considered assumptions (An) and dependencies (Dn) for each threat and the resulting 2nd (IIn) and 3rd (IIIn) order effects. The correlation of threat to likelihood and impact variable are shown in Table 1. By utilizing both of these techniques together, risk levels called consequential risk factors (CRF) were able to be generated.

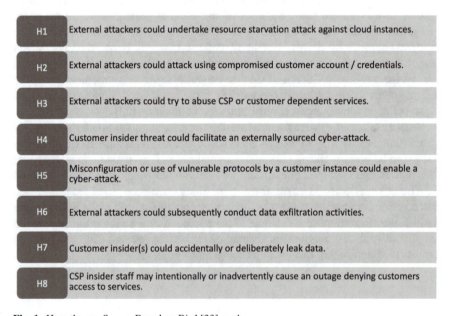

H1	External attackers could undertake resource starvation attack against cloud instances.
H2	External attackers could attack using compromised customer account / credentials.
H3	External attackers could try to abuse CSP or customer dependent services.
H4	Customer insider threat could facilitate an externally sourced cyber-attack.
H5	Misconfiguration or use of vulnerable protocols by a customer instance could enable a cyber-attack.
H6	External attackers could subsequently conduct data exfiltration activities.
H7	Customer insider(s) could accidentally or deliberately leak data.
H8	CSP insider staff may intentionally or inadvertently cause an outage denying customers access to services.

Fig. 1 Hypotheses. *Source* Based on Bird [23], p. 4

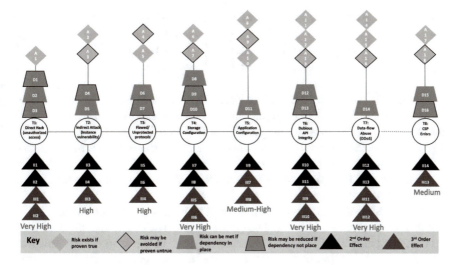

Fig. 2 String of Pearls technique

Table 1 Likelihood and impacts

Threats	Likelihood score	Impact score
T1: Direct hack (unauthorized access)	More probable	Very high
T2: Indirect hack (instance vulnerability)	Probable	High
T3: Flawed/unprotected protocols	More probable	High
T4: Storage configuration	Probable	Very high
T5: Application configuration	Probable	Medium-high
T6: Dubious API integrity	Probable	Very high
T7: Data-flow abuse—Distributed denial-of-service (DDoS)	Less probable	Very high
T8: CSP Errors	Infrequent	Medium

Source Based on Bird [23], p. 4–5 and Bird [25], p. 3)

4.2 Risk Model

Lowder has stated that technical risk consists of a function of pertinent variables [26]. Bodungen et al. [27] have stated that there is a relationship between threat event, likelihood, impact, and consequence in order to establish a risk rating. Therefore, it was asserted that a mathematical function [28] could be defined as a rule-set and used to calculate contributory variables in order to establish risk. The resulting mathematical function in Eq. (1) [23] was used to define this rule.

$$\begin{pmatrix} a1 \\ a2 \\ a3 \\ a4 \end{pmatrix} + \begin{pmatrix} b1 \\ b2 \\ b3 \\ b4 \end{pmatrix} + \begin{pmatrix} c1 \\ c2 \\ c3 \\ c4 \end{pmatrix} = \begin{pmatrix} a1 + b1 + c1 \\ a2 + b2 + c2 \\ a3 + b3 + c3 \\ a4 + b4 + c4 \end{pmatrix} = R \qquad (1)$$

Equation (1) comprises the following variables: an = SC, bn = likelihood and cn = impact where R = CRF. Therefore, matrix addition was used where the sum of those values in each category can quantify the CRF weightings. The rule detailed in Fig. 3 [25] was thereby produced and used as a mechanism to transpose quantitative values into qualitative values for the CF.

This then enabled a defense-in-depth model of risk attribution to be formed for the CF. Subsequently, a consensus model of values was then used to discern CRFs in order of SF importance and SC ranking. But by going through this process, a supposition was made that it would be erroneous to consider that an individual CRF relating to an individual SF would unilaterally equate to that level of risk. This was ratified by the Code Spaces example. It would be all too easy to speculate how the attackers gained access to the Code Spaces AWS account, but supposition indicates that the attack was either enacted through a compromised password that may have been phished during a precursory DDoS attack or a brute force dictionary attack [3]; however, the underlying root-cause was undoubtedly the use of weak passwords and certainly the lack of MFA.

By working through this scenario, it was asserted that one or more SFs could together present a risk when improperly implemented or are flawed, misconfigured, or overlooked. Consequently, a sequence of evolutionary stages was conducted that built upon the Consensus Model. An arrangement of SFs called Collectives became apparent that each comprised a number of CSGs that themselves contained various SFs. This arrangement enabled a defense-in-depth model to be formed for the CF shown in Fig. 4; categories are shown on the left-hand side, the Collective layer

Rule			
Security Context	Likelihood	Impact	Consequential Risk Factor weighting
Significant (4)	More Probable (4)	Very High (4)	Aggregator (<=12)
Serious (3)	Probable (3)	High (3)	Contributor (<=9)
Strong (2)	Less Probable (2)	Medium-High (2)	Facilitator (<=6)
Some (1)	Infrequent (1)	Medium (1)	Catalyst (<=3)

Fig. 3 Rule. *Source* Cited in Bird [25], p. 4

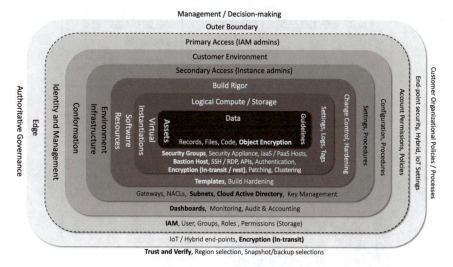

Fig. 4 Cloud defense-in-depth collectives. *Source* Based on Bird [25], p. 4. © David Bird

purpose is articulated across the top, examples of security enforcing function (SEF) dependencies per layer are represented on the right-hand side and abridged examples of SFs for each Collective are shown across the bottom; for the latter the Primary SFs are emboldened while Secondary SFs are in standard font. The resulting percentages for each CRF type are shown in Fig. 5 using the scale of highest to lowest risk already articulated in Fig. 3.

4.3 Risk Reduction Model

The public cloud is based on the premise that customers can build their own environments using infrastructure-as-code. This necessitates a need for reliable coding during template creation. Hence, configuration measures, settings, and policies have been identified as essential SEF risk mitigation considerations. Examples of configuration options for the Virtual Instantiation collective's Primary and Secondary SFs are listed in the third column of Table 2.

Based on a wider analysis of SEF categories available for the SFs across the other Collectives it was determined that it would be possible to define a method for calculating residual risk; especially, since the baseline CRF scores had already been devised in Fig. 3 where a risk drop could be quantified as being more than or equal to three from a scale of up to and including 12. Based on this analysis, a risk mitigation approach was then evolved for the CF to assess the decrease in risk per Collective through counter-measure implementation.

Defence-in-Depth Collectives	Conceptual Risk Factor %	
Edge	Aggregator	4.2
	Contributor	2.8
Identity and Management	Aggregator	7.0
Conformation	Aggregator	1.5
	Contributor	9.0
	Facilitator	1.5
Environment Infrastructure	Aggregator	8.7
	Contributor	13.0
Software Resources	Contributor	3.0
	Facilitator	3.0
Virtual Instantiations	Aggregator	17.3
	Contributor	7.2
	Catalyst	2.8
Data	Aggregator	2.0
Authoritative Governance	Facilitator	4.25
	Catalyst	12.75

Fig. 5 Consequential risk factor types and percentages. *Source* Based on Bird [25], p. 4

Mathematical Approach

A principle here is risk reductions can be calculated based on the selection of adequate pre-defined SEF criteria. Therefore, it is reasoned that mathematical expressions can be used to:

- Qualify the rigor of SEF implementation through selected mitigation categories, and
- Once applied, quantify the remaining risk level of each common CRF type between multiple and related CSGs ($CSG_{(a-n)}$) within a Collective (in this example it is $Collective_A$).

The linkages between the two are shown in Fig. 6 and this reasoning has been rationalized into two rules shown in Eqs. (2) and (3).

$$Collective_A = CSG_a \Rightarrow CSG_n \tag{2}$$

$$\begin{aligned} &\because Primary_{SF} \propto Primary_{CRF} \\ &Primary_{SEF} \propto Primary_{CRF\_Reduction} \\ &\because Secondary_{SF} \propto Secondary_{CRF} \\ &Secondary_{SEF} \propto Secondary_{CRF\_Reduction} \end{aligned} \tag{3}$$

Table 2 Virtual instantiations collective of security enforcing functions by security factor

Defense-in-depth Collectives	Security factor examples	Examples of security enforcing function implementations
Virtual Instantiations	**Security groups**	Set ingress rule (source/destination IP, protocol allow–deny) Set egress rule (source/destination IP, protocol allow–deny)
	Security appliance	Assign appliance/Web application firewall rules
	Tenants (IaaS)	Set change-state logging Set data-flow logging Set meaningful meta-data tags
	Containers (PaaS)	Set container registry Set container engine dependencies
	IoT (PaaS)	Set queues Allocate publish/subscribe topics
	Bastion Host	Apply SSH/RDP to Bastion Host
	RDP/SSH	Assign Public-key cryptography key-pairs for TLS 1.2+
	REST APIs	Assign certificates/Pre-shared keys/Security tokens Set permissions
	Authentication	Disable password-only access Enable MFA
	Encryption (in-transit)	Manage TLS 1.2 + centrally
	Encryption (at-rest)	Attach persistent storage Enable block encryption
	Patching	Patch application Patch operating system
	Virtual machine (VM) clustering	Set VM pool Set scaling rules Set scaling thresholds

Relation theory and difference mathematics were used to develop a set of risk reduction formulae in order to establish a theoretical risk reduction model. This enabled a correlation to be defined between (a) the SEF implementation(s) for each SF and (b) the respective CRF and subsequent risk reductions for each CRF category.

Relation Formulae

Adopting set mathematics, relations [28] between the Common Primary SF ($Common_{a\,PSF}$) and related Primary SEFs ($Common_{a\,PSEF}$) and equivalent secondary controls are shown in Fig. 7; where the implementation measures (I) fulfill the SEF functionality. This relationship is represented in Eqs. (4) and (5).

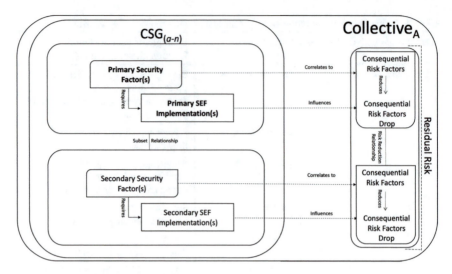

Fig. 6 Conceptual framework linkages to reduction

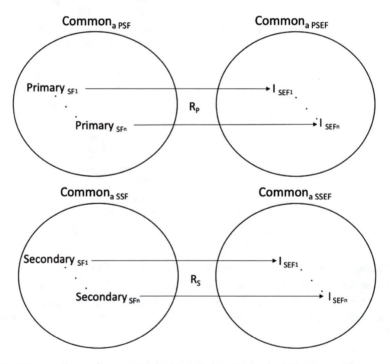

Fig. 7 Primary and secondary security function relations to security enforcing functions

$$\text{Common}_{a\,PSF} \in \{\text{Primary}_{SF1}, \ldots, \text{Primary}_{SFn}\}$$
$$\text{Common}_{a\,PSEF} \in \{I_{SEF1}, \ldots, I_{SEFn}\}$$
$$\therefore R_P \subseteq (\text{Primary}_{SFn} \times \text{Primary}_{SEFn}) \tag{4}$$

$$\text{Common}_{a\,SSF} \in \{\text{Secondary}_{SF1}, \ldots, \text{Secondary}_{SEFn}\}$$
$$\text{Common}_{a\,SSEF} \in \{I_{SEF1}, \ldots, I_{SEF_n}\}$$
$$\therefore R_s \subseteq (\text{Secondary}_{SFn} \times \text{Secondary}_{SEFn}) \tag{5}$$

$$\therefore \text{Common CSG}_{a\,CRF\,Drop} \sim \overset{n}{\Sigma}(R_P : R_s) \tag{6}$$

Hence, the risk reductions called CRF Drops can be derived based on the Common SEFs of related CSGs within a CRF category. Through the application of Common SEFs, it was deduced from the schematic in Fig. 6 based on their sub-set relationship that the CRF Drop level per CSG (Common CSG$_{a\ CRF\ Drop}$) is a ratio of primary and secondary counter-measures. By implication, within a collective the cumulative effect of applying risk mitigation categories per CRF is expressed in Eq. (6).

Difference Formula
The connection between CRF and CRF Drop is denoted as a range and represented diagrammatically in Fig. 8. By applying requisite SEF counter-measures to mitigate the CRFs for each Collective (in this example Collective$_A$), the risk reduction could be defined by using difference mathematics as shown in Eq. (7). The risk reduction is thereby surmised to be the sum of the difference between one or more CRF Drops per CRF for one or more related CSGs (from a range of a to n), in this case within Collective$_A$. This is formalized by Eq. (8) showing the collective risk reduction per CRF type.

$$\text{Difference} = \text{Minuend} - \text{Subtrahend} \tag{7}$$

$$\therefore \text{Collective}_A \text{ Residual Risk per CRF} = \sum \left\{ \begin{bmatrix} \text{CSG}_{a\,CRF} \\ \vdots \\ \text{CSG}_{n\,CRF} \end{bmatrix} - \begin{bmatrix} \text{CSG}_{a\,CRF\,Drop} \\ \vdots \\ \text{CSG}_{n\,CRF\,Drop} \end{bmatrix} \right\} \tag{8}$$

Hence, it is asserted that this method is repeatable and can be used to ascertain the resultant risk reductions per Collective across the entire defense-in-depth model.

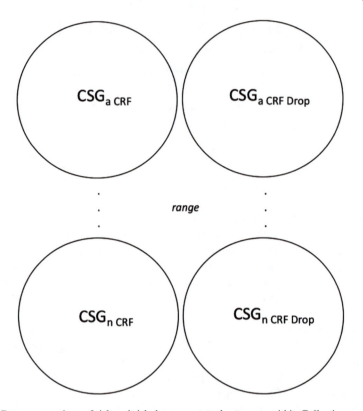

Fig. 8 Range connexions of risk and risk drop per control-set group within Collective$_A$

5 Conclusion

In 2014, a study revealed that there was a diversification of cyberattacks taking advantage of publicly known software vulnerabilities [29]. The implementation of adequate security controls is a particularly poignant topic after the spate of customer-centric cloud data exposures in 2017. However, Webber [30] has declared that 'the more controls there are, the more ways there are to slip up.' So proportionate and relevant counter-measure deployments are needed.

This research has re-imagined risk assessment approaches and goes beyond the methods used in matrices normally attributed to compliance-mapping approaches. The CF was developed from tasks and deliverables using unbiased data analysis and subjective analytical interpretations. Key proponents of the CF are the SC, likelihood, and impact ratings that enabled the establishment of maximum CRF values based on the threats; where:

- The probability of weakness or vulnerability occurrence are realized, and
- Risks are improperly mitigated due to misconfiguration oversights or implementation gaps.

This approach contextualized maximum CRFs that were determined by understanding dependency relationships with other sub-systems within the cloud system of systems. The theory behind the proposed method of calculating residual risk has been devised based on the associated CSGs per Collective and the SEF categories per SF. Therefore, based on a normalized set of SEF selection options, it is thereby argued that it is feasible to establish residual risk levels for each common CRF type across the related CSGs of each Collective.

6 Further Work

More work is required to fine tune the details relating to the SEF implementation categories across the SFs for all Collectives hitherto discussed in this paper. For example, the configuration of security groups will need to consider the finer details of assigning ingress and egress policies to include protocol, source/destination IP and not consider unfettered policies such as 'any-any' for inbound or outbound traffic.

It is argued that the amalgamation of principles in this paper provides the foundational elements of a risk tool. By using such a tool, it could be possible to quantify and qualify the rigor of counter-measure implementation used by cloud customers; enabling them to practically work out and comprehend the residual risks through their choice of SEF implementations. Additionally, the tool could also be used to assist cloud customers in re-thinking their cloud environment architecture and their selection of appropriate SEF criteria within their 'shared security responsibility' and encourage a more informed counter-measure selection process.

Acknowledgements Table 1 and Figs. 3, 4, and 5 have been adapted from Bird [23, 25].

References

1. Pistorious, M.: The Quick Guide to Cloud Computing and Cyber Security. StreetLib, USA (2015)
2. EY.: Building trust in the cloud: creating confidence in your cloud ecosystem, insights on governance, risk and compliance. EYGM Limited, USA (2014)
3. The attack that forced Code Spaces out of business—what went wrong? https://www.itgovernance.co.uk/blog/the-attack-that-forced-code-spaces-out-of-business-what-went-wrong/. Last accessed 3/1/2019
4. Old AWS API key led to search provider's cloud security breach. http://searchcloudsecurity.techtarget.com/news/2240224543/Old-AWS-API-key-led-to-search-providers-cloud-security-breach. Last accessed 19/1/2019
5. Deloitte hit by cyber-attack revealing clients' secret emails. https://www.theguardian.com/business/2017/sep/25/deloitte-hit-by-cyber-attack-revealing-clients-secret-emails. last accessed 3/1/2019

6. Keeping Britain safe: how GCHQ's new cyber security agency will protect us from hackers. http://www.wired.co.uk/article/national-centre-cyber-security-ian-levy. Last accessed 3/1/2019
7. Everything you need to know about the NIS Directive. https://blog.infinigate.co.uk/everything-you-need-to-know-nis-directive-network-information-security. Last accessed 19/1/2019
8. Understanding the risks of cloud computing. http://searchcloudsecurity.techtarget.com/tip/Understanding-the-risks-of-cloud-computing. Last accessed 19/1/2019
9. End complacency and help address cyber crime threat, NCA tells business. http://www.computerweekly.com/news/450412817/End-complacency-and-help-address-cyber-crime-threat-NCA-tells-business. Last accessed 19/1/2019
10. Public snapshots pose significant Amazon EBS Security Risks. https://searchaws.techtarget.com/tip/Public-snapshots-pose-significant-Amazon-EBS-security-risks. Last accessed 12/1/2019
11. McAfee says cloud security not as bad as we feared…it's much worse. https://www.theregister.co.uk/2018/10/30/mcafee_cloud_security_terrible/. Last accessed 3/1/2019
12. Amazon S3 users exposing sensitive data, study finds. http://m.crn.com/news/security/240151857/amazon-s3-users-exposing-sensitive-data-study-finds.htm?itc=xbodyrobwes. Last accessed 19 January 2019
13. Default AWS S3 encryption walls off vulnerable customer data. http://searchaws.techtarget.com/news/450429898/Default-S3-encryption-walls-off-vulnerable-customer-data/. Last accessed 3/1/2019
14. Study: Lax Security Enforcement Behind Rise in Amazon S3 Exposures. https://awsinsider.net/articles/2017/10/11/redlock-lax-cloud-security.aspx. Last accessed 3/1/2019
15. Ahmed, N., Albakri, S., Idris, N., Samy, G., Shanmugam, B.: Traditional security risk assessment methods in cloud computing environment: usability analysis. Jurnal Teknologi **73**(2), 483–495 (Penerbit UTM Press, Malaysia) (2015)
16. Risk Matrices Failures. https://www.causalcapital.club/single-post/2019/01/09/Risk-Matrices-Failures. Last accessed 12/1/2019
17. Nurse, J., Radanliev, P., Creese, S., De Roure, D.: If you can't understand it, you can't properly assess it! The reality of assessing security risks in IoT systems. In: PETRAS Living in the Internet of Things Conference Proceedings (2018). https://doi.org/10.1049/cp.2018.0001
18. Booz Allen stock plummets on word of federal government probe. https://www.cnbc.com/amp/2017/06/15/booz-allen-stock-plummets-on-word-of-federal-government-probe.html. Last accessed 3/1/2019
19. AWS and the General Data Protection Regulation (GDPR). https://aws.amazon.com/blogs/security/aws-and-the-general-data-protection-regulation/. Last accessed 19/1/2019
20. A shared responsibility. https://www.bcs.org/content/conWebDoc/58147. Last accessed 1 February 2019
21. Alexander, D., Amanda, F., Sutton, D., Taylor, A.: Information Security Management Principles, 2nd edn. BCS Learning and Development Ltd, UK (2013)
22. Information Security Forum: Security Architecture Workshop. Information Security Forum Limited, USA (2006)
23. Bird, D.: Information Security risk considerations for the processing of IoT sourced data in the Public Cloud. In: PETRAS Living in the Internet of Things Conference Proceedings, pp. 1–10. IEEE Xplore, USA (2018). https://doi.org/10.1049/cp.2018.0040
24. Hoffman, B.: Red Teaming: Transform Your Business by Thinking Like the Enemy, 1st edn. Piatkus, UK (2017)
25. Bird, D.: A conceptual framework to identify cyber risks associated with the use of public cloud computing. In: 11th International Conference on Security of Information and Networks Proceedings, pp. 1–4. ACM International, USA (2018). https://doi.org/10.1145/3264437.3264466
26. Why the 'Risk = Threats x Vulnerabilities x Impact' Formula is Mathematical Nonsense. https://www.bloginfosec.com/2010/08/23/why-the-risk-threats-x-vulnerabilities-x-impact-formula-is-mathematical-nonsense/. Last accessed 3/1/2019

27. Bodungen, C., Singer, B., Shbeeb, A., Hilt, S., Wilhoit, K.: Hacking Exposed Industrial Control Systems: ICS and SCADA Security Secrets & Solutions. McGraw Hill Education, USA (2017)
28. Function (mathematics). https://simple.wikipedia.org/wiki/Function_(mathematics). Last accessed 19/1/2019
29. Reasons We Need to Boost Cybersecurity Focus in 2019. https://securityaffairs.co/wordpress/80080/security/6-reasons-boost-cybersecurity.html. Last accessed 19/1/2019
30. A New ISO Standard for Cloud Computing. http://privacylawblog.fieldfisher.com/2014/a-new-iso-standard-for-cloud-computing/. Last accessed 3/1/2019

Securing Manufacturing Intelligence for the Industrial Internet of Things

Hussain Al-Aqrabi⬡, Richard Hill⬡, Phil Lane and Hamza Aagela

Abstract Widespread interest in the emerging area of predictive analytics is driving the manufacturing industry to explore new approaches to the collection and management of data through Industrial Internet of Things (IIoT) devices. Analytics processing for Business Intelligence (BI) is an intensive task, presenting both a competitive advantage as well as a security vulnerability in terms of the potential for losing Intellectual property (IP). This article explores two approaches to securing BI in the manufacturing domain. Simulation results indicate that a Unified Threat Management (UTM) model is simpler to maintain and has less potential vulnerabilities than a distributed security model. Conversely, a distributed model of security out-performs the UTM model and offers more scope for the use of existing hardware resources. In conclusion, a hybrid security model is proposed where security controls are segregated into a multi-cloud architecture.

Keywords Business Intelligence · Security · Industrial Internet of Things

1 Introduction

The Internet of things (IoT) [1–6] and the Industrial Internet of Things (IIoT) [7–10] are a collection of enabling technologies that can provide data-driven visibility of physical systems. Industrial organizations, particularly those in the manufacturing

H. Al-Aqrabi · R. Hill (✉) · P. Lane · H. Aagela
Department of Computer Science, University of Huddersfield, Huddersfield, UK
e-mail: r.hill@hud.ac.uk

H. Al-Aqrabi
e-mail: h.al-aqrabi@hud.ac.uk

P. Lane
e-mail: p.lane@hud.ac.uk

H. Aagela
e-mail: hamza.aagela@hud.ac.uk

© Springer Nature Singapore Pte Ltd. 2020
X.-S. Yang et al. (eds.), *Fourth International Congress on Information and Communication Technology*, Advances in Intelligent Systems and Computing 1027, https://doi.org/10.1007/978-981-32-9343-4_21

industry, make use of myriad material handling and conversion processes that require data to control at both micro- and macro-levels.

At the micro level, data is an intrinsic part of a control loop to assure quality and repeatability of the process. At the macro level, data is used to coordinate the logistics of material movements between individual processes and manufacturing units, for local, national and even international distribution. As the costs associated with technology constantly reduce, the availability of hardware platforms and Software-as-a-service (SaaS) becomes more commonplace, more and more enterprises are installing IIoT devices to increase the awareness of their operations [11].

The installation of new manufacturing plant often includes the capability to extract data about operations in real time. However, older plant does not have such capabilities, and from a return on investment (ROI) perspective, there is little justification to replace plant that is functioning satisfactorily.

Thus, there is scope for the retro-fitting of IIoT devices to plant and machinery, in order to gain valuable insight into any potential process optimizations that might occur as a result of having improved visibility of process data. Once there is an IIoT infrastructure in place, where the process data can be monitored, collected and analyzed, there is a range of opportunities to visualize process data, as well as the macro level whereby collections, or even entire supply chains can be monitored.

One phenomenon that becomes apparent with the installation of IIoT is the considerable increase in volume of fast-moving data that has to be captured, transported, stored and managed. Ultimately, to realize the full potential of such data requires analysis. The addition of modelling and forecasting operations, using the data, gives rise to emerging interest in the field of *predictive analytics*.

The adoption of an analytics platform from a remote cloud service does mean; however, that process data from the enterprise must be transported outside of the boundaries of the organization, a prospect that is viewed with some skepticism. There are two issues at play here. First, organizations wish to retain full control of their data, and this can only be assured by keeping data on the premises. Second, whilst an enterprise may yet not yielded any value from the analysis of the raw data, it is conceivable that the analyzed sensor data, stored in a cloud which is off the premises, is an obvious target for cyber attack, and therefore this introduces a vulnerability into the security of the organization's operations [12–18]. Raw data that has been processed, analyzed and used to construct models of the business, is the key source of knowledge for *Business Intelligence*.

Traditionally this knowledge has been retained tacitly by human machine operators and managers, but the adoption of IIoT equipment to sense and gather this data means that valuable Intellectual property (IP), that is the core of any organization's competitive advantage, can be managed just like any other asset. The act of collating knowledge brings significant possibilities into being [19, 20], whilst also increasing the risks of the knowledge being lost or stolen [21, 22].

Whilst a malicious employee might share the 'secrets' of a process that they manage, it would be difficult to envisage how a larger security breach could be coordinated. Conversely, a malicious attempt to attack a remote server might provide a back door into a system that can be copied, disrupted, or even monitored covertly.

In the light of these challenges, this work explores how the security of IIoT-centric BI might be augmented to increase trust between physical plant within an organization's premises, with the remote analytics capability that is hosted remotely. In particular, we propose, simulate and evaluate two models that offer significant improvements in security. These improvements will be of interest to enterprises who cannot yet justify the wholesale management of IIoT analytics management on premises by mitigating some of the additional risks presented by established off-premises solutions.

2 Business Intelligence Services

The infrastructure of BI is a complex collection of functions that inter-operate to consume, share and process data for reporting purposes. Data sources are varied and include operations data from enterprise systems and sales order processing, as well as process data for the coordination of operations activities [23]. The adoption of IIoT gives enterprises the ability to include in-process data from manufacturing operations, increasing the scope and depth of business reporting that is possible. In general, transaction data is stored in relational databases, data marts and data warehouses, which is available for query by the BI system [24].

As such, BI is an enhanced interface that facilitates human interaction with the organizations' data, leading to judgements, decisions and actions around operational interventions. This important role of BI elevates its status beyond that of traditional IT reporting infrastructure, and thus creates a greater dependency between its users and the business.

Such is the dynamic nature of business, especially in the fast-moving manufacturing industry, that the requirements placed upon data usage will invariably change depending upon the needs of the business, whether it be from the external environment, internal disruptions within the physical systems, or the aspirations of the business executive.

Each of these transactions and human interactions are opportunities for vulnerabilities either through human error or nefariously. Attempts to model complex, multi-agency systems [25] have enabled greater understanding and communication of complex human interactions with sensitive personal, in this case health data. As a consequence, a corpus of security challenges lie in the procedural and human interactions with the BI system [24]. The work described in this article is focused on the technical security challenges and solutions when BI is hosted off the premises.

2.1 BI Architecture

BI processes are often composed from orchestrated, collaborating services, which are consumed by different users. In the broader domain of businesses that collect and consume data from IIoT systems, and who have opted to utilize BI as a service, each of the corresponding component services that constitute the overall BI application may, in fact, request aggregated data from multiple cloud systems, each of which may be in a different security realm.

These internal services have to be executed dynamically at runtime, and therefore the requisite authorization and permissions need to be in place to allow such actions to happen.

Cloud-based systems are inherently heterogeneous in nature which necessitates the automation of authorisation wherever possible, to facilitate the timely and seamless delivery of BI services. It is a considerable technical challenge to enable this secure collaboration.

In order to address this security challenge, an authentication framework is required to establish trust amongst BI service instances and users by distributing a common session secret to all participants of a session. We have addressed this challenge by designing and implementing a secure multiparty authentication framework for dynamic interaction, for the scenario where members of different security realms express a need to access orchestrated services [18].

This framework exploits the relationship of trust between session members in different security realms, to enable a user to obtain security credentials that access cloud resources in a remote realm. The mechanism assists cloud session users to authenticate their session membership, thereby improving the performance of authentication processes within multiparty sessions.

We see applicability of this framework beyond multiple cloud infrastructure, to that of any scenario, where multiple security realms have the potential to exist, such as the emerging Industrial Internet of Things (IIoT) within the manufacturing industry, commonly perceived as an essential component of Industry 4.0 [7–9, 11].

2.2   BI Security Challenges and Controls

As a business builds its BI capability, the complexity of new BI workflows increases as workers begins to ask more insightful queries of their data. As such, it can be very challenging for information and security architects to be able to design the necessary safeguards that will enable business users to pose queries, without hindering the potential for more complicated inquiry at a future date.

Data and information that are useful exist within all aspects of an enterprise system, including, but not limited to: hardware and network systems; data marts and warehouses together with associated metadata repositories; OLAP servers; the data presentation and application services layers, as well as the appropriate layers of authentication [12–14].

Using the basic principle that data is secure by default and only shared on a 'need-to-know' basis, the authentication services apply to each generator or collector of data within the system. It follows that the security controls are bound to the repositories that hold both temporal and permanent data [26]. Extract, transform and load (ETL) functions within data marts require this data in order to perform the correct operations on the pertinent data at the correct time. The security controls apply both to system level and human stakeholders, depending upon the nature of the interaction with the business system.

Multi-dimensional models for the secure management of objects within data warehouses have been proposed [15, 27, 28], particularly for Online analytical processing (OLAP) queries. Using Object Security Constraint Language (OSCL) to extend the Unified Modelling Language (UML), a multi-dimensional model is described. Such is the arrangement of the data repositories, together with the tight coupling of security control objects to the repositories that a hierarchical organization of the security objects also emerges.

Whilst the cohesive approach to integrating secure control objects into the multi-dimensional data model is effective when the data domain owner is known, instances of temporal data may not have clearly identifiable owners [16], and therefore the argument for this approach is less convincing. During the course of querying data for a particular purpose, it is likely that a particular BI user requires access to data that it is not the owner of. One example is of data streams from shopfloor manufacturing processes that have been fused with relevant logistics data. In this particular instance (that is very commonplace), the data that is required does not necessarily have sufficient metadata to describe the constraints under which that data can be processed or even viewed [23].

In this scenario, we need to ensure that the controls that apply to intermediate repositories of transient data will require greater security controls, than if the data had been extracted from its source, where the domain owner of both the owner and the requester can be ascertained through conventional authentication mechanisms [17]. One of the trade-offs of a cohesive coupling of security objects to multi-dimensional arrangements of repositories is that there is an increased cost for ETL transactions in terms of performance overhead. This is further complicated as additional levels of abstraction are considered (for other purposes), such as the rules engines required for OLAP processes, that are usually separated from underlying data warehouses [19, 29].

There comes a point at which additional expenditure on infrastructure is not feasible. This point may be reached quicker with the adoption of IIoT technologies that can rapidly overload existing network architectures with extra data transport, a significant part of which is directly incurred as a by-product of the additional authentication transactions for the additional physical objects, and the associated queries that become possible as a result [30–32]. This is also in addition to essential infrastructure services that invariably increase as more equipment is added. Figure 1 illustrates one attempt to cope with this challenge by federating clusters of data warehouses. Each data warehouse is separated by a network switch and associated data storage, and enables data within multiple warehouses to be striped or *depersonalized*, and queried as one object [20].

The framework [20] assigns federated repositories to separate groups of administrators. The process of striping the data enables an element of obfuscation in that any data within one resource is meaningless without its relevant contextual information. The striped deployment of relational databases onto a cloud could possess elastic characteristics, and therefore any additional objects, such as for the implementation of security controls at object-level, could be accommodated in an architecture that scales when necessary [21].

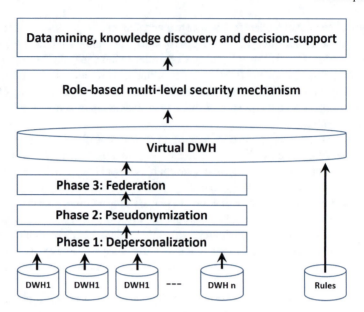

Fig. 1 Distributed security for a system of federated data warehouses [20]

2.3 OLAP Security

OLAP is a prevalent part of BI implementations, and it provides a means by which reporting can be performed against a multi-dimensional repository. Views of the data, referred to as 'cubes', enable visualization applications such as dashboards to present data to business users.

An OLAP presentation contains at least a services and a user's cube, each of which has associated with it a set of security controls that protect the cubes from unauthorized access to data [33].

OLAP functionality permits more complex operations to be performed on data, in a controlled way, than would normally not be permissible, nor practical with a traditional management information system [27]. The granular control of this extended functionality is administered by a series of OLAP operations, themselves being constrained by a Multi-Dimensional Security Constraint Language (MDSCL) [24]. Figure 2 illustrates the security controls to restrict view operations.

Business users can access controls via instances of cubes, typically as part of a query modelling process carried out by the users [27] or system administrators.

Since OLAP organization facilitates the creation and customization of different views of the data, it is important to ensure that inadvertent access is not granted to unauthorized users, in order that the privacy of both users, users' access to data, and the data itself is preserved.

The Secure Distributed-OLAP aggregation protocol (SDO) is a means by which such privacy can be assured. Using SDO, the owners of views of multi-dimensional

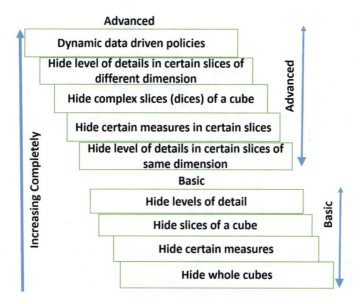

Fig. 2 Distributed security controls in an OLAP cube comprising viewing and operations restrictions [27]

queries can be secured and logged. This enables the control and prevention of unauthorized access as described above, and manages the data that is held about a user, their permissions and the details about their interactions with data objects [28].

These 'privacy metrics' [33] will be hierarchical in structure [29], and reflect detailed signatures of users' interactions, in relation to the sensitivity of the data, as well as the originating privacy criteria of the repositories [33], through their interactions with the various cubes. A variation of this proposal [22] provides the introduction of further controls (and overhead) to manage constraints on the OLAP cubes and underlying repositories. By the augmentation of Intrusion Detection and Prevention Systems (IDPS) into the scheme, requests made to repositories can be marshalled as an additional layer of protection against unauthorized access. Similarly, Lightweight Directory Access Protocol (LDAP) is a means by which access to services on a network can be managed through an authentication mechanism. At a lower level, Secure Socket Layer (SSL) protocols can also be utilized to ensure that interactions within objects are transacted via an encrypted exchange [22].

This has been a considerable challenge for the research community for some time now. There appears to be a preference for security mechanisms that are distributed across systems, with individualized controls at object level. In contrast, the Unified Threat Management framework (UTM) approach considers the system as a whole entity that must be protected as such and requires centralized governance. However, the emergence of IIoT, and the awareness that systems need to be designed and built to scale elastically in future, is also generating enthusiasm for cloud-provisioned resources and approaches to software development such as Microservices Architectures (MA) [34]. We have attempted to approach the modelling of this situation by recognizing that a manufacturing organization will want to adopt IIoT devices in a fluid and scalable way and have adopted good practice for the modelling of networks that may include mobile sensors and other entities as advocated by [35].

We describe the modelling and simulation of both a UTM approach versus a distributed approach to security, in the context of the manufacturing domain where IIoT devices are included as part of the system. Since there is a need for rapid allocation, de-allocation and re-allocation of resources, we have adopted a cloud-centric model, though this does not in itself necessitate an off-the-premises solution.

3   Description of the Models

Since the UTM approach is centralized, it is logical to group security controls into the following layers: network, transport, session and presentation. For the distributed approach, application layer security is appropriate, utilizing each of the components within the infrastructure. Both scenarios have been modelled in OPNET. Model A uses a UTM cloud within a demilitarized zone (DMZ) to contain the security components that are required to govern and marshall the whole system. In contrast, Model B has no requirement for a UTM cloud as the security components have been distributed at object level, hence all connection requests are made directly to the object that is to be invoked, via appropriate object-specific security controls. Both approaches are illustrated in Fig. 3.

OPNET allows the creation of bespoke application configurations, which includes the ability to assign them to different resources (in this case servers) according to the role that is played within the overall system. For example, the OLAP_DASHBOARDS, DW_DM and OLAP_VIEWS are all represented by OPNET *profiles* and are assigned within the BI application and database cloud in Model A (UTM model). The remainder of the controls reside within the UTM cloud in the DMZ.

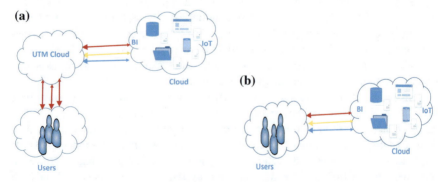

Fig. 3   a) Unified Threat Management (UTM) model; b) Distributed security model. Adapted from [36]

Since the UTM cloud (and DMZ) is not present in Model B (distributed security model), the entirety of the roles are undertaken by the BI application (via OLAP) and database (in this case a data warehouse) servers within the BI cloud. Therefore, Model B has the role of security distributed across all the application and database servers, embedding the marshalling at object level. Model A's centralization of security roles into a separate security domain contrasts with the distributed approach of Model B.

3.1   Model A: Unified Threat Management Approach

Figure 4 illustrates the BI system which is composed of two server arrays. The BI on cloud (comprising of application servers and database servers) has been modelled employing four different arrays. There is an array of five OLAP servers connected to cloud switch 2, an array of four database servers connected to cloud switch 1, another array of four database servers connected to cloud switch 3 and an array of four high-end Cisco switches (Cisco 6509 routing switches). The two sets of database server arrays have been presented to model the concept of splitting records between two hardware clusters.

Each of the databases that store the temporal and permanent data are represented by the DW_DM servers. OLAP applications are hosted on server arrays, including both the applications and the view cubes. Server connectivity is provided via simulated Cisco 7000 switches.

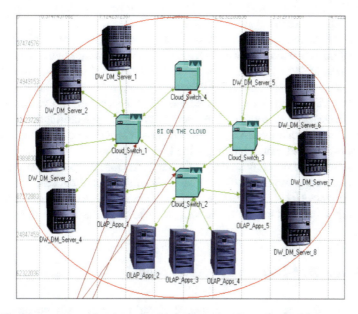

Fig. 4 The BI server arrays forming a cloud infrastructure. Adapted from [36]

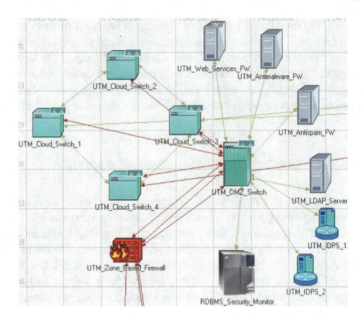

Fig. 5 Security components within the Unified Threat Management cloud. Adapted from [36]

Figure 5 illustrates the security components contained within the DMZ. Each user session is directed through a zone-based firewall, which isolates all user sessions and prevents them from accessing the cloud switches directly without the appropriate authority. All user sessions are forwarded to the UTM switch.

Source and destination rules for each client and server have been created to inform the simulation as to which client and server are the target for a particular request. This feature within OPNET had been employed to create complex traffic flows between clients (representing users) and the relevant cloud servers. The destination servers for the clients (in the extranet domain of BI users) are the servers with various security roles in this DMZ, and the cloud OLAP servers are in turn the destination servers for these DMZ-based security servers. The traffic is now strictly rule-based, an approach that is an established method amongst the body of research work in this area. The RDBMS security monitor holds all the rules deployed for ETL processes, and others are regular security servers. The LDAP server is deployed as a reference database of all user accounts. Hence, the RDBMS security monitor has a destination rule pointing towards this server and this server, in turn points towards the cloud servers.

3.2 Model B: Distributed Security Model

Model B is presented in Fig. 3. As described earlier, the security components are integrated with BI servers in the cloud-based arrays, and all client requests are made directly upon the cloud switches. Since the centralized UTM cloud is absent in Model B, the necessary security components and services have been embedded within the OLAP, database and data warehouse servers as follows: BI_DW_DM: the data warehouse and data marts servers; BI_Application: OLAP apps servers; BI_Security_UTM: all security services are enabled on each server. This represents the concept of distributing embedded security controls, as advocated by the research community. The application destination preferences have been directed towards the OLAP application servers, which in turn have their destination preferences as the data warehouse and data mart servers.

4 Results

In this section, the simulation results of the two approach are presented, together with a discussion about the relative efficacy of each approach to security.

4.1 *Performance*

For Model A, the average response time to database queries across the entire network exceeds 20 seconds, and the average HTTP object response time can take up to 3 seconds. TCP delays are in excess of 20 seconds, with TCP segment delays of 4 seconds. The TCP retransmission count exceeded 1000 on two occasions (Fig. 6).

All the user traffic is being routed through the security services servers in the UTM cloud. The users are connected to a zone-based firewall, which routes their

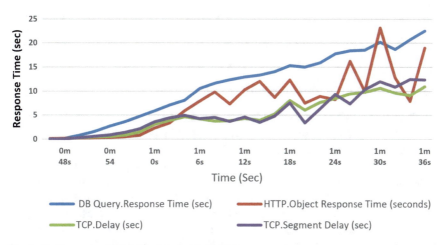

Fig. 6 Performance of Model B. Adapted from [36]

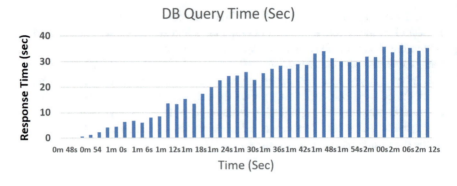

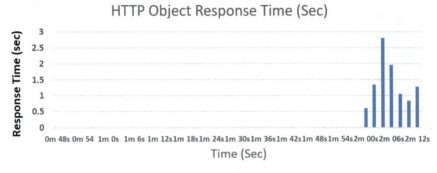

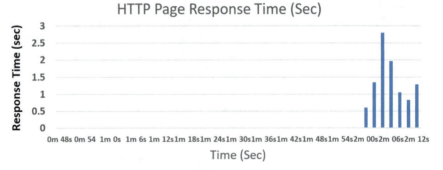

Fig. 7 Performance of Model B. Adapted from [36]

traffic to the security servers. In this arrangement, the capacity may not be an issue, but the routing of traffic through the servers placed in the demilitarized zone (DMZ) is adding a delay component onto all of the servers.

The performance report of Model B is presented in Fig. 7. In this model, the performance is within the expected limits and there are no TCP delays, TCP segment delays and TCP re-transmissions. We observed that the performance of database query response and HTML page/object responses is superior to Model A and is satisfactory from a user's perspective. The response times returned in this model are as expected from a cloud hosted BI application. Removal of the UTM cloud layer has ensured that all the session bottlenecks in the network are eliminated.

The load of the security sessions is distributed amongst all servers in the OLAP and data warehouse arrays given that the databases serving the security applications will also be partitioned and spread across the servers. It is conceivable that a UTM cloud owner may increase the number of significantly large sized arrays of servers, and then partition all security-related databases to produce a massively parallel system [24]. In such a scenario, a UTM might be the first choice for a system architect. Innovative methods are emerging for the meta scheduling of multi-cloud instances, in a bid to counter the increased complexity of balancing resource efficiency against the need to provide an acceptable level of service [37, 38]. However, whilst the performance at the UTM cloud will improve, network and session layer congestion cannot be reduced. This is because the system will be a cascade of two large clouds and the cascade itself will be a bottleneck.

This limitation will exist when two clouds are interconnected and the user sessions will be allowed to pass through one cloud to the server arrays of another. Performance will improve when the users are allowed to connect to two clouds independently for different purposes and there is no inter-cloud cascade. However, such a system will not be effective from security point of view, because the user sessions need to pass through a common checkpoint before being allowed to access application resources. We must also consider that the demands being placed upon BI applications are continuing to stretch performance as users derive new insight from innovative applications, such as automated object-tracking from video stream data, which is finding many applications in the manufacturing domain.

4.2 Security Efficacy

Model A incorporates both a UTM and DMZ, which are well-proven security concepts, whereas the distributed approach of Model B is yet to be proven to the same degree. To summarize:

1. For Model A, each user is marshalled to the BI application via the DMZ. In Model B, users connect directly to the BI application components.
2. Cyber attacks into a Model A arrangement are faced with only one secure entry point to overcome. Again, for Model B there are multiple potential vulnerabilities that will increase as more components and devices are added to the system over time;
3. Security updates and maintenance is a constant requirement that will be managed via the UTM cloud in Model A. For Model B, this results in a very challenging environment, whereby the security responsibilities of components are individualized.

However, Model A suffers from performance-related issues. Close scrutiny of the simulation results shows that UTM_DB_ACT_MON is generating maximum traffic, a cause of the performance degradation. The Security-as-a-service (SECaaS) approach via a UMT cloud is attractive to adopt as user sessions pass through the UTM zone-based firewall, whereas user sessions directly hit the cloud servers in the distributed security model (Model B). In addition, security administration is much easier than the distributed model.

However, the issue of security vulnerability through the multiple entry points within Model B could be addressed by the use of a cloud server to prevent an adversarial attack, by marshalling and containing the attack, whilst maintaining the integrity of the overall system.

Thus, we feel that a hybrid configuration of models would offer considerable potential, balancing performance requirements without jeopardizing security efficacy. All application security controls can be embedded (Model B approach), whereas the network, transport and session security can be incorporated in the DMZ (Model A approach). This segregation of security roles will offer the potential of delivering the users' requirements.

5   Conclusions and Future Directions

The performance of Model B is superior to that of Model A, because there is an additional overhead of a DMZ cloud in Model A with rule-based traffic forwarded through multiple hops of servers. Every hop added in-between causes a delay that is reflected in the simulation results. In contrast, there is only one hop between users and cloud servers in Model B, and there are no forwarding rules (because every server acts as a security node). For Model B, the performance is within the expected limits and there are no TCP delays, TCP segment delays and TCP re-transmissions. We observed that the performance of database query response, HTML page, and object responses, is improved over Model A, and is satisfactory from the users perspective. The response times returned in this model are as expected from a cloud hosting of BI. Elimination of the UTM cloud layer has ensured that all of the session bottlenecks in the network are eliminated.

Hence, a combined model with application security embedded in the BI cloud, with the rest deployed in the UTM cloud, may be an optimum solution, both in terms of performance and effectiveness of security controls. In a mixed model, the security controls will be technically sound with appropriate positioning of security layers, better security processes, clearly defined roles and accountabilities (application security and network security), and better effectiveness of controls.

Additional investigation is required to assess how such a hybrid model of BI security in the cloud can be implemented and is the mandate for this research going forward. In particular, we are investigating how traditional security components (like those shown in the Model A DMZ) can be implemented in the form of cloud arrays in distributed architectures.

References

1. Ashton, K.: That "Internet of Things" thing. RFiD J. (2009)
2. Gubbi, J., Buyya, R., Marusic, S., Palaniswami, M.: Internet of Things (IoT): a vision, architectural elements, and future directions. Future Gener. Comput. Syst. **29**(7), 1645–1660 (2013)
3. Chen, F., Deng, P., Wan, J., Zhang, D., Vasilakos, A., Rong, X.: Data mining for the Internet of things: literature review and challenges. Int. J. Distrib. Sens. Netw. **11**(8) (2015)
4. Gartner: Gartner Says 6.4 Billion Connected Things? Will Be in Use in 2016, Up 30 Percent From 2015?, Gartner website. http://www.gartner.com/newsroom/id/3165317. 10th Nov (2015)
5. Ikram, A., Anjum, A., Hill, R., Antonopoulos, N., Liu, L., Sotiriadis, S.: Approaching the Internet of things (IoT): a modelling, analysis and abstraction framework. Concurrency Comput. Pract. Experience **27**(8), 1966–1984 (2015)
6. Sundmaeker, H., Guillemin, P., Friess, P., Woelfflé, S.: Vision and challenges for realising the Internet of Things. Cluster of European Research Projects on the Internet of Things, CERP IoT (2010)
7. PWC: Industry 4.0: building the digital enterprise. http://www.pwc.com/industry40 (2016)
8. Wang, S., Wan, J., Zhang, D., Li, D. Zhang, C.: Towards the smart factory for Industrie 4.0: a self-organized multi-agent system assisted with big data based feedback and coordination, Comput. Netw. **101**, 158–168 (2016)
9. Wang, S., Wan, J., Li, D., Zhang, C.: Implementing smart factory of industrie 4.0: an outlook. Int. J. Distrib. Sens. Netw. **12**, 1 (2016)
10. Liu, Q., Wan, J., Zhou, K.: Cloud manufacturing service system for industrial-cluster-oriented application. J. Internet Technol. **15**(3), 373–380 (2014)
11. Hill, M., Devitt, J., Anjum, A., Ali, M.: Towards In-Transit Analytics for Industry 4.0. FCST2017. IEEE Computer Society (2017)
12. Katic, N., Quirchmayr, G., Schiefer, J., Stolba, M., Tjoa, A.M.: A Prototype Model for Data Warehouse Security Based on Metadata. University of Vienna, pp. 1–9 (2006)
13. Rosenthal, R., Sciore, E.: View security as the basis for data warehouse security. In: Jeusfeld, M., Shu, H., Staudt, M., Vossen, G. (eds.) Proceedings of the International Workshop on Design and Management of Data Warehouses (DMDW'2000), pp. 1–8. Stockholm, Sweden, June 5–6 (2000)
14. Kadan, A.: Security Management of Intelligent Technologies in Business Intelligence Systems. Yanka Kupala State University of Grodno, pp. 1–3 (2012)
15. Fernandez-Medinaa, E., Trujillo, J., Villarroel, R., Piattini, M.: Developing secure data warehouses with a UML extension. Inform. Syst. **32**, 826–856 (2007)
16. Ahmad, S., Ahmad, R.: An Improved Security Framework for Data Warehouse: A Hybrid Approach. IEEE, pp. 1586–1590
17. Brankovic, L., Estivill-Castro, V.: Privacy issues in knowledge discovery and data mining. Department of Computer Science and Software Engineering, The University of Newcastle, Australia, pp. 1–12 (2000)
18. Al-Aqrabi, H., Liu, L., Hill, R., Antonopoulos, N.: Taking the business intelligence to the clouds. In: 9th International Conference on Embedded Software and Systems (HPCC-ICESS), pp. 953–958. IEEE Computer Society, Liverpool (2012)
19. Blanco, C., Prez-Castillo, R., Hernndez, A., Fernndez-Medina, E., Trujillo, J.: Towards a modernization process for secure data warehouses. In: Pedersen, T.B., Mohania, M.K., Tjoa, A.M. (eds.) LNCS, vol. 5691, pp. 24–35. Springer, Heidelberg (2009)
20. Stobla, N., Banek, M., Tjoa, A. M.: The Security Issue of Federated Data Warehouses in the Area of Evidence-Based Medicine. Vienna University of Technology and University of Zagreb, pp. 1–11
21. Abadi, D. J.: Data management in the cloud: limitations and opportunities. Data Eng. 32(1), 3–11 (IEEE Computer Society) (2009)

22. Fienberg, S.E.: Privacy and confidentiality in an e-Commerce world: data mining, data ware-housing, matching and disclosure limitation. Statistical Sci. **21**(2), 143–154 (Institute of math-ematical Statistics) (2006)
23. Al-Aqrabi, H., Liu, L., Hill, R., Antonopoulos, N.: A multi-layer hierarchical inter-cloud con-nectivity model for sequential packet inspection of tenant sessions accessing BI as a service. In: Proceedings of (CSS) and (ICESS), pp. 137–144. IEEE, France, Paris (2014)
24. Al-Aqrabi, H., Liu, L., Hill, H., Cui, L., Li, J.: Faceted search in business intelligence on the cloud. In: Proceedings of GREENCOM-ITHINGS-CPSCOM, pp. 842–849. IEEE, China, Beijing (2013)
25. Beer, M., Huang, W., Hill, R.: Designing community care systems with AUML. In: Proceed-ings of the International Conference on Computer, Communication and Control Technologies (CCCT '03) and the 9th International Conference on Information Systems Analysis and Syn-thesis (ISAS '03), Orlando, USA, July 31–August 2, IEEE Computer Society, pp. 247–253 (2003)
26. Farhan, M.S., Marie, M.E., El-Fangary, L.M., Helmy, Y.K.: Transforming conceptual model into logical model for temporal data warehouse security: a case study. Int. J. Adv. Comput. Sci. Appl. **3**(3), 115–122 (2012)
27. Priebe, T., Pernul, G.: Towards OLAP security design survey and research issues. In: Proceed-ings of Third ACM International Workshop on Data Warehousing and OLAP (DOLAP 2000), pp. 33–40, 10 Nov 2000. ACM, McLean, VA, USA (2000)
28. Cuzzocrea, A., Bertino, E., Sacca, D.: Towards a theory for privacy preserving distributed OLAP. In: PAIS12, pp. 1-6, 30 March 2012. ACM, Berlin, Germany (2012)
29. Agrawal, R., Srikant, R., Thomas, D.: Privacy preserving OLAP. In: SIGMOD 2005, pp. 1–12, 14–16 June 2005. ACM, Baltimore, Maryland, USA (2005)
30. Shu, Z., Wan, J., Zhang, D., Li, D.: Cloud-integrated cyber-physical systems for complex industrial applications. Mobile Netw. Appl., 1–14 (2015)
31. Li, X., Li, D., Wan, J., Vasilakos, A., Lai, C., Wang, S.: A review of industrial wireless networks in the context of industry 4.0. Wireless Netw., 1–19 (2015)
32. Baker, C., Anjum, A., Hill, R., Bessis, N., Liaquat Kiani, S.: Improving cloud datacentre scalability, agility and performance using OpenFlow. In: Proceedings of the 4th International Conference on Intelligent Networking and Collaborative Systems (INCoS), IEEE Computer Society (2012)
33. Wang, L., Jajodia, S., Wijesekera, D.: Securing OLAP data cubes against privacy breaches. In: Proceedings of the 2004 IEEE Symposium on Security and Privacy (S&P04), pp. 1–15. IEEE Computer Society (2004)
34. Shadija, D., Rezai, M., Hill, R.: Towards an understanding of microservices. In: 23rd Interna-tional Conference on Automation and Computing (ICAC). IEEE Computer Society, Hudders-field, UK (2017). https://doi.org/10.23919/IConAC.2017.8082018
35. Garcia-Campos, J.M., Reina, D.G., Toral, S.L., Bessis, N., Barrero, F., Asimakopoulou, E., Hill, R.: Performance evaluation of reactive routing protocols for VANETs in urban scenar-ios following good simulation practices, In: 2015 9th International Conference on Innovative Mobile and Internet Services in Ubiquitous Computing (IMIS), Santa Cantarina, Brazil, 8–10 July, pp. 1–8. ISBN: 978-1-4799-8872-3 (2015). https://doi.org/10.1109/IMIS.2015.5
36. A1-Aqrabi, H., Liu, L., Hill, R., Antonopoulos, N.: Business intelligence security on the clouds: challenges solutions and future directions. In: Proceedings of 7th International Symposium on Service Oriented System Engineering, pp. 137–144. 25–28 March 2013
37. Sotiriadis, S., Bessis, N., Antonopoulos, N., Hill, R.: Meta-scheduling algorithms for managing inter-cloud interoperability. Int. J. High Perform. Comput. Netw. **7**(2), 156–172 (2013)
38. Sotiriadis, S., Bessis, N., Anjum, A., Buyya, R.: An inter-cloud meta-scheduling (icms) simu-lation framework: Architecture and evaluation. IEEE Trans. Serv. Comput. (2015)

The Private Sector's Role in e-Government from a Legal Perspective

Kaisa Lõhmus, Katrin Nyman-Metcalf, Rozha K. Ahmed,
Ingrid Pappel and Dirk Draheim

Abstract The private sector constantly develops new innovative technologies and governments are always eager to adopt emerging technological possibilities, in order to respond better to citizens needs. While this promises benefits, it can also bring uncertainty from a legal perspective. The implications of emerging technologies and how these can be regulated effectively are of great importance. The aim of this paper is to answer, how the legal framework should be set up in order to shape the private sectors role and, in particular, how the responsibility between the state and private enterprises should be divided. In service of this, we conducted six interviews with high-level stakeholders and experts from state agencies as well as enterprises and triangulated the insights with the relevant regulations and secondary literature. The analysis shows that, currently, public procurement laws play a pivotal role – the private sectors role is crucially determined in public development plans and when it comes to public procurement, the public sector critically relies on the know-how that is connected to delivered products and services.

Keywords e-Government · e-Governance · Private sector · Ict and society · Ict reforms · Ict acts · Legal framework for e-Governance · Ict policies · Best practices in e-Government

K. Lõhmus · Rozha K. Ahmed (✉) · I. Pappel · D. Draheim
Information Systems Group, Tallinn University of Technology, Akadeemia tee 15a, 12618
Tallinn, Estonia
e-mail: rozha.ahmed@taltech.ee

K. Lõhmus
e-mail: kaisa.lohmus@taltech.ee

D. Draheim
e-mail: dirk.draheim@taltech.ee
URL: http://is.taltech.ee

K. Nyman-Metcalf
e-Governance Academy, Rotermanni 8, 10111 Tallinn, Estonia
e-mail: katrin.nyman-metcalf@ega.ee
URL: https://ega.ee/people/katrin-nyman-metcalf/

© Springer Nature Singapore Pte Ltd. 2020
X.-S. Yang et al. (eds.), *Fourth International Congress on Information
and Communication Technology*, Advances in Intelligent Systems
and Computing 1027, https://doi.org/10.1007/978-981-32-9343-4_22

1 Introduction

The private sector constantly develops new innovative technologies and governments are eager to adopt emerging technological possibilities, in order to respond better to citizens needs. While this presents benefits, it can also bring uncertainty from a legal perspective. The implications of emerging technologies and how these can be regulated effectively are of great importance. The aim of this research is to analyze how should the legal framework to be see up in order to ensure private sectors role and responsibilities in the context of an e-Government. Estonia, as an advanced digital society, has been particularly successful in the way it took up eID [1–3] and digital signing [4] and in its use of e-voting, e.g., as early as in 2011, 24.3% of voters cast their votes electronically in the general elections. The implementation of new technology is greatly supported by the legal framework, however, there it still needs more understanding of how and what should be changed in the *status quo*.

Non-state actors, e.g. companies, have an intermediary role and the ability to influence new technologies. ICT enterprises such as Internet Service Providers or search engines, become proxy regulators for the interests of others. Furthermore, supranational organizations, such as EU or UN, who have the global reach, often take the lead in areas like privacy and data privacy in particular. This is an important aspect to consider, as markets for new technologies are worldwide, therefore, the ability for nation states to effectively regulate new technologies is limited.

Speaking in broad term, regulating is usually viewed as a political decision. If a social problem is important enough, there comes a political response to warrant a coordinated response. Regulation could also be used to create a new social practice which generates the desired social effect. That happens in a situation where a social effect is wanted but there is no social practice to produce those effects. Focusing on the regulation, the social practice or the social effects, it is the meaning attached to practice and effect that provides the justification for regulation. For instance, competition laws (also referred to as antitrust laws) are traditionally conceived as regulation of the marketplace to ensure private conduct does not suppress free trade. [5] pointed that such laws prohibit business behaviour which has the objective or the effect of preventing or restricting competition.

Fundamentally, smart governments are relying on databases, software and devices, which are often a competence of private sector. Hence, as it will be presented by this research, competition plays has an important part to play.It is clear that, the citizen wants a high quality and affordable supply of public services, and private companies often want to engage in fair competition to provide such services of general interest. Therefore, inclusion of competition in such case is indispensable. States have laws that seek to safeguard and foster market competition and for that it is possible to establish competition policies that urge regulatory bodies to achieve public policy objectives in ways that are compatible with keeping markets competitive [6].

The objective of this research is to anlyze how should the legal framework be set up in order to ensure private sectors role and responsibilities in a context of e-Government. Moreover, the ultimate goal is to gain more insight on how efficient are todays norms and regulations while including private sector in an e-state. When todays systems flaws are detected, suggestions of improvement for the future will be possible.

The paper will begin with a problem statement and research objective to present the used research methodology in Sect. 2, that will be followed by the description of findings and analysis in Sect. 3. In Sect. 4, discussion will be presented. Finally, the paper will be finished with a conclusion.

2 Problem Statement and Research Objective

When it comes to services of general interest, the citizens want a high quality and affordable supply of public services. Often, such e-services are designed, developed and maintained with the private sector. However, there is no deep understanding from a legal perspective the actual role of the private sector in an e-Government and whether the states should regulate such co-operation.

The governance system often relies on infrastructure usually provided by the private sector. While the state as a force remains important for making sure the availability of critical frameworks, market forces are increasingly exerting influence over infrastructure. Inglehart and Welzel [7] have associated this to linking free markets to self-expression, but an important difference is that the market should set the rules in keeping control. Furthermore [8, 9] has created a model of regulation where he identified four regulatory modalities - law, social norms, architecture or design, and markets. He states that market forces encourage to facilitate online commerce and as it develops, it fundamentally transforms its regulability. Mueller [10, 11] has pointed out that new technologies distribute control and together with liberalization of the telecommunications sector, technology decentralizes participation and ensures that decision-making processes are no longer aligned very closely with state authorities. In addition [12] states that many of the initiatives emanate from the private sector, therefore, private sector might be the one to lead the innovation. Certain is that e-Government, supports administrative processes, improves the quality of public services, and increases efficiency of public sector. This is the way to advance the modernization of public administration in the EU.

Considering the above explanation and based on the existing literature review this research seeks to answer the following research question:

- *What is current consideration of the private sectors role in an e-state and how should it be regulated from a legal perspective?* Along with two sub-questions (both in the context of an e-state):
  - *How can competition be viewed as a reason for including or excluding the private sector?*
  - *How should the responsibility between the state and private enterprises be divided?*

In the following sections, the research evaluates current state of consideration regarding private sectors role, and what further development needed of such role from legal perspective. In addition to explaining how competition policy affect including or excluding private sector, based on the states procurement policy. The current research also provides an explanation of responsibility between a procuring authority and a supplier, that are typically determined in a written contract.

2.1 Research Methodology

A qualitative research method was used to collect relevant data. Triangulation of expert interviews and thematic analysis of relevant public documentation was used and the patterns from the data collection are compared to the theoretical framework in order to conduct the analysis.

The current paper generates data from public documents combined with semi-structured interviews with experts. Thematic analysis is used for analyzing documents. Documentation was publicly available and the sample of documents were chosen in order to examine how private sectors role is currently stated in main strategic documents and regulations.

2.1.1 Document Analysis

According to [13], as the European Commission aims for a digital society, including smarter cities, improving access to e-Government and digital skills. Moreover, Digital Single Market strategy, supports the member states on availability and take-up of e-Government services at European Union level. Speaking broadly, regulations are legal acts that are automatically and uniformly applied to all EU countries. Hence, the regulation (EU) No 910/2014 of the European Parliament and of the Council of 23 July 2014, on electronic identification and trust services for electronic transactions in the internal market and repealing Directive 1999/93/EC [14] was chosen. In addition, EUs Public procurement strategy was chosen to examine how private sectors role is stated in todays public procurement development documents on an international level. The main reason on choosing Digital Agenda 2020 for Estonia [15], was to have further data about how nation state has approached on the role of private sector. Hence, the state is one of the pathfinders in e-Government and therefore a suitable sample to examine.

2.1.2 Interviews

In February and March 2018, six interviews were conducted in Estonian, as interviewers were native speakers of Estonian, and audio recorded by face-to-face meetings. The sample of experts was chosen based on the professional background and experience, from the e-state and legal field. The sample was composed by two policy makers from Estonian Ministry of Economic Affairs and Communications; representative of ITL; a lawyer with previous experience in competition law, public procurement law and IT law; and representatives from two IT enterprises, Tieto Estonia AS and AS Datel.

Interviews were semi-structured or informal, interviewer used verbal interchange with experts to bring forward information by asking questions [16]. The questionnaire was divided into three main thematic blocks, which were chosen based on the theoretical framework and relevant literature. The collected data was coded thematically.

3 Research Findings and Analysis

The findings are divided in thematic parts, which are looking for answers for the main research question and sub-questions.

3.1 The Private Sector's Role and the Regulatory Framework

Experts pointed out that the main regulatory framework followed by states is public procurement laws. They set the limits and commands on co-operation with the private sector.

3.1.1 Status Quo in Private Sector Inclusion

Experts interviewed pointed out that the most common form is an open procurement, where any interested party is allowed to submit a tender. It allows procuring technology development components or buying in full service, in addition to renting workforce. In recent years, the EU has taken a stronger approach on the IT field. In 2014 the EU took steps for public authorities to procure in a more flexible way. For that purpose new generation of public procurement directives were adopted, such as; (Directive 2014/23/EU, Directive 2014/24/EU, and Directive 2014/25/EU). Furthermore, in 2016, EU also launched a Public procurement package, which aims at sustainable and balanced public procurement in EU member states.

Todays legal norms is more about what role public or private sector sees for each other. An example from Estonia, most of the development has gone through private sector services as they form innovative partner and knowledge exporter. Estonian Ministry of Economic Affairs and Communications [15] states that the existence of a competent and innovative partner and service provider, e.g. a competitive ICT sector, is important for the development of public sector ICT solutions and economy in general.

Experts from public sector stated that in addition to developing or maintaining IT systems, inclusion could also mean policy making, participating in an community collaboration or including private sector to creation of a strategy. On an European level, eIDAS regulation states that member states are free to decide whether to involve the private sector in the provision of means of e-Government. Estonian Ministry of Economic Affairs and Communications [15] also points out that information society should be developed in cooperation with the public, private and third sector of all other relevant parties, but it is not clarified what exactly counts as inclusion. Thus, including private sector in a provision of e-state is rather a suggestion, not stated by a rule of law. Even though, both experts from public sector and private sector emphasized on the importance of community co-operation. Therefore, good legislation process requires including stakeholders. Hence, this raises a place for consideration, whether procurement laws are sufficient, since there is no precise role for private sector, or other regulation to be put in place.

As an example, Estonian IT enterprises and public sector representatives have formed an organizing body, that covers working groups. Working groups are directed to solve any matters raised by participating parties, and developing good customs and instructions on how to establish better procurement frameworks and inclusion policies. Organizing body and its working groups gather on a regular basis in order to discuss and find solutions to problems concerning both sectors and the ICT field.

As mentioned before, current legislation do not insist including private sector, and community co-operation is also not regulated in any way. All parties make an effort to carry out ideas developed in an organizing body, but as a principle, decisions are not legally binding nor can they be in contradiction with any other existing regulations. Nevertheless, such co-operation form has become the main gateway for public and private sector to collaborate.

3.1.2 Alternatives for Creating or Modifying Regulation

Market forces encourage to use ICT possibilities and as those possibilities develop, they fundamentally also transform ICT regulability. Hence, there is possibility that modification of existing regulations and norms, and/or create new legislation needed. Normative theory of regulation states that law needs to be set in a broader context that takes full account of the variety of norms [12]. Therefore, while regulating new technologies, e-Government or private sectors role, there must be clear understanding how different legislation exist together.

Experts did not favour creating any new regulation, which would be addressed directly on e-Government and inclusion of private sector. In addition to that, According to the Normative theory of regulation, a possible risk in creating or modifying regulation always entail expenses [16]. Not regulating at all was also mentioned to be as too idealistic idea. But, experts from public sector stated that, if there is a common understanding of the end result and shared responsibility, the roles of public and private sector become clear as well.

However, experts from both, private and public sector, pointed out that in practice, when contractual relationship between public and private sector as a service provider occur, the end-value or goal should be mutual. Both should be partners, forming one chain. That should be taken into account while modifying existing (procurement) regulation. Experts pointed out that regulations must not exclude the usage of e-solutions and law must be technology neutral, without any distinctions. There should not be a separate legal area for IT, while a specific regulation from business point of view needed. Such norm would also include how public management should be regulated through a context of IT. As an example [15] emphasized that a principle of e-Government helps public sector to become better customer and private sector to be a better provider. Hence, the goal is to create more co-operation between private and public sector, it is policy, which needs to adapt, not norms or regulations. European Comission [13] also states that stepping up the involvement of private sectors businesses in service delivery leads to reducing red tape, easing use and lowering delivery costs. In addition, interviewed experts as a compromise suggested that, every domain should be prepared to develop its own digital subjects that private sector fits in it, instead of creating specific regulation to determine a role of private sector in sectoral development plans. And, ultimately they would be binded and concluded in one whole information society development plan which would therefore clearly state how private sector is included. On the other hand, when it comes to domains development plan, they are often incomplete or do not exist at all and even if it is stated in domains development plan that private sector must be included and co-operation is necessary, practice shows that in reality development plans are interpreted differently. Therefore, community co-operation was again mentioned for a lack of integral strategic vision.

In addition, public sector experts visualized a situation where public sector is procuring development service from private sector, leaving them the know-how and the certain product. Private sector would be responsible for the quality of the service and public sector would be free to choose the provider. Such approach would also be more sustainable, since only one state as a customer is not enough for a private sector.

Another point stated by private sector experts was the total life cycle cost of purchases. Public sector is using public money, therefore all expenses must be cost-efficient and well throughout. Therefore, the cheapest offer of short-term is unfortunately chosen. Hence, development of an e-service might include saving, but in a long-term maintenance and modification expenses are much higher than initially proposed. Therefore, the total life cycle cost of purchases should be considered from the beginning and also stated in procurement legislation. Same principle is emphasized at the EU level, where in order to enhance innovation in public procurement, all procurement procedures should take into account the total life cycle cost of purchases and therefore know the long-term financial benefits. However, this is only possible when a state is not dependent on external financing, i.e. European funds. Ideally, the state is allowed to use national budget, which allows to choose more flexible ways how or what to obtain, with ensuring transparency and the procurement has to meet certain rules.

Therefore, the goal of public and private sectors co-operations should evolves on a running basis. Being more flexible with norms and regulations, helps defining the role of private sector, even if the co-operations lacks a written contract right from the beginning.

3.2 Impact on Competition

This section is analyzing how can competition be viewed as a reason for including or excluding private sector in a context of e-state and in such case, what is the impact on competition.

Based on the interviews, there was a clear understanding of not establishing any strict competition regulations directed specifically on a certain field, such as e-Government. However, a mandatory promotion and development of e-Government was encouraged on the state level. In a context of e-Government, development decision and competition promote can be made by the state. For example; making procurement legislation more flexible and hence including private sector even further. As mentioned before, public sector is the force that drives innovation and private sector is producing solutions which are ordered by a state. Hence, creating other competition regulations is not needed, but establishing competition policies helps to achieve public policy objectives for keeping markets competitive. The states procurement policy has an important role on stimulating private sectors competitiveness. Therefore, smart customer state concept was emphasized by most of the experts. In addition [15] also states that public sector should be a smart customer. public procurement's should give freedom for offering innovative solutions in order to contribute of the ICT sector. Further more, the government itself is an important player in the market and hence it affects the market both by creating rules and purchasing products [8, 9]. Public sectors IT procurement, as a purchase with important market power, directly affects competence and competitiveness of private sector. Estonian Ministry of Economic Affairs and Communications [15] also points that the existence of competent and innovative ICT sector is vital for the development of public sector ICT solutions and the economy in general.

Todays procurement regulations and competition laws are flexible. Therefore, experts are more after on guidance, mainly through secondary legislation and guidance notes, on how to apply existing regulation. In that sense (competition) regulations should be more specific and in detail, meaning that they should be more widely communicated and promoted, in order to ensure consistent know-how through-out public sector.

3.2.1 Position of Small and Medium-Sized IT Enterprises

Maximizing consumer welfare must be the goal of competition law, but in many cases regulation strongly gives advantage to certain groups [17, 18]. In a context of e-Government and competition, it means that regulating could also give advantages to some parties included, e.g. certain IT enterprises. According to experts and based on EUs Public procurement strategy, competition regulation often leaves small and medium-sized enterprises (SMEs) in disadvantage. EUs Public procurement strategy points out that currently SMEs win only 45 percent of the value of public contracts, as often they simply do not have a capacity to offer. That is not consistent with Chicago school theory, which explain that politics honor passionately expressed preferences [16]. In case of SMEs, the preferences are more homogeneous than in larger IT enterprises, hence they have an advantage in that for the yield per member of the group is greater. However, experts pointed out that the procurement laws, as the main regulatory framework followed, do not favour SMEs, as they are said to be too strict and also entailing too heavy administrative burden. Generally, big IT enterprises, also named as IT factories by some experts, have adequate amount of resources to deal with bureaucracy. As todays legal norms role of private sector not cleared, that might be interpreted as an disadvantage for all IT enterprises, being in more difficult situation. Following all regulatory norms is not executable for small IT enterprises, but any competitive advantages are mainly discouraged by the experts and the same was pointed out in a context of e-state. Therefore, it should be the accountability of public authorities to foster an environment where SMEs can compete successfully in the market.

Currently, the main solution for including small IT companies is collaboration with bigger IT enterprises. Therefore, two companies can co-operate to match a procurement. Experts from private sector stated that often SMEs have very specific niche, which is not offered by big IT enterprises. Therefore, co-operation and willing for collaboration is important also from international point of view. Partnership is a direction supported by the EU as well, the Unions public procurement rules encourage innovation partnerships, which allow the combination of research and public requirements, with specific rules in competition and research and development. Estonia also has examples where procurement's are made with a prerequisite supply consortium. It means that there might be the main provider, but procurement conditions state that at least two or three other companies must be included. Furthermore, current practice also includes situations where small sized companies come up with an idea, present it to policy makers and if its approved, the idea becomes alive. Estonian Ministry of Economic Affairs and Communications [15] also points out pilot projects to test and implement innovative solutions and technologies, including in the private sector in the case of services of general interest, as one of the action lines.

Another interesting idea proposed by private sector experts, for helping SMEs, would be prioritizing IT systems and registers. There are IT systems and registers with various sizes and with different importance, with different impact on society. But, they would give IT enterprises different ways to obtain experience. If a state makes a decision to distinguish registers and divide them on categories, based on their impact for example, it would give small IT companies a chance to learn. IT systems and registers with large effect on society means strict procurement laws and heavy administrative burden, which is manageable for big IT companies. IT systems and registers with little impact and responsibility entails more room to learn, therefore smaller IT enterprises have a chance to get experience. A compromise would be establishing so called IT test lab, which allows to make concessions within procurement regulations, specifically from the procurement point of view. There would be development projects that every IT company is allowed to try. If necessary, benefits would be included, but only the best companies would receive a long-term contract. As a result, the amount of bureaucracy would reduce and all companies would equally be allowed to show their competence.

3.2.2   Larger Role of State-Owned IT Centers

Experts pointed out that recently there has been a shift towards the state-owned IT centers. They are not only administrators of IT systems, but also involve interior development units. Hence, services that could be promoted by private sector are now developed internally by public state-owned IT centers. Trend is also that state-owned IT centers buy specialists from private sector for a specific work assignment. After the assignment, the specialist might not return to private sector, as a result, the state is actively participating on labor market competition. Experts from private sector emphasized that, private sector and ICT companies are strongly against such practice, as it restricts sectors own development competence. Moreover, keeps the knowledge within public sector, making it extremely difficult to export. A neo-classical model of competition states that production and distribution of services in competitive free markets promotes rationality, stability, and equal welfare [19]. Therefore, the state being a player in the free market would be tolerated. However, experts claimed that it is killing competition when a state is interfering to service offering. They pointed out that there should be a regulation stating that a state itself cannot offer a product or a service, in a situation where private sector has capability to provide those services. Hence, banning the state from doing some things would be an innovative solution in many countries and an interesting thing that legislation can actually do.

Another point by experts from private sector was, if there is a (political) will to offer a service by a state, in any case, financial resources are found. According to [20] who point out political willingness and a political decision-making as a powerful tool. Meaning that regulating competition more is also usually viewed as a political decision. Supportive politicallegal environment has the power to adjust a potential norm.

Experts also mentioned that when it comes to the private sector, they have to offer services which are in demand. Because, they are not spending resources if there is no possibility of making profit. Moreover, private sector has to constantly compete with other providers while continuously adapting with the needs of end-user. On the other hand, the state offered services do not have to compete, nor do they have to be constantly modified and adjusted, which is a normal situation when competing with other suppliers. Estonian Ministry of Economic Affairs and Communications [15] also emphasizes constantly redesigning information systems, otherwise their administration costs will soar. Continuous modification and adjustment are important aspects, but when a state is a supplier itself, whose main priority is not exporting the know-how, nor constantly developing competence. There is a strong possibility pointed out by the experts that state-offered services might not be with as good quality and as cost effective compared to private sector services.

As a solution, experts suggested that, state-owned IT centers to be intermediary force, they should be a link between a contracting authority and a provider. However, as mentioned before, current situation is state-owned IT centers compete against private sector, while recruiting specialists and developing themselves. Therefore, they have become from intermediary force to provider of services. As a results, state-owned IT centers also compete with each other, lowering the price of the service even more. Experts from private sector emphasized that private companies are not capable to offer a service with similar price as public IT centers, thus again, competition is killed.

3.3 Division of Responsibility

This section is analyzing how the responsibility should be divided between a state and a supplier. Responsibilities of a procuring authority and a supplier are typically determined in a written contract and in a broader context regulated with procurement laws. In some states it is stated that a responsibility must be divided between a contracting authority and a supplier. However, in practice, is often written contract states that the private sector carries most of the risks. Contractual penalties are often used as a tool to for security. Public procurement contracts demand guarantees for objectivity and responsibility.

Experts interviewed suggested that more emphasis should be put on sharing risks and responsibility. It was highlighted by the experts that contractual penalties are definitely not the way to go forward, as they do not improve co-operation between sectors, but rather hinder it. They create a risk that private sector deliberately refrains from co-operation with public sector. Another important factor is that a general risk tolerance, especially from public sectors point of view, must be higher. A low risk tolerance, especially from public sectors point of view, is wider than just e-Government, as it is closely connected with public opinion. Experts emphasized that with an e-Government, there is always a possibility that risks become reality, whether the public is willing to accept it or not. Therefore, it is necessary to say it publicly from the beginning that a procured project might not work out, but there is a cross-sectorial common vision, to develop a service which would benefit citizens.

Moving on to political responsibility, experts emphasized that responsibility for the existence of a service cannot be delegated from a state to private sector. According to Normative intervention theory of regulation, regulating is usually viewed as a political and ideological decisions [20]. If a social problem is important enough, a political responses coordinated. Un-supportive or supportive politicallegal environment has the ability to push through or vice-versa, put a lid on a potential norm.

3.3.1 Alternatives for Regulating Responsibility

The more transparent and prepared procurements and contracts are, the better is the end-result. Responsibility could be regulated through written contracts, but there must be a definite guarantee that responsibilities and risks are shared in practice as well, not only on paper. Experts emphasized that public sector must have a vision and a willingness to finance innovation and at the same time share risks. In contractual relation, co-operation and partnership between parties, and sharing risks should be balanced. Some experts pointed out that public sector should take responsibility on behalf of society, and this cannot be defined without a written legal norm. Public procurement can also promote social responsibility. When public sector links social responsibility to public procurement processes, socially responsible procuring becomes a rule, not exception. Experts suggested community responsibility as an idea to follow. Another way to better enhance responsibility, experts suggested to promote wider sharing of good practices, meaning that successful e-Government practices, which have been created in public-private co-operation, should be gathered in one database. Public procurement contracts demand confidentiality, however, know-how about which practices are successful and which mistakes not to make is relevant to all parties included in e-Government.

3.3.2 Regulating New Technologies

An expert from public sector pointed out that when a state is dealing with emerging technology, there is no rush to regulate something, but there has to be a strict position on how new technologies are dealt with.

The expert made an example from cloud computing, and how legal framework can be a driver or an obstacle. States have different opinions about restoring data outside the country. If a state takes direction to own server resources outside of its own territory, to be used for operate services in addition for data backup, it is important that such resources would still be under states control. However, government level cloud computing is an innovative idea and some would say risky as well. Also, there should be a certain position made by the state, how such technology is to be treated and whether there is a need to regulate. According to [15], innovative technologies such as cloud computing should constantly be analyzed and piloted on a states level, not regulated immediately.

An important point was made by an expert with a long-term legal background - with emerging new 43 technologies, it is important to agree on what is technologically feasible, and socially acceptable. As an example, e-state can be evolved with no privacy at all, but the question is whether this is also a state citizens want to live in. Individuals have objectives, such as personal freedom, that are affected by regulations and technology [21]. The state can drive innovation or restrict it. According to [20], the purpose of regulation is to achieve a different social effect or to ensure that people behave in the desired and familiar manner. In addition, boundaries must be well justified, as protecting some social norms, may hinder technical innovation.

4 Discussion

Generally, Private sectors role in a context of an e-Government is not stated in relevant strategic documents, nor in public procurement laws, which are the main regulatory framework followed. As, private sector companies are the suppliers and export knowledge to public sector. While, legal perspective, if there are no legally established guarantees for continuous co-operation, companies dont have any certainty. Furthermore, if private sectors competitiveness is hindered, economic situation in general is jeopardized.

The goal is to create more defined co-operation between private and public sector, and determine the roles and responsibilities of both parties. Service provider and contracting authority should be partners, forming one chain, and having mutual goal. For that, policy, not norms or regulations, need to adapt. As creating new regulations might hinder the innovation, the states, as contracting authorities, set the direction of innovation. This direction should not be to create any new legislation, which would be addressed directly on e-Government and inclusion of private sector. When a state is a smart customer, the needs are clear, procurement's aim to order something innovative. On the other hand, if public sector do not order innovative solutions, private sector is also forced to offer outdated ideas, which in turn ends up restraining private sectors competence and competitiveness as well. It is rather interesting that dominating opinion among experts is skepticism towards the need to regulate private sectors role more. Internationally this is not always the attitude. EU has put increasingly more emphasis on data protection regulation, cyber security laws and payment directives. Moreover, raising EU standards might help the Union as a whole or states which are not digitalized enough. However, in states, which are already successfully using technological possibilities in the context of an e-state, there is a risk that EU rules might hinder not favor digital innovation. In the future, the divide between highly innovative countries and rather conservative ones become more visible. ICT, as a horizontal field, affects other domains as well. Hence, Europe has given the direction that member states should set more control over technology and how it affects everyday lives. Additionally, as a compromise for not creating any specific regulation, sectoral development plans could be used for forming a basis to communicate the role of private sector, and procurement regulation would be main legal framework. When public procurement takes place, public sector procure from private sector, as they have the know-how about their certain products. This results in not giving any advantages to state-owned IT centers. Therefore, public ICT procurement enhances competence and competitiveness of private sector. Public sectors IT procurement is a purchase with an important market power, as it affects private sectors competence and sets the direction of innovation. Moreover, states should not establish any strict competition regulations directed specifically on an e-Government. Instead, take responsibility of mandatory promotion and development of e-Government.

In addition, there should be a norm to define a guideline that at least three percent out of all public sector procurement should be directed to innovative solutions.

Existing procurement laws are for some SMEs too strict and entail heavy administrative burden, but any competitive advantages are mainly discouraged in a context of e-state. Therefore, public authorities should foster an environment where SMEs can compete successfully in the market. SMEs could be included more when collaborating with bigger ICT enterprises through innovation partnerships or procurements with prerequisite of supply consortium. However, state-owned IT centers have shifted to become one of potential providers of public e-services. they are administrators of IT systems in addition to involvement in interior development units. On the other hand, as private sector is capable to offer e-services which benefit citizens, Therefore, the state is generally only responsible for ensuring the availability of e-services, rather than interfering in a well functioning market. Participation of state-owned ICT centers in the market directly restricts private sectors development competence. Legally, radical solution would be banning state from being a provider. State-owned IT centers would be an intermediary force, and would be a link between a contracting authority and a provider from private sector. Responsibilities of contractual parties regulated with procurement laws, clarified in a written contract. As analysis showed, current way of regulating responsibility seems to be acceptable, meaning that additional new legal framework needed. However, current legal landscape do not emphasize enough sharing risks and responsibilities. In some states, procurement laws may refer to sharing risks, but in a written contract the reality is often different. Contractual relations should be based on co-operation and partnership, with higher risk tolerance.

E-projects are often with high market value and with a major impact on society. In a context of e-Government, if initial risks become reality, it affects large amount of citizens. In the long run, that leads to a crucial status regarding the publics willingness to accept failed e-projects. Thus, it is necessary to emphasize from the start that an e-state itself is somehow a pilot project with an aim to bring greater good. In turn, states own a will to experiment and wider public is more indulgent when it comes to failing. When it comes to emerging new technologies, the ability to anticipate technological changes and flexibility to adapt is vital. Therefore, no rush to regulate everything new. The state has to be willing to take risks and, create a strict position on how new technologies are dealt with, new Technologies could be piloted and analyzed. When a state has taken a position to regulate and set boundaries on some innovative technical solution, it automatically sets limits to private sectors innovation. It is important to agree on what is technologically feasible and at socially acceptable at the same time. Again, when there is a willingness to regulate quickly, there has to be a understanding that when wanting to protect some social norms, technical innovation is automatically hindered.

5 Conclusion

Common understanding is that legal norms and regulations do not set the standards on what is exact role of the private sector and their co-operation, in a context of e-state. However, the direction should not be into creating new legislation.

Solution would be using sectoral development plans as a form to communicate the role of private sector. Procurement regulation would be main legal framework. Private sector is responsible for the quality of the service. Therefore, public sectors procurement enhances directly competence and competitiveness of private sector. In addition there could be a norm, stating that out of all public sector procurements at least three percent should be directed to innovative solutions, which help to develop e-state. There has been a shift towards state-owned IT centers. Their participation in the market restricts private sectors development competence. A solution would be legally banning the state-owned IT centers, from being a supplier. From a legal perspective there is no need to create any new framework for stating responsibility. Todays legal solutions such as procurement laws, responsibilities determined in a written contract, with no enough emphasis on sharing risks and responsibilities. Instead, the states mentality in procuring, should be changed and the contractual relations be based on partnership and shared risks. Moreover, certain position has to be taken by state on how new technologies are dealt with, set the direction and boundaries on innovative technical solution. This automatically sets limits to (private sectors) innovation.

References

1. Tsap, V., Pappel, I., Draheim, D.: Key success factors in introducing national e-identification systems. In: Dang, T.K., Wagner, R., Küng, J., Thoai, N., Takizawa, M., Neuhold, E.J. (eds.) Future Data and Security Engineering, pp. 455–471. Springer International Publishing, Cham (2017)
2. Republic of Estonia: Electronic Identification and Trust Services for Electronic Transactions Act. https://www.riigiteataja.ee/en/eli/527102016001/
3. Republic of Estonia: Identity Documents Act https://www.riigiteataja.ee/en/eli/521062017003/
4. Pappel, I., Pappel, I., Tepandi, J., Draheim, D.: Systematic digital signing in estonian e-government processes, pp. 31–51. Special Issue on Data and Security Engineering, Transactions on Large-Scale Data- and Knowledge-Centered Systems XXXVI (2017)
5. Huffman, M.: Bridging the divide? theories for integrating competition law and consumer protection. Eur. Compet. J. 6(1), 7–45 (2010)
6. Aydin, U., Bthe, T.: Competition law & policy in developing countries: explaining variations in outcomes; exploring possibilities and limits. Law & Contemp. Probl. 79(4), 1–36 (2016)
7. Inglehart, R., Welzel, C.: Changing mass priorities: the link between modernization and democracy. Perspect. Polit. 8, 551–567 (2010)
8. Lessig, L.: Code and Other Laws of Cyberspace. Basic Books, New York (1999)
9. Lessig, L.: Code. Basic Books, New York (2006)
10. Mueller, M.: Networks and States: The Global Politics of Internet Governance. MIT Press (2010)
11. Mueller, M.L., Drake, W.J.: Networks and States: The Global Politics of Internet Governance (Information Revolution and Global Politics). MIT Press (2013)
12. Brownsword, R.: In the year 2061: from law to technological management. Law, Innov. Technol. 7(1), 1–51 (2015)
13. European Comission: EU eGovernment Action Plan 2016-2020 – Accelerating the digital transformation of government. Communication from the Commission to the European Parliament,

the Council, the European Economic and Social Committee and the Committee of the Regions. COM(2016) 179 final (April 2016) 1–12

14. European Union: Regulation (EU) No 910/2014 of the European Parliament and of the Council of 23 July 2014 on electronic identification and trust services for electronic transactions in the internal market and repealing Directive 1999/93/EC

15. Estonian Ministry of Economic Affairs and Communications: Digital Agenda 2020 for Estonia (2013)

16. den Hertog, J.: General theories of regulation. In: Elgar, E. (ed.) Encyclopedia of Law and Economics. Utrecht University, pp. 223–270 (1999)

17. Tsoulfidis, L.: Classical vs. neoclassical conceptions of competition. Technical Report Discussion Paper No. 11/2011, University of Macedonia (2011)

18. Stigler, G.: The theory of economic regulation. Bell J. Econ. Manag. Sci. **2**(1), 3–21 (1971)

19. Plummer, P.S., Sheppard, E.: Modeling spatial price competition: Marxian versus neoclassical approaches. Ann. Assoc. Am. Geogr. **88**(4) (1998)

20. Sheehy, B., Feaver, D.P.: Designing effective regulation: a normative theory. UNSW Law J. **38**(1), 392–425 (2011). November

21. Joskow, P.L., Noll, R.C.: Regulation in theory and practice: an overview. In: Fromm, G., (ed.) Studies in Public Regulation. The MIT Press (1981)

Toward an Effective Identification of Tweet Related to Meningitis Based on Supervised Machine Learning

Thierry Roger Bayala, Sadouanouan Malo and Atsushi Togashi

Abstract Epidemic surveillance requires a rapid collection and integration of data and events related to the disease. Adequate measures, including education and awareness, must be rapidly taken to reduce the disastrous consequences of the disease. However, developing countries, especially those in West Africa, face a lack of real-time data collection and analysis system. This situation delays the analysis of risk and decision making. The aim of this research is to contribute to the surveillance of the meningitis epidemic based on Twitter datasets. The approach, we adopted in this research is divided into two parts. The first part consisted of investigating different methods to convert the tweet data into numerical data that will be used in machine-learning algorithms for the classification tasks. The second step is to evaluate these approaches using different algorithms and compare their performance in term of training time, accuracy, F1-score, and recall. As a result, we found that the SVM machine algorithm performed good with 0.98 of accuracy using the TF-IDF embedding approach while the ANN algorithm performed good with accuracy of 0.95 using the skip-gram embedding model.

Keywords Natural language processing · Twitter · Machine learning · Meningitis · Neural network · Word embeddings · Support vector machine · Logistic regression · Random forest

T. R. Bayala (✉) · A. Togashi
Miyagi University, Sendai, Miyagi 981-3298, Japan
e-mail: thierrybayala@gmail.com

A. Togashi
e-mail: togashi@myu.ac.jp

S. Malo
Université Polytechnique de Bobo-Dioulasso, 01 BP 1091 Bobo, Burkina Faso
e-mail: sadouanouan@yahoo.fr

1 Introduction

Twitter is a microblogging platform where many interconnected users share frequently information among themselves. This platform allow users to interact through short messages by posting their opinion about a current situation such as presidential election, stock market, and healthcare issues that occurs in their surroundings. This diversity of topics that are tackled everyday has made Twitter an interesting source of data used by researchers to predict events. In the recent years, Twitter data has allowed to conduct successfully a lot of researches in fields of epidemic disease prevention such as the epidemic of influenza [1–4].

In this paper, we tackled the problem of meningitis which is an infectious disease affecting many people in West African countries every year. The reasons we decided to work on this disease are described as follow:

- Meningitis affects mostly young people. In 2017, a report relating elaborated by United Nations Department of Economic and Social Affairs estimated the percentage of African population who suffer from meningitis between 0- and 14-years old is about 41% [5].
- There is almost no system for data collection and analysis to prevent the outbreak of meningitis epidemic in real time.
- The ability of the disease to create serious damages such as hearing loss and sometimes mental disorders in a short period after the exhibition requires to be diagnosed earlier to reduce the disastrous damages [6].
- There is a need to discover if the country faces a situation of the epidemic to limit the spreading over the country through sensitization of the population.

To reach our goal, we trained a classifier that can be able to identify the label of a given tweet mentioning the keyword "meningitis." Since we approached this problem as a supervised machine learning problem, we proceed to the annotation of our dataset based on the features we defined as meningitis dictionary. Most of the existing machine-learning algorithm deals with numerical data. Therefore, we trained a word embedding model that consists of converting textual data(tweet) into numerical data to convert our tweet into a numerical vector. We used the embedding model to train several classifiers with the target labels "*Infection, Concern, Vaccine, Campaign*, and *News*." We investigated the support vector machine (SVM), artificial neural network (ANN), logistic regression (LR), and random forest (RF) models and compared their results.

2 Related Work

Some existing work has been carried out to tackle the problem of meningitis. In general, most of these researches deal with clinical data that take time to confirm if the patient is affected by meningitis. Some others used climate data to predict whether there is a risk of meningitis epidemic. In this section, we present some of the researches done in the field of meningitis as follow:

In 2006, E. C. Savory et al. proposed an approach to evaluate a risk of epidemic by collecting and analyzing data reported by WHO's surveillance system from 2000 to 2004, PubMed database, ProMED-mail, and the International Disasters DataBase (OFDA/CRED). In their research, the authors predicted the location of the epidemic based on humidity data analysis. As a result, a total of 71 meningitis epidemics was reported in 25 countries from January 2000 to April 2004 affecting 721 (22%) of the 3281 continental African districts [7].

In 2017, Basil Benduri Kaburi et al. experimented another approach in Ghana that collected data from some sub-municipal health centers in Ghana trough a questionnaire. In this paper, there were three possibilities of result depending on the clinical conditions of the patient: Suspected meningitis case, Probable meningitis case, and Confirmed meningitis case. The authors combined the clinical test results and the symptoms exhibited by the patient to predict if the patient is affected or not. They reached a predictive positive value of 100% from 2010 to 2014 and 63.3% in 2015 [8].

In their research published in 2015, the authors Clément Lingani et al. proposed a method that analyzed the data from the weekly surveillance bulletins and the World Health Organization database. They were able to report 341,562 suspected and confirmed cases over 10-years study [9].

While the previous research used the reporting system data for their analysis, we used the Twitter data source for the reason that tweets are publicly accessible in real time and in huge volume.

3 Methodology

3.1 Data Collection

We collected historical data using the hand chosen keyword "meningitis." We were able to collect 373,765 tweets from 2009 to 2014. Figure 1 shows the number of tweets mentioning the keyword "meningitis" with a significant peak in 2012–2013 corresponding to a period when Nigeria has seen an increased rate of meningitis in some of its states. Each entity of tweet is composed of two features: *text* and *date*.

3.2 Data Pre-processing

Twitter data contain a lot of information that affects negatively the quality of data. The presence of this information affects the performance of our system in terms of time and memory consumption. The most frequent unnecessary words we found in

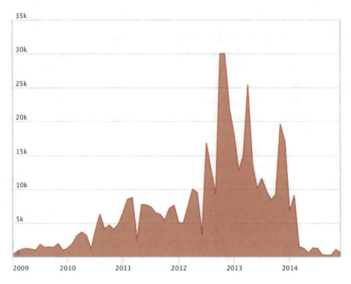

Fig. 1 Tweets mentioning the keywords meningitis from 2009 to 2014

our dataset are URLs, emoji, hashtags, apostrophes, punctuations, and stop words. Our pre-processing tasks removed this information from the collected data.

3.3   Training Data

The training process defined a label for each tweet which can be Infection, Concern, Vaccine, Campaign, or News. We assigned one of these labels to our tweets based on a list of features we defined as meningitis dictionary. We asserted that the label of a tweet corresponds to the category name and a tweet must have only a feature belonging to one of these categories. For the feature's selection, we were inspired by the work done by Lam et al. in their research related to influenza tracking through Twitter data [10]. We extended the list of categories by adding the categories News and Campaign. We also extended the list of features presented in the paper publish by Lam et al. The following Table 1 presents the list of features we considered in our dictionary.

We used this training data to train our classifiers. Before the training step, we proceeded to the transformation of the tweet into numerical data. This process is so-called word or document embeddings. One of the earlier works done by G. Salton et al. in 1975 was information retrieval system with the main idea to represent a document as a vector in a vector space [11]. The result of this allows the progress in information retrieval which is used in many research engines. Recently, this concept has been taken up by other researchers to improve it for natural language processing tasks [12, 13].

Table 1 Features related to meningitis

Features	List of terms
Infection	Contracted meningitis, contract meningitis, contracting meningitis, get meningitis, got meningitis, getting meningitis, have meningitis, having meningitis, had meningitis, has meningitis, catch meningitis, caught meningitis, infect meningitis, infected meningitis, recovering meningitis, recovered meningitis
Concern	Worried, afraid, scared, fear, worry, nervous, dread, dreaded, terrified, panicked, tormented, wondering
Vaccine	Meningitis vaccine, meningitis vaccines, meningitis shot, meningitis shots
Campaign	Raised money, raised funds, raised fund, collected funds, collected fund, collected money, fund raising, support, supported, campaign, raise funds, raise fund, collecting money, donation
News	Meningitis outbreak, meningitis alert, meningitis killed, outbreak of
Negation	No sign of, not had, not have, rules him out, rules her out, etc.
Self	Me, i, myself, we, us
Other experiencers	You, he, she, they, relative parents (son, aunt, brother, etc.), some common English names (John, Zax, etc.)

There exist several approaches of document embedding, and they can be divided into two groups: The embedding based on count (EBC) and the embedding based on prediction (EBP). The EBC group's concept is based on counting words in the document. Some examples of the algorithm are TF-IDF, count vector, and co-occurrence matrix.

Different from EBC group, the EBP group is composed of two algorithms: The CBOW or continuous bag of word and the skip-gram algorithms. Both of these algorithms use an artificial neural network to generate the vector representation of a word or document. However, the approaches used by the algorithms are different. While the main concept of Skip-gram model is to predict the surrounding word called *context* given a corpus; the CBOW's main idea is to predict a word given a context. These algorithms were introduced by Mikolov et al. in their paper "*Distributed Representations of Words and Phrases and their Compositionality*" [14].

In this paper, we used the approaches of TD-IDF and skip-gram models for the reason that the TF-IDF approach emphasizes the features compared others EBC algorithms and skip-gram model gave us a better a representation of word in terms of similarity.

3.4 Classification

For the classification, we used different algorithms such as support vector machine (SVM), artificial neural network, random forest, and the logistic regression.

Table 2 Accuracy, F1-score, and recall using TF-IDF embedding model

		Accuracy	F-1 score	Recall
Infection	SVM	1.00	1.00	1.00
	ANN	0.99	0.99	1.00
	Random forest	0.59	0.74	0.98
	Logistic regression	0.98	0.99	0.99
Concern	SVM	1.00	0.99	0.98
	ANN	1.00	0.98	0.96
	Random forest	0.00	0.00	0.00
	Logistic regression	1.00	0.71	0.55
Vaccine	SVM	1.00	1.00	1.00
	ANN	1.00	1.00	1.00
	Random forest	0.97	0.98	0.99
	Logistic regression	1.00	1.00	1.00
Campaign	SVM	1.00	1.00	1.00
	ANN	1.00	1.00	1.00
	Random forest	1.00	0.16	0.09
	Logistic regression	1.00	1.00	0.99
News	SVM	1.00	1.00	1.00
	ANN	0.98	0.97	0.99
	Random forest	0.97	0.76	0.67
	Logistic regression	0.99	0.99	0.99

4 Result

After having trained the different algorithms, we compared them on the training time, the accuracy, F1-score, and the recall. We found out that using the TF-IDF approach to represent our tweet take long time compare to the skip-gram. This situation is due to the fact that the size of the document vector was huge using the TF-IDF compare to the vector size generated by the skip-gram model. However, even if the training time was higher using the TF-IDF model, the classifiers accuracy was better than the skip-gram. Tables 2 and 3 are the result of each algorithm using the TF-IDF and skip-gram embedding approaches.

5 Discussion

The good accuracy we obtained after training our model can be explained by the quality and the quantity of the dataset. In fact, we spent roughly 60% of this work cleaning and normalizing the tweets. This step allows us to get a good quality of the

Table 3 Accuracy, F1-score, recall using skip-gram embedding model

		Accuracy	F1- score	Recall
Infection	SVM	0.95	0.97	0.99
	ANN	0.96	0.97	0.97
	Random forest	0.82	0.86	0.90
	Logistic regression	0.95	0.97	0.98
Concern	SVM	0.0	0.0	0.0
	ANN	0.31	0.29	0.27
	Random forest	0.0	0.0	0.0
	Logistic regression	0.0	0.0	0.0
Vaccine	SVM	0.99	0.99	0.99
	ANN	0.99	0.99	0.99
	Random forest	0.93	0.92	0.91
	Logistic regression	0.99	0.99	0.99
Campaign	SVM	0.99	0.99	0.99
	ANN	0.99	0.99	0.99
	Random forest	0.95	0.85	0.86
	Logistic regression	0.99	0.99	0.98
News	SVM	0.99	0.98	0.98
	ANN	0.99	0.99	0.99
	Random forest	0.84	0.87	0.87
	Logistic regression	0.98	0.98	0.98

datasets which contribute to increasing the accuracy of the model. Using the approach of skip-gram, we were able to have a rich representation of the different tweets in a vector space that contribute to learn very faster the different models. However, the embedding model of TF-IDF performed a good accuracy. The keywords used in our vocabulary were too small and they have been used to label the tweet. The variation of word is very small, and therefore, allowed the TF-IDF to emphasize the features. This situation explains the reason why we obtain a good accuracy during the classification test. The result relies on the number of features than the words similarities, therefore a tweet can be misclassified if the user uses a synonym of the features presented in Table 1.

6 Future Work

Every year, meningitis affects many people in Africa. From Senegal to Ethiopia including Burkina Faso, these countries are well known as meningitis belt. Our future work will consist to evaluate our model on the dataset from these countries.

However, depending on the counties, the main languages spoken vary from English to French. Since, in this paper, we deal with English language, we will consider the French language in the future work. We intend to use an ontology implemented by Cédric Béré et al. as vocabulary to extend the list of features we used in this work and for the tweet annotation [15].

Acknowledgements This research would have been impossible without the support of the Japanese International Cooperation Agency in short JICA that allowed me to pursue a Master thesis at Miyagi University. I would like to thank JICA for their financial support. I would also like to thank the CEA-MITIC, which is a research institution in ICT-based in Sénegal that supported me financially to participate in this conference.

References

1. Mowery, J.: Twitter influenza surveillance: quantifying seasonal misdiagnosis patterns and their impact on surveillance estimates. Online J. Public Health Inform. (2016)
2. Paul, M., Dredze, M., Broniatowski, D., Generous, N.: Worldwide influenza surveillance through twitter. In: AAAI (2015)
3. Culotta, A: Towards detecting influenza epidemics by analyzing twitter messages. In: KDD Workshop on Social Media Analytics (2010)
4. Broniatowski, D.A., Michael J.P., Dredze, M.: National and local influenza surveillance through twitter: an analysis of the 2012–2013 influenza epidemic. PLoS ONE (2013)
5. United nations department of economic and social affairs/population division, world population prospects, p. 17 (2017)
6. Logan, S.A.E., MacMahon E.: Viral meningitis. BMJ (2008)
7. Savory, E.C., Cuevas, L.E., Yassin, M.A., Hart, C.A., Molesworth, A.M.: Thomson MC Evaluation of the meningitis epidemics risk model in Africa. Epidemiol. Infect. **134**, 1047–1051 (2006)
8. Kaburi, B.B., Kubio, C., Kenu, E., Ameme, D.K. et al.: Evaluation of bacterial meningitis surveillance data of the northern region. Ghana, 2010–2015, Pan Afr. Med. J. (2017)
9. Lingani, C., Bergeron-Caron, C., Stuart, J.M., et al.: Meningococcal meningitis surveillance in the African meningitis belt, 2004–2013. Clin. Infect. Dis. (2015)
10. Lamb, A., Paul, M.J., Dredze, M.: Separating fact from fear: tracking flu infections on twitter. In: North American Chapter of the Association for Computational Linguistics (NAACL) (2013)
11. Salton, G., Wong, A., Yang, C.: A vector space model for automatic indexing, communications of the ACM **18** (1975)
12. Mikolov, T., Chen, K., Corrado, G., Dean, J.: Efficient estimation of word representations in vector space. arXiv preprint arXiv:1301.3781 (2013)
13. Bengio, Y., Schwenk, H., Senecal, J.-S., Morin, F., Gauvain, J.-L,: Neural Probabilistic Language Models, Innovations in Machine Learning. Springer (2006)
14. Mikolov, T., Sutskever, I., Chen, K., Corrado, G., Dean, J.: Distributed representations of words and phrases and their compositionality. In: Proceedings of NIPS 2013 (2013)
15. Béré, W.R.C., Camara, G., Malo, S., Lo, M., Ouaro, S.: Towards meningitis ontology for the annotation of text corpora. In: Kebe, M.F., Gueye, C., Ndiaye, A. (Eds.) Innovation and interdisciplinary solutions for underserved areas. CNRIA 2017, InterSol 2017. Lecture Notes of the Institute for Computer Sciences, Social Informatics and Telecommunications Engineering, vol. 204. Springer, Cham (2018)

The Effect of Data Transmission and Storage Security Between Device–Cloudlet Communication

Nhlakanipho C. Fakude, Ayoturi T. Akinola and Mathew O. Adigun

Abstract With the popularity of mobile cloud computing and edge computing, the deployment of cloudlet technology in these modern computing paradigms has been affected by security issues. A cloudlet helps mobile devices in accessing the cloud by enhancing the processing time, lowering latency, and enabling better service time but it does not provide enough security assurance to its consumers. However, several studies have been conducted in this field and most of the proposed models mainly focus on either data transmission or data storage between the communication of mobile devices and cloudlets. In this study, we proposed a security model which addresses both transmission and storage of data, by protecting the information that is shared between the mobile devices and the cloudlet resources which is invariably linked to the cloud. The study considered two scenarios for the security model: Three-tier architecture and a Three-tier enhanced with Edge Orchestrator (EO) architecture. The security model used an encryption algorithm called hybrid cryptosystem to provide an efficient security solution which secures transmission and storage of data by addressing three security parameters: confidentiality, data integrity, and non-repudiation. EdgeCloudSim was used to implement and evaluate the proposed security model. The results obtained showed that the proposed security model between mobile devices and cloudlets communication performs better in Three-tier enhanced with EO architecture when considering security and performance metrics such as processing time, service time, and network delay.

Keywords Cloudlet · Security model · Hybrid cryptosystem · Edge orchestrator · Three-tier architecture

N. C. Fakude (✉) · A. T. Akinola · M. O. Adigun
Department of Computer Science, University of Zululand, Private Bag X1001,
Kwadlangezwa 3886, South Africa
e-mail: ncfakude30@gmail.com

© Springer Nature Singapore Pte Ltd. 2020
X.-S. Yang et al. (eds.), *Fourth International Congress on Information and Communication Technology*, Advances in Intelligent Systems and Computing 1027, https://doi.org/10.1007/978-981-32-9343-4_24

1 Introduction

Edge computing is a distributed computing paradigm in which computing is largely or completely performed on distributed device nodes know as smart devices or edge devices. This is typically against the norms of computing processes taking place in a centralized cloud computing. A cloudlet thus becomes a mobility-enhanced small-scale cloud data center that is located at the edge of the Internet. The cloudlet technology aims to add support to mobile applications that are interactive and intensive to resources, such as an augmented reality application which requires less end-to-end latencies [1]. However, literature has opened up several challenges that are confronting the use of cloudlet technology today among which are the concern about the security of the transmitted data between the end-users and the cloudlet [2]. Organizations utilizing edge computing allow end-users-cloudlets transmission and store confidential data to the cloud; hence, these organizations suffer from the lack of proper and efficient security measures for securing users data. Authors emphasized the importance of confidentiality, data integrity, and non-repudiation as a yardstick for any system to be considered secured [3, 4]. This was born out of the fact that several malicious programs often gain unauthorized access to the transmitted data, thereby resulting in loss of control of the users' data [5]. In the context of a cloudlet, there is need for a strong security measure on the communication between the end-users and the cloudlet; hence, this is the challenge in cloudlet deployment that will be addressed in this paper [6].

In this work, hybrid (encryption algorithm) cryptosystem which is the combination of both the symmetric and asymmetric encryption algorithms will be used on the model [7]. The hybrid encryption algorithm will address how confidentiality, data integrity, and non-repudiation are achieved. The security model will be evaluated for performance using four applications: augmented reality, health, infotainment, and heavy computation applications.

The rest of the paper is organized as follows: Section 2 presented a brief description of the related work. A brief overview of the implemented model was presented in Sect. 3. Section 4 discussed the background of the architectures (scenarios) that were used in the security model implementation. Section 5 contains the experiment and results. Finally, the conclusions and the future works were provided in Sect. 6.

2 Related Work

In [8], the authors identified mobile cloud computing (MCC) security challenges by the use of 3G/LTE and proposed a secure MCC model based on the cloudlet concept where the mobile users can directly connect to the cloud. The proposed cloudlet-based model can be used in many applications where security is required. The simulation results of their model showed that it is more efficient and reliable than other mobile cloud computing models that do not use a cloudlet. But the proposed work did not provide a well-secured communication for cloudlet-based model to guarantee users' confidence toward using the cloudlet.

In the article [9], the authors described a security challenge which explained that mobile users lose control of their information while utilizing cloudlet technology. Hence, a data security protocol was proposed for a distributed cloud architecture having cloudlet integrated with the base station. Their protocol does not only protect data from any unauthorized user but also prevented unnecessary exposure. However, they lacked the security on transmission which this paper seeks to address.

In [10], the authors discussed the sharing of critical medical data and the challenging security measures. Thus, the authors built a novel healthcare system utilizing the flexibility of the cloudlet platform. The number theory research unit (NTRU) method was used to encrypt data collected from wearable devices. But the trust model used does not cater for non-medical data (e.g. videos, images, etc.). Thus, the considered literature has not fully addressed the challenge of secured transmission and storage in a cloudlet deployed environment.

Moreover, In [11], the authors proposed a security solution by using hybrid encryption algorithm. The authors enhanced RSA asymmetric encryption algorithm by increasing the length of the key so that it can generate big primes. The authors further merged AES symmetric encryption algorithm with the enhanced RSA algorithm. But this study only considered lightweight data on the cloud storage service. The study did not address the security of scalable data and also the deployment of the security solution to a cloudlet.

In the mentioned studies, the authors designed security models that address either data transmission or data storage security and the security models did not address clear and direct integration into a cloudlet environment. Hence, all these studies lack the design of a model that addresses both transmission and storage of data in a cloudlet environment where mobile devices can communicate with the cloudlet using encryption algorithms that are publicly proven to be secured. In this study, a security model is designed using hybrid encryption algorithm which has proven to be a secure communication and data encryption tool between any communicating entities based on our results.

3 Proposed Security Model

This section provides a brief overview of the proposed security model, which will secure data in transit and at rest while utilizing hybrid encryption algorithm. The notations in Table 1 were used in defining the encryption process flow in Figs. 1, 2, and 3. The proposed model includes the following components which are depicted in Fig. 1.

- Client (Mobile devices)
- Cloudlet
- Cloud.

Table 1 Security model notations

Notation	Description
PKc	Public key of client
PrKc	Private key of client
PKcl	Public key of cloudlet
PrKcl	Private key of cloudlet
Hash	The result of a hashing process

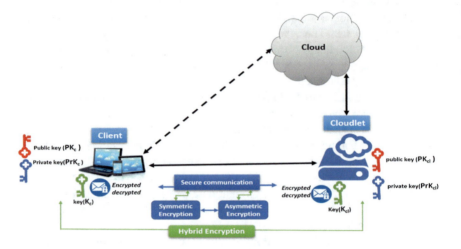

Fig. 1 Client-Cloudlet security model

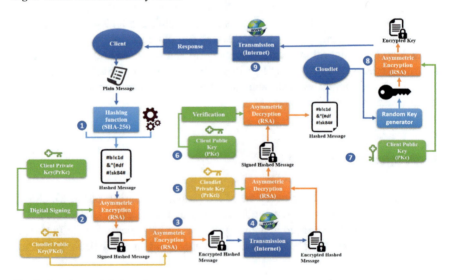

Fig. 2 Security model phase 1 processes

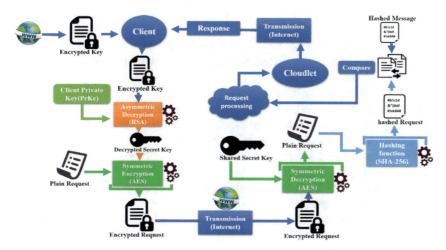

Fig. 3 Security model phase 2 processes

The cloudlet acts as an intermediary for the client and the cloud. Moreover, the client–cloudlet communication consists of a trusted management system which enables only eligible clients to access the information/data. This is achieved by the use of hybrid cryptosystem which is the combination of both the asymmetric and the symmetric encryption algorithms. The symmetric encryption algorithm used in this study is the Advanced Encryption Standard (AES) [5, 6, 11]. This encryption algorithm was used because it enables fast encryption and decryption processing which is ideal for mobile devices as these have less processing power. The asymmetric encryption algorithm used in this study is Rivest–Shamir–Adleman (RSA) [5, 6, 11]. This algorithm allows devices with limited resources to relatively perform key exchange processing efficiently.

Having understood from the literature that a secured system should address the following security properties: confidentiality, data integrity, and non-repudiation [3, 4], thus our study proposes a model that address all these major properties. The processes are being described below and in addition, the description of each of the security properties was shown.

3.1 Data Transmission

In Fig. 2, the client establishes a connection with the cloudlet in the following manner:

Message Signing and Verification

- The client will first generate a *hash* from the data to be transmitted using an SHA-256 hashing algorithm, and then encrypt the *hash* with the private key (**PrKc**) using RSA to get a ciphertext which is a signed message (*hash*).
- The signed message (*hash*) will then be encrypted with the cloudlet's public key (**PKcl**) to get a ciphertext which is the encrypted *hash*.
- The encrypted *hash* is then sent to the cloudlet by the client over the Internet.
- The client waits for a response from the cloudlet.
- Once the encrypted *hash* is received, the cloudlet will decrypt it using its public key (**PKcl**) to get a ciphertext.
- The ciphertext will be decrypted using the client's public key (**PKc**) to get the signed message (*hash*).
- The cloudlet will validate the authenticity of the message by using the client's public key (**PKc**).
- This *hash* is kept for later use.

Key Exchange (RSA)

- Once the cloudlet has verified the message and is certain that the *hash* was sent by the client, the cloudlet will generate a random password (key).
- The random password (key) is then encrypted using the verified client's public key (**PKc**) to get a ciphertext.
- The ciphertext is then sent as a response over the Internet to the client who is waiting.
- Once the encrypted password (key) is received, the client will decrypt it using the private key (**PrKc**) to get a plain password (key).
- Data Encryption (AES).

Data Encryption (AES)

- Now, that the client is certain that it is communicating with the cloudlet and also has the password sent by the cloudlet, data encryption using AES takes place.
- The encrypted data (ciphertext) is then sent to the cloudlet.
- Once the cloudlet receives the encrypted data, it can easily decrypt it using the shared password (key) to get the actual data.
- Encryption is the process of transforming a message using a key so that anyone viewing the message without the key view the message.

Confidentiality

Confidentiality is achieved by the key exchange (RSA) process and the data encryption (AES). Basically, confidentiality is achieved by the encryption of messages using a secret key so that anyone viewing the message without the key cannot determine its contents.

Data Integrity

After the above processes both entities (client and cloudlet) have a secure communication channel, but to verify that the data sent by the client has not been tampered with, the cloudlet will firstly generate a hash from the decrypted data to get hash2. Since the cloudlet kept the hash received from the client, the cloudlet will compare the hash with hash2 to validate that the data has not be altered.

Non-repudiation

In the message signing and verification processes, the cloudlet verifies the authenticity of the client. Thus, the hash has been signed with the client's private key (PrKc), it can only be decrypted with the client's public key (PKc); hence, the cloudlet is sure that the message was sent by the client and the client cannot deny that they have sent the message.

The same procedure will be taken in order for cloudlet to establish a secure connection with the client.

3.2 Data Storage

The transmitted data is stored on the cloud via the cloudlet. The following steps will take place for the storage of encrypted data:

- The cloudlet will send the encrypted data to the cloud for storage using the same transmission process as described in the data transmission section.
- On receiving the response, the cloudlet will pass it to the client.

The reason for not allowing the client to send a request directly to the cloud is to maximize the performance, i.e. it is important to delegate the cloudlet to perform the complex tasks and return less heavy responses to the client. Thus, the hybrid processes described in the data transmission section above will increase the confidence in security measures between the communicating entities. However, the average time of deploying hybrid encryption algorithm is slightly higher than symmetric encryption algorithm; this is because both symmetric and asymmetric encryption algorithms are used simultaneously, hence the time required to compute both is higher than symmetric encryption algorithm alone [11].

4 Overview of Architectures

A Three-tier architecture in edge computing is a scenario in which a local tier of mobile devices(nodes) and a middle tier(cloudlets) of nearby computing nodes usually located at the edge of the network where there is a limited amount of resources, and a remote tier of distant cloud servers, which typically have infinite resources. This architecture has the benefits of computation offloading from mobile devices

Table 2 Application specifications

Parameters	Augmented reality	Health	Heavy comp	Infotainment
Usage percentage (%)	30	20	20	30
Prob cloud selection	20	20	40	15
Delay sensitivity (s)	0.90	0.05	0.15	0.5
Active period (s)	45	10	60	15
Idle period (s)	15	20	60	45
Data upload (KB)	1500	1250	2500	25
Data download (KB)	25	20	250	2000

to external servers while limiting the use of the cloud whose higher latency could negatively impact the user experience.

In this study, EdgeCloudSim simulator was used to implement and evaluate the performance of the security model [12]. Two scenarios were used: Three-tier architecture and Three-tier architecture enhanced with edge orchestrator, and the specifications are described below. Four applications: augmented reality app, health app, infotainment app, and heavy computation app were also used to evaluate the performance of the model on different environments. The specifications of the applications are described in Table 2. Moreover, the client–cloudlet communication consists of a trust management system which enables only eligible clients to access the information/data.

This is achieved by the use of hybrid cryptosystem which is the combination of both the asymmetric and the symmetric encryption algorithms. To demonstrate the capabilities of the proposed security model, EdgeCloudSim is used because it allows an experimental setup based on different edge architectures and the simulations show the effect of the computational and networking system parameters on the results of the demonstrated model [12].

The security model has been implemented on both the Three-tier architecture (Fig. 4) and the Three-tier enhanced with edge orchestrator (Fig. 5). The edge orchestrator holds an entire overview of the network between the clients and the cloudlets including the cloud in the system [12]. Hence, it is the decision maker of the whole system, i.e. it decides how and where to handle incoming client requests. Therefore, the security model was implemented on this platform. The specifications of the simulation parameters are depicted in Table 2.

5 Experiments and Results

To evaluate the performance of the security model, augmented reality, health, infotainment, and heavy computation apps were deployed into the edge computing setups; Three-tier architecture and Three-tier enhanced with edge orchestrator. The performance metrics considered were: processing time, service time, and network delay.

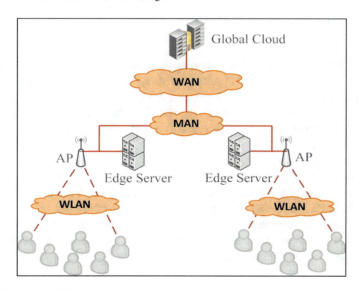

Fig. 4 Three-tier architecture

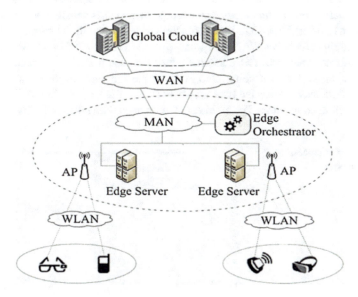

Fig. 5 Three-tier with EO

5.1 Applications

As mentioned in the above sections, four applications are used to evaluate the performance of the security. The four applications are specified in Table 2. From the table, the applications are chosen based on requirements of resources. The heavy computational application requires a lot of resources and the health application requires fewer computation resources as compared to the other applications.

Processing Time
The average performance time is the combination of the processing times of augmented reality, health, infotainment, and heavy computation applications. From Fig. 6, the simulator shows that three-tier with edge orchestrator processes better than the three-tier architecture. As mentioned in the previous section, better processing time is achieved because this architecture is able to distribute work to other cloudlets that are free; hence, there is less processing on the cloudlets.

Service Time
In Fig. 7, the average service time performance is shown with respect to the average task size parameter. In this study, the average task size is selected as 250 million instructions (MI) and 4000 MI. Since both the Three-tier and the Three-tier enhanced with EO architectures has WAN delay, they provide a relatively low performance but Three-tier enhanced with EO performs better because it is able to handle a high number of tasks due to it being able to distribute the work to different cloudlets (servers). The average service time increases due to the increase in mobile devices which result in congestion. The architecture enhanced with edge orchestrator is able

Fig. 6 Average processing time on cloudlet

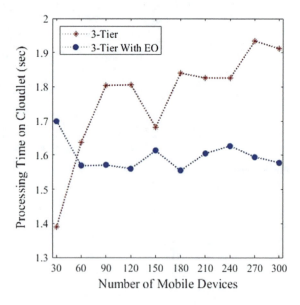

Fig. 7 Average service time on cloudlet

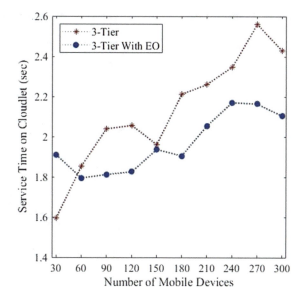

to handle the congestion because it can distribute the client requests to the cloudlets, so the service time is not increasing as compared to the three-tier architecture.

Network Delay Time
In Fig. 6, the average network delay includes both the WAN delay and the LAN delay. If the number of mobile devices is low, the three-tier architecture provides better delay performance because there is no congestion in the network. In the Three-tier enhanced with edge orchestrator architecture, the delay grows faster than the Three-tier architecture due to work distribution. Since each and every cloudlet is connected using WAN the distribution of tasks between the cloudlets results in an increasing network delay time (Fig. 8).

6 Conclusions and Future Work

In this paper, we proposed a security model which secures data transmission and data storage between mobile devices and cloudlets. The model takes the advantages of the hybrid encryption algorithm which combines the asymmetric and symmetric encryption algorithms. The security model addressed the three basic security properties (confidentiality, data integrity, and non-repudiation) in detail and from the simulation results, this work demonstrated that in an edge computing environment, the three security properties can be achieved while improving the performance of a secured edge computing setup. Hence, our security model showed that a secured edge computing environment is better implemented in an architecture enhanced with an edge orchestrator.

Fig. 8 Average network delay

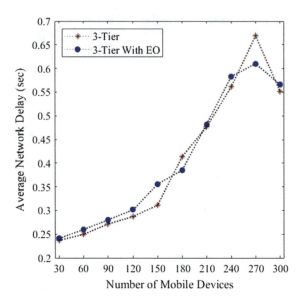

In the future work, it would be interesting to add availability as one of the security properties. Availability enables mobile devices can be able to access data from anywhere in the network, i.e. the cloudlets should interact with one another in order to hold an overview of the network, hence making them access any data from the other cloudlets. This can be achieved by hand-off. Hand-off will enable the mobile devices to move to any position while having the same state on the network.

References

1. Schneider, M., Rambach, J., Stricker, D.: Augmented reality based on edge computing using the example of remote live support. In: 2017 IEEE International Conference on Industrial Technology, vol. March, pp. 1277–1282 (2017)
2. Tawalbeh, L., Jararweh, Y., Dosari, F.: Large scale cloudlets deployment for efficient mobile cloud computing. J. Netw. **10**(1), 70–76 (2015)
3. Brauer, K.: Authentication and Security Aspects in an International Multi-user Network, pp. 3–5 (2011)
4. Hansen, E.: Analysis of design for implementing confidentiality, integrity, authentication, and non–repudiation solutions. SANS Inst. 2003, no. Security 401, pp. 1–39 (2003)
5. Zhou, Z., Huang, D.: Efficient and secure data storage operations for mobile cloud computing. In: 8th International Conference on Network and Service Management (CNSM), pp. 37–45 (2012)
6. Gupta, S., Sharma, J.: A hybrid encryption algorithm based on RSA and Diffie-Hellman. In: 2012 IEEE International Conference on Computational Intelligence and Computing Research (ICCIC 2012), pp. 5–8 (2012)
7. Sharma, M., Sharma, V.: A hybrid cryptosystem approach for file security by using merging mechanism. In: Proceedings 2016 2nd International Conference on Applied and Theoretical Computing and Communication Technology (iCATccT 2016), pp. 713–717 (2017)

8. Tawalbeh, L.A., Bakheder, W., Mehmood, R., Song, H.: Cloudlet-based mobile cloud computing for healthcare applications. no. Mcc (2016)

9. Jindal, M., Dave, M.: Data security protocol for cloudlet based architecture. In: IEEE Icraie, pp. 8–12 (2014)

10. Chen, M., Qian, Y., Chen, J., Hwang, K., Mao, S., Hu, L.: Privacy protection and intrusion avoidance for cloudlet-based medical data sharing. IEEE Trans. Cloud Comput. **XX**(c), 1–1 (2016)

11. Liang, C., Ye, N., Malekian, R., Wang, R.: The hybrid encryption algorithm of lightweight data in cloud storage. In: 2nd International Symposium on Agent, Multi-Agent Systems and Robotics (ISAMSR 2016), August, pp. 160–166 (2017)

12. Sonmez, C., Ozgovde, A., Ersoy, C.: EdgeCloudSim: an environment for performance evaluation of edge computing systems. EdgeCloudSim An Environ. Perform. Eval. Edge Comput. Syst. pp. 39–44 (2017)

Slang-Based Text Sentiment Analysis in Instagram

Elton Shah Aly and Dustin Terence van der Haar

Abstract A large amount of user-generated content on social media has led to the pursuit of quickly and accurately mining through data and gathering useful insights. Text sentiment analysis has become a necessary tool in classifying user opinions within Web generated content. Due to the various ways, opinions can be conveyed, performing text sentiment analysis in specific domains becomes a difficult task. With an even greater degree of difficulty added when slang or colloquialisms are used. There is a great deal of research into investigating various classifiers in a traditional natural language processing setting each with their own merits and demerits. In this paper, we present a slang-based dictionary classifier with the objective of determining the sentiment of Instagram comments within the context of fashion, or more specifically sports shoes, and compare it with the performance of other classifiers such as a Naive Bayes, J48, lexicon and random forest. The dataset used for the benchmark was created from popular fashion Instagram accounts. Overall, the random forest classifier yields the best results with an accuracy of 88%, precision of 84% and a recall of 88%.

Keywords Sentiment recognition · Natural language processing · Machine learning

1 Introduction

Opinion mining, also known as text sentiment analysis, has many practical applications. A few examples include election forecasting, product pricing and perhaps the most relevant to this paper competitive intelligence [1]. The average consumers

E. S. Aly · D. T. van der Haar (✉)
Academy of Computer Science and Software Engineering, University of Johannesburg, Gauteng, South Africa
e-mail: dvanderhaar@uj.ac.za

E. S. Aly
e-mail: eltonaly@gmail.com

© Springer Nature Singapore Pte Ltd. 2020
X.-S. Yang et al. (eds.), *Fourth International Congress on Information and Communication Technology*, Advances in Intelligent Systems and Computing 1027, https://doi.org/10.1007/978-981-32-9343-4_25

purchasing process is guided by various variables, some of which organisations cannot control, and these include culture, social, personal and psychological [2]. The purpose of this paper is in determining if comments gathered from Instagram on multiple images in the environment of fashion expresses a subjective opinion and how strongly, positive or negative, the expressed subjective opinion is towards the brand and the sneaker, the image represents.

The focus is on basketball brands dealing mainly with sneakers and the comments of sneaker fans, more commonly known as sneakerheads, taking into account their culture and way of speaking to determine their interest or disinterest in the different sneakers released by different sneaker brands through the use of text sentiment analysis. By implementing a slang-based dictionary classifier and other classifiers such as a Naive Bayes classifier, we attempt to determine the sentiment of comments, whilst also comparing the results of each classifier.

Nike, Adidas, Under Armour and other companies have signed various lucrative deals with fan favourite basketball players such as Michael Jordan (considered one of the greatest ever to play the game) with a 110 million dollar Nike deal. Illustrating how much of an influence the culture of basketball has on its fans and their purchasing patterns when it comes to shoes. For the average consumer, sneakers are used for a specific sporting reason [3]. Sneakerheads or rather individuals with a great love for wearing, collecting and often reselling sneakers are the opposite. To them, sneakers are used as a fashion statement [3]. Most sneakerheads are identified as being big sports fans often falling under the category of long-time basketball fans [3]. The word sneaker is slang for athletic shoes used for fashion as opposed to sport. Throughout this paper, athletic shoes will be referred to as sneakers.

2 Problem Background

In recent times, the use of the Internet has become a crucial and ever so prominent part of our lives. Social networks, blogs and forums are a major part of the Internet. The immense popularity of social media has created an environment where its users now generate a large amounts of content daily, which can be of value in many contexts [4]. The generation of such volumes of user-generated content has sparked research into extracting and understanding this content in an attempt to better understand a users biases, moods, interests and more regarding specific topics [5].

Such content can now be used to keep track of and determine peoples opinions about other people (presidential candidates), products and services, thus making it a valuable asset to organisations and academics [4]. As a result, sentiment analysis on certain platforms such as Twitter has become a well-studied area since its creation with both the academic and business worlds spending a great number of resources to understand the opinions of Twitter users [6].

2.1 Text Sentiment Analysis

Sentiment analysis or opinion mining is a field of study related to the gleaning and evaluating of text in order to determine the emotions and opinions within said text [5]. Sentiment analysis is done in multiple languages with most works focusing on English [5]. Some languages are more complex than others thus posing a greater challenge. A good example is the Arabic language: it has no capital letters, a high number of words have more than one meaning and Arabic consists of many variants regarding its typographic form [5].

Most text sentiment analysis techniques can be classified into two approaches, namely supervised and unsupervised [5]. For the supervised approach, a dataset is separated into a test set and a training set. The machine-learning algorithm then learns from the separated training data and builds a model which it then compares with the test data to determine its accuracy [7]. A model is the implementation of an algorithm for recognising a pattern [8]. Singh et al. in [4] show the typical machine-learning process which often yields higher accuracy than unsupervised sentiment analysis, but it requires a large dataset, expertise, is resource intensive and time-consuming [8]. Naive Bayes classifier, maximum entropy classifier and support vector machine (SVM) are the most explored machine-learning classifiers for machine-learning approaches in text sentiment analysis [1].

The second approach, lexicon-based (unsupervised) approach determines the polarity of a word or sentence based on a pre-defined dictionary, i.e. lexicon [7]. In this lexicon, each word is associated with a value. These values can be either be between +1, −1 or 0 for positive, negative and neutral, respectively, or they can be ranges such that a word with a value of +4 is much more positive than a word with the value of +1 [7]. The lexicon is created either manually or automatically. The polarity of the text/corpus is determined by adding the score of each word which is derived from the lexicon. This approach does not do well in different domains but is considered practical [7]. Due to the constantly changing English language, the unstructured way people communicate in different environments, and the different cultures, many approaches have emerged to perform text sentiment analysis [4].

2.2 Related Work

In Naidu et al. sentiment analysis is achieved using the Telugu SentiWordNet [9]. The authors use a two-step sentiment analysis process. The first step being subjectivity analysis, where a sentence is classified as either objective or subjective. The objective sentences are seen as neutral. Secondly, the subjective sentence is classified as either positive or negative.

Abdelhameed uses SentiWordNet as a tool for sentiment analysis. SentiWordNet is a sentiment lexicon, which is some form of a dictionary or language model. It has four files each containing positive, negative, neutral and ambiguous words. The words in each file are further split into five parts of speech, adjective, noun, verb, adverb and unknown. The neutral words file are used for subjectivity and objectivity classification, and the positive and negative files for sentiment classification [5]. The proposed system obtained an accuracy of 74% for subjectivity and objectivity classification and 81% for sentiment classification in the news environment. The authors concluded that the result is completely based on the quality of the SentiWordNet [5]. The greater the volume of words covered by the lexicon, the greater the outcome.

In [7] a similar lexicon-based approach is used to determine sentiment in an Arabic language, but with further pre-processing, which normalises certain characters that Internet users use. This approach also noted that the accuracy of the results increases with the growth of the lexicon, but the percentage of growth starts to decline the bigger the lexicon grows [5].

Finally, in [10] the authors implement a Naive Bayes and SVM classifier and compare their results with that of a lexicon-based approach implemented using SentiStrength. The Lexicon approach is easy to implement but is often very domain specific and thus fails when tried in a different domain [10]. Overall the supervised approaches performed better than that of the unsupervised approach [10]. However, all of the above work does not explore the impact of these results when colloquialism or slang is included in the text or corpus.

3   Experiment Setup

The Instagram API is used to pull sneaker images and their respective comments from Instagram. All comments including training and testing data are gathered from a popular Instagram profile known as Sneakernews. The API limits the number of comments the system can pull to 150 at a time, and due to a policy update, further restrictions are placed on certain media, such as the images. Furthermore, the API only grants access to profiles of which the username and password can be supplied, so a test profile was created on which comments from Sneakernews were posted. The system was tested on a total of 398 comments and trained each supervised classifier (Naive Bayes, J48 decisions tree and random forest model) with a total of 803 comments. These comments were manually chosen and gathered from Instagram as there are no openly available datasets specific to slang and basketball sneakers (based on research done).

For the lexicon-based classifier, a combination of dictionaries is used. A word list was downloaded from https://github.com/ms8r/mpqa/blob/master/subjclues.tff and then further adjusted to suit the domain better to form 8222 words each labelled with their polarity (negative, positive) and subjectivity strength. Along with this, a bigram dictionary of 118 bigrams, a trigram dictionary of 85 trigrams and a slang dictionary of 162 words were used in combination with the list mentioned above to determine sentiment. Finally, a dictionary of common design words (such as colourways, sole, and laces), common buying words and common brands are used to gather some insights about the text.

Keeping the lexicons up-to-date with relevant slang words proved to be a challenge as new slang words emerged quite often. The same time and effort were dedicated to gathering training data and the pulling of comments from the private Sneakernews profile, to the created test profile for testing purposes as both processes are done manually.

The built system is shown in Fig. 1. Initially, the training data goes through some pre-processing which consists of the removal of stop words, removal of punctuation and emoticons, normalisation and spell checking. Stop words consist of words that do not add much or any meaning to the comment. Punctuation and emoticons are removed to reduce some of the complexity in text processing. The process of normalisation takes exaggerated words such as "helllllloooooo" or "baaaaaaadd" and changes them to "hello" and "bad" respectively. Finally, a spell check, which takes slang into account, is carried out due to the context.

The processed training data is then transformed from text to a set of vectors in order to train the Naive Bayes, J48 decision tree and the random forest models. Word2Vec is used to change each comment to a set of three vectors and StringToWordVector changes each comment to a set of vectors whose size depends on the length of the comment. The output of the Word2Vec process is then passed through a min–max normalisation process which maps the vector values between a maximum and minimum. The normalised values of the Word2Vec are used to train the Naive Bayes model, and the result of the StringToWordVector is used to train the J48 and random forest models.

Next, the Instagram comments go through the same pre-processing as the training data, and some feature extraction follows. Feature extraction consists of parts of speech, unigrams, bigrams, trigrams and the transformation of the comments from text to vectors using Word2Vec and StringToWordVector. Finally, the comments are classified as being either positive or negative based on the score given by the lexicon classifier and the probabilities given by the Naive, J48 and random forest classifiers.

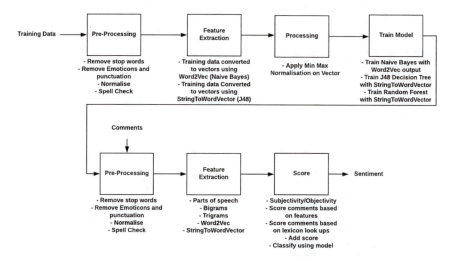

Fig. 1 The model for achieving slang-based text sentiment analysis for Instagram comments

Apaches OpenNLP is used for POS tagging, Apache Spark machine-learning library is used to implement the Naive Bayes classifier and Weka (collection of machine-learning algorithms) is used to implement the J48 decision tree and random forest.

4 Benchmark and Results

The performance of the four classifiers is compared using precision, recall, specificity and accuracy, along with each respective ROC curve. A classifiers performance is often determined by: specificity (TN/(TN + FP)), sensitivity also known as recall (TP/(TP + FN)), precision (TP/(TP + FP)) and the negative predictive value (TN/(TN + FN)) [11]. Specificity measures the false positive rate (FPR), sensitivity measures the true positive rate (TPR), precision measures how often does the classifier correctly predict a positive and the negative predictive value measures how often does it predict a negative. The receiver operating characteristics(ROC) analysis is based on the relationship between the specificity (FPR) and the sensitivity (TPR) of a binary classifier [12]. The decision threshold of a classifier is the value above or below predicted scores are considered positive or negative [11]. Decreasing the decision threshold yields an increase in the false and positive rates. Each FPR and TPR for that specific decision threshold then forms a point to plot the curve where TPR is the y-coordinate and FPR the x-coordinate [12]. This curve can then be used to determine a threshold with the minimum misclassification and allows for comparison with other classifiers by facilitating the highlighting of regions.

The lexicon classifier was the first to be implemented. In addition to being compared with the results of the other classifiers, it was also compared with a lexicon version without slang dictionaries. The classifier with the use of slang dictionaries classified 56% of the comments as negative and 44% as positive and without the use of slang dictionaries the classifier classified 84% as negative and 16% as positive. The classifier without the use of slang dictionaries tends to classify more comments as negative. This result is to be expected as many slang words, which are considered positive are considered negative in standard English.

Overall the random forest tree has the highest performance with the lexicon classifier in a close second, the decision tree in third and the Naive Bayes last as shown by Table 1 and Fig. 2. The lexicon-based classifier is the simplest to adjust as it often

Table 1 A table comparing the evaluation metrics for slang-based text sentiment analysis on the Instagram dataset

Classifier	Precision (%)	Recall (%)	Specificity (%)	Accuracy (%)
Naive bayes	0	0	100	42
J48	66	93	34	68
Lexicon	85	64	84	72
Random forest	84	88	77	83

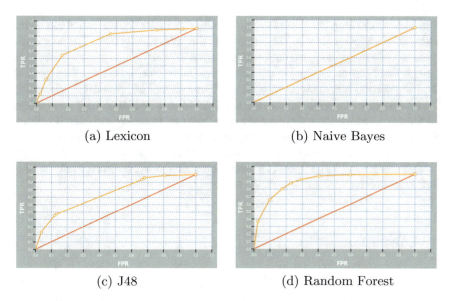

(a) Lexicon (b) Naive Bayes

(c) J48 (d) Random Forest

Fig. 2 A comparison of all the study's ROC curves

required a simple addition, removal and/or adjustment of rules for the dictionaries. The other classifiers require more attention to detail, and the results after tinkering can often be surprising or non-existent.

A classifier can achieve a high recall (also known as sensitivity) by always returning positive, a high specificity by always returning negative and a high precision by returning positive only on what it is most certain of [13]. These characteristics are why more than just accuracy is taken into account. Evaluating Table 1 shows that the Naive classifier, with a specificity of 100%, always classifies a comment as negative and only achieves an accuracy of 42% as a result of 42% of the comments being negative. In a case where the dataset contained mostly negatives, it would yield a high accuracy even though the classifier is extremely poor overall. The lexicon and the random forest classifiers have a more balanced set of metrics and as a result, achieve more reasonable accuracy. With this being said the lexicon still leans more to classifying comments as negative, whereas the random forest has a better balance between sensitivity (recall) and specificity. Lastly, the J48 classifier has a high recall, meaning that it has a high tendency to classify comments as positive and as a result comments classified as negative are likely to be a true negative.

The Naive, random forest and J48 models are all trained on the same dataset. The decision tree models yielded better results and trained much quicker than the Naive model. In an attempt to solve the poor performance of the Naive Bayes, more training data was added, multiple training/tests splits were tested and multiple Word2Vec vector sizes were tested. These steps showed no improvement in the results and led to the conclusion that the Word2Vec is the issue as it is the only major difference between the training of the Naive and the decision tree model. To remedy this, instead of using the Word2Vec as a feature extractor for the Naive model the StringToWordVector is used. The results after this were equally poor. The poor result is a result of manipulating the output of the StringToWordVector to feature vectors of a single fixed size as required to train the Naive Bayes model. Alternatively, pruning the J48 decision tree showed an accuracy improvement of around 4%. Increasing the number of trees in the random forest model from an initial of 10–15 and a final 20 showed a significant improvement in the classifiers accuracy and overall metrics. Finally, first additions to the slang dictionaries resulted in a significant improvement in the lexicon classifier which then began to decrease as the number of words in the dictionaries grew. The decrease is a result of certain clashes between words causing further ambiguity.

5 Conclusion

Four classifiers are implemented to classify heavily slang-based comments and evaluated. The lexicon classifier and the random forest yielded the best results and improvements after adjustments. The negative of the lexicon classifier being that with its dictionaries being so finely tuned for this specific environment, and it would perform poorly in any other environment. The J48 and random forest classifiers required the least effort to implement, and model training is relatively fast. The Naive classifier showed to be the hardest of the four to implement and yielded the worst results. The poor performance of the J48 and Naive Bayes model is likely due to a small set of training data and poor feature extraction mechanisms.

Granted suitable results are achieved, and these classifiers could prove beneficial to sneaker companies in gauging their online popularity within the young demographic. Evaluating the success of the design of a sneaker, may also be facilitated with the results from this system of classifiers.

References

1. Sun, S., Luo, C., Chen, J.: A review of natural language processing techniques for opinion mining systems. Inf. Fusion **36**, 10–25 (2017)
2. Sub, K., Adamu, E., Afewerk, F.: Factors affecting consumers shoe preference: the case of addis (2011)
3. McCracken, A., Dong, H., Murphy, C., Hoyt, M., Niehm, L.: The stories that come with the shoe: a qualitative study of male sneaker collector motivations, experiences, and identities (2016)
4. Singh, J., Singh, G., Singh, R.: A review of sentiment analysis techniques for opinionated web text. CSI Trans. ICT **4**(2–4), 241–247 (2016)
5. Abdelhameed, H.J., Muñoz-Hern'andez, S.: Emotion and opinion retrieval from social media in arabic language: Survey. In: 2017 Joint International Conference on Information and Communication Technologies for Education and Training and International Conference on Computing in Arabic (ICCA-TICET), pp. 1–8. IEEE (2017)

6. Arslan, Y., Birturk, A., Djumabaev, B., Küçük, D.: Real-time lexicon-based sentiment analysis experiments on twitter with a mild (more information, less data) approach. In: 2017 IEEE International Conference on, Big Data (Big Data), pp. 1892–1897. IEEE (2017)
7. Abdulla, N.A., Ahmed, N.A., Shehab, M.A., Al-Ayyoub, M.: Arabic sentiment analysis: Lexicon-based and corpus-based. In: 2013 IEEE Jordan Conference on Applied Electrical Engineering and Computing Technologies (AEECT), pp. 1–6. IEEE (2013)
8. Chappell, D.: Introducing azure machine learning. Sponsored by Microsoft Corporation, A Guide for Technical Professionals (2015)
9. Naidu, R., Bharti, S.K., Babu, K.S., Mohapatra, R.K.: Sentiment analysis using telugu senti-wordnet (2017)
10. Williams, G., Mahmoud, A.: Analyzing, classifying, and interpreting emotions in software users' tweets. In: Proceedings of the 2nd International Workshop on Emotion Awareness in Software Engineering, pp. 2–7. IEEE Press (2017)
11. Sammut, C., Webb, G.I.: Encyclopedia of Machine Learning and Data Mining. Springer Publishing Company, Incorporated (2017)
12. Flach, P.A.: Roc analysis. In: Encyclopedia of Machine Learning. Springer, pp. 869–875 (2011)
13. Powers, D.M.: Evaluation: from precision, recall and f-measure to roc, informedness, markedness and correlation (2011)

A Temporal Cognitive Model of the Influence of Methylphenidate (Ritalin) on Test Anxiety

Ilse Lelieveld, Gert-Jan Storre and S. Sahand Mohammadi Ziabari

Abstract This paper presents the connection between exam or test anxiety and the illicit use of Ritalin (methylphenidate). Ritalin is used by students as a cognitive enhancer and thus to increase their learning and memory abilities. However, a side effect of Ritalin has altered fear reactions to stimuli. Thus, this paper aims to gain insight in the mechanisms of Ritalin on the anxiety system. The created network visualizes the overlapping systems of Ritalin and anxiety and possible outcomes of the use of Ritalin around exam weeks. First, the exam is noticed, with the induced (normal) stress response on having this exam. The wish to increase learning abilities during exam weeks surfaces, and Ritalin is administered as a result of cognitive decision making. Finally, the model shows how Ritalin can decrease test anxiety.

Keywords Integrative temporal-causal network model · Biological · Affective · Cognitive · Stress · Caffeine

1 Introduction

In the past few years, an increase in the use of Ritalin is observed [17]. Ritalin or methylphenidate (MPH) is normally used by persons who have attention or concentration deficits to suppress attention deficiency, improve mood and decrease hyperactivity. However, Ritalin can also increase learning and concentration abilities in healthy persons [22].

I. Lelieveld · G.-J. Storre · S. Sahand Mohammadi Ziabari (✉)
Behavioural Informatics Group, Vrije Universiteit Amsterdam, Amsterdam,
The Netherlands
e-mail: sahandmohammadiziabari@gmail.com

I. Lelieveld
e-mail: Ilse.lelieveld@xs4all.nl

G.-J. Storre
e-mail: gjstorre@gmail.com

© Springer Nature Singapore Pte Ltd. 2020   331
X.-S. Yang et al. (eds.), *Fourth International Congress on Information
and Communication Technology*, Advances in Intelligent Systems
and Computing 1027, https://doi.org/10.1007/978-981-32-9343-4_26

In the past few years, the number of these illicit or non-prescribed Ritalin users has increased, and it is estimated that between the 5 and 35% of the students in the US, Canada and Australia has used Ritalin during their studies to enhance their cognitive abilities [40]. In the Netherlands, the IVM found 25% of the students has used or is using Ritalin to enhance their study abilities, which is a four-fold increase within a decade [5]. Additionally, Ritalin is used as party drug; however, this is less common than the use for study purposes [5].

Even though Ritalin seems to have a positive effect on learning and concentration abilities, the use of Ritalin has side effects. These include common side effects, such as headaches, nausea and sleep problems, but also less common, more serious problems: nervousness, depression and anxiety disorders. In the US, the number of emergency service visits for illicit Ritalin use had increased three times between 2005 and 2010 for persons between the 18–25 years [24]. This indicates a trend towards an increase in non-medical Ritalin use for young adults.

Since the age group between 18 and 25 is likely to study, with the current pressure of gaining high grades for exams and finishing a study on time, the likelihood of exam anxiety increases [39]. Supporting this, 33% of adolescents (15–20 years old) has experienced test anxiety [39]. Test anxiety could be caused by lack of preparation, fear of negative feedback or previous bad experiences. Furthermore, test anxiety can cause concentration problems, bad experiences, failure on exam, hereby increasing the chance on test anxiety during an exam [33]. Thus, the increase in test anxiety could explain why Ritalin is so popular within that age group [29], since Ritalin reduces the concentration and memory problems that exam anxiety would normally increase.

However, since there is little research done on what kind of effect non-medical Ritalin use has on the pathway that involves test anxiety, this research will look into the effect of methylphenidate on test anxiety, by creating simulations of a network model of the anxiety and methylphenidate pathway. In Sect. 2, the biological and neurological principles concerning test anxiety and the mechanisms of methylphenidate are explained. Section 3 discusses the network model based on the principles of Sect. 2. In Sect. 4, the results are shown and discussed, followed by a discussion in Sect. 5.

2   Underlying Biological and Neurological Principles

This research is based on the effects of methylphenidate and its function on the conditioned fear pathway. The conditioned fear pathway can be induced by factors such as stress and trauma and is strengthened by conditioning based on external stimuli that enhance the fear or anxiety stimuli [8]. In this case, the conditioned fear is exam anxiety. By anticipating a situation as stressful, an anxiety response can be triggered in the individual [8]. Anticipation of a stressful situation can induce the ventral tegmental area (VTA) to synthesize dopamine (DA) which plays a major role in acquiring conditioned fear [12]. The major role VTA plays in our network will be discussed in this chapter.

The VTA is a brain structure that consists of neurons that produce and transmit DA to a variety of brain structures which use dopamine for fear acquisition or regulation [27]. Targets of the VTA which are included in this research are the amygdala, the nucleus accumbens (Nacc), prefrontal cortex (PFC), the locus coeruleus (LC) and the hippocampus [11, 27]. The transmission of DA to the amygdala has an important function as the amygdala structure is responsible for the creation and expression of fear [14, 26]. The amygdala sends out DA to the ventral side of the hippocampus as well as the PFC and the Nacc. The linkage between the hippocampus and the amygdala is important for fear memory as the ventral side of the hippocampus works together with the amygdala to memorize current fear stimuli to a conditioned state [6]. The signal that the amygdala sends to the PFC is to induce fear expression but also for fear modulation [31]. In addition to this, the hippocampus is also directly stimulated by the VTA by DA transmission stimulation long-term memory [21]. The other part of the hippocampus, the dorsal part, is the part that is responsible for extinction of a fear memory and reduces activity of the ventral part of the hippocampus [23, 38].

The ventral part is thus inhibited by the dorsal part of the hippocampus and is excited by the amygdala. The ventral part of the hippocampus itself can excite the amygdala, the prefrontal cortex and the dorsal part of the hippocampus. The inhibiting function of the dorsal part of the hippocampus can be inhibited by the ventral part of the hippocampus [38]. The ventral part of the hippocampus can excite the amygdala in case of fear renewal after the previous conditioned fear stimuli were extinct [16]. In addition, it also has an Hebbian learning connection with the PFC where it modulates fear memory extinction [25]. The dorsal part of the hippocampus is stimulated by Norepinephrine (NE) which is synthesized by the LC [10, 18].

The LC plays an important role in both DA and NE synthesizes [18]. It activates the VTA which stimulates the VTA to synthesize DA [28]. The VTA also has a connection with the LC as earlier discussed, which activates the LC to make NE [28]. This NE then activates the dorsal part of the hippocampus as earlier discussed.

The VTA has two targets left in this model: the Nacc and the PFC. The Nacc is stimulated by the dopamine VTA that transmits to the Nacc [7, 20]. The Nacc is an important factor for the emotions that can modulate fear as the activation of Nacc is correlated with experiencing positive beliefs and feelings [37]. The Nacc can be inhibited by dopamine transmission from the amygdala but also acts as a modulator for the transmission from the amygdala [32]. The Nacc is also targeted by the ventral hippocampus which seems to be an inhibitor of the Nacc as malfunctioning ventral hippocampus makes a rat less susceptible to depression [4]. The Nacc stimulates the PFC with dopamine where PFC has a negative feedback to the Nacc reducing the available DA [15]. A major part of the model is the PFC and its functions. The PFC is also activated by the VTA directly with DA which triggers the memory extinction of learned fear and anxiety [1]. The PFC has a function in emotion regulation where it inhibits both the Nacc but also the amygdala in its fear response [19]. It also inhibits the LC which then makes less DA and NE [28]. On the other hand, the NE the PFC receives induces attention [3].

The effects of methylphenidate are caused by the blockage of dopamine trans-porters (DAT) and norepinephrine transporters (NET). The blockage of these trans-porters means inhibited reuptake of norepinephrine and dopamine into the synaptic cells resulting in an increase of free dopamine and norepinephrine in the synaptic cleft [36]. As can be seen in the model, methylphenidate (MPH) targets the VTA and LC which trigger more available dopamine and norepinephrine to target areas in the brain. This results in more DA in the PFC which increases attention [9]. In addition, it also leads to more NE available in the LC and thus the PFC which also triggers attention but also more NE for the dorsal hippocampus [18]. This abundance of NE in the hippocampus seems to reduce fear and anxiety [13], but also for the amygdala and the Nacc [41]. This seems to modulate fear which is in correlation with previous research [30].

3 The Temporal-Causal Network Model

Conceptual representation

This study shows a conceptual representation of a temporal-causal network model (Fig. 1) for the effect of MPH on test anxiety. The conceptual model presents states, connections between them and the impact on each other. These states can be activated on different times. Three important measures are taken into account in the conceptual representation. First of all, all connections have different *impact strengths,* since this resembles real life. Additionally, *states can receive multiple impacts*, which require a way to combine these multiple impacts and to calculate the total impact. Lastly, *speed of change* of a state is taken into account to represent the timing of the impacts. These three measures will be stated in the model as follows [34]:

- **Strength of a connection** $\omega_{X,Y}$: Each connection from a state X to a state Y has a *connection weight value* $\omega_{X,Y}$ representing the strength of the connection, often between 0 and 1, but sometimes also below 0 (negative effect) or above 1.
- **Combining multiple impacts on a state** $c_Y(..)$: For each state, a *combination function* $c_Y(..)$ is chosen to combine the causal impacts of other states on state Y. Combination functions are available in different forms and approaches. It specifies how impacts on a state are aggregated.
- **Speed of change of a state** η_Y: For each state Y, a speed factor η_Y is used to represent how fast a state is changing upon causal impact.

Figure 1 depicts the conceptual representation of the model. Table 1 contains explanations and information concerning Fig. 1.

Explanation of states

In this subsection, the states of the conceptual representation (Fig. 1) will be explained. The states ws_f and ws_c are the world states for fear and *contextual stimulus* c (the exam). ss_f and ss_c are the sensory states for fear and the contextual stimulus.

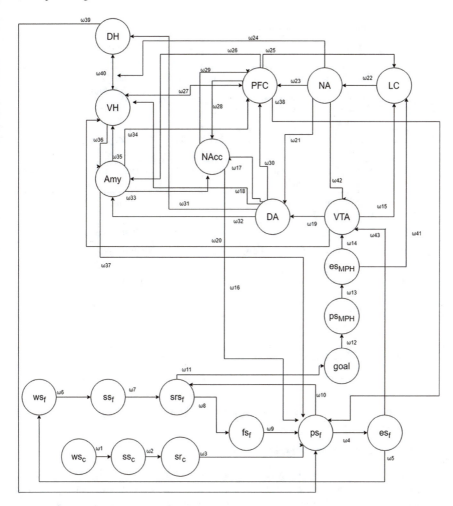

Fig. 1 Conceptual representation of the adaptive temporal-causal network model

The sensory representations for fear and the contextual stimulus are represented by sr_f and sr_c. fs_f represents feeling state of fear. Additionally, the preparation state (ps_f) for feeling fear is influenced by the feeling state of fear and the sensory representation of the contextual stimulus. The state ps_f influences the execution state of fear (es_f), which in its turn influences the world state of fear again (ws_f).

Finally, the sensory representation state leads to the goal of wanting to improve concentration in order to improve learning abilities to decrease fear. The goal leads to the preparation state for taking MPH (ps_{MPH}) and eventually the execution state of taking MPH (es_{MPH}). The states that follow on taking MPH are brain areas and neurotransmitters that are involved with the anxiety and fear pathway. MPH interacts with these areas and transmitters as well and can change the

Table 1 Explanation of the states in the model

X_1	ws$_f$	World state, fear	X_{11}	ps$_{MPH}$	Preparation state taking methylphenidate
X_2	ss$_f$	Sensor state, anxiety	X_{12}	es$_{MPH}$	Execution state taking methylphenidate
X_3	ws$_c$	World state contextual stimulus (exam)	X_{13}	LC	Locus coeruleus, brain area
X_4	ss$_c$	Sensory state contextual stimulus	X_{14}	NA	Norepinephrine, neurotransmitter/hormone
X_5	sr$_f$	Sensory representation contextual stimulus	X_{15}	PFC	Prefrontal cortex, brain area
X_6	srs$_c$	Sensory representation state, releasing anxiety	X_{16}	DA	Dopamine, neurotransmitter
X_7	fs$_f$	Feeling state, feeling fear	X_{17}	Nacc	Nucleus accumbens, brain area
X_8	ps$_f$	Preparation state for emotion, feeling fear	X_{18}	VTA	Ventral tegmental area, brain area
X_9	es$_f$	Execution state of extreme emotion, conditioned fear for exam	X_{19}	Amy	Amygdala, brain area
X_{10}	Goal	Improved learning	X_{20}	VH	Ventral hippocampus, brain area
X_{21}	DH	Dorsal hippocampus, brain area			

effect of the pathway. The interactions between brain areas and neurotransmitters are explained in Chap. 2.

Explanation connection weights

The connections in the lower part of the model are as follows: The states ss$_c$ and ss$_f$ have incoming connections from ws$_c$ and ws$_f$ with weights ω_1 and ω_6. This is followed by a connection to the sr$_c$ and sr$_f$ with connection weights ω_2 and ω_7. sr$_f$ continues to fs$_f$ with connection weight ω_8. The fear preparation state (ps$_f$) receives input from both fs$_f$ and sr$_c$ with connections ω_9 and ω_3, respectively. The preparation has a body loop back to srs$_f$ with connection ω_{10}. Additionally, it receives a input from the amygdala (ω_{37}), nucleus accumbens (ω_{16}), dorsal hippocampus (ω_{39}) and the prefrontal cortex (ω_{38}). The latter three are inhibiting inputs. The preparation state is connected towards es$_f$ via ω_4, after which in its turn loops back to ws$_f$ with connecting ω_5. The goal has an incoming connection weight from the sensory representation of fear via ω_{11} after which connections ω_{12} and ω_{13} lead to the preparation and execution state of taking MPH.

The upper part of the model resembles the brain parts involved in fear and anxiety. The VTA has three incoming connections: from the es$_f$, es$_{MPH}$ and from NA, respectively, ω_{43}, ω_{14} and ω_{42}. Dopamine gets incoming connections from the VTA via ω_{19} and NA via ω_{21}. The LC has three incoming connections from the VTA (ω_{15}), es$_{MPH}$

(ω_{41}) and the PFC (ω_{25}). Norepinephrine has a connection from the LC (ω_{22}). The PFC has multiple incoming connections from NA (ω_{23}), DA (ω_{30}), the amygdala (ω_{34}), VH (ω_{27}) and the Nacc via ω_{29}. The Nacc in its turn also receives connection ω_{28} from the PFC. Additionally, Nacc receives input from the amygdala via connection weight ω_{33} and from dopamine via ω_{17}. The amygdala has three entering connections: ω_{32} from dopamine, ω_{36} from the VH and ω_{26} from the PFC. The VH has an incoming connection of the amygdala as well, with connection weight ω_{35}. Furthermore, the VH has entering connections from the VTA (ω_{20}), DA (ω_{18}), the PFC (ω_{27}) and the DH (ω_{40}). Finally, the DH has entering connections from DA (ω_{31}) and VH (ω_{40}). ω_{40} represents a Hebbian learning function, which is inhibited by ω_{24} from the PFC.

Numerical representation

The conceptual representation was also put into a numerical representation. The numerical representation is as follows [34]:

- at each time point t each state Y in the model has a real number value in the interval [0,1], denoted by $Y(t)$
- at each time point t each state X connected to state Y has an impaxt on Y defined as **impact$_{X,Y}$(t)** $= \underline{\omega}_{X,Y} X(t)$ where $\omega_{X,Y}$ is the weight of the connections from X to Y
- The aggregated impact of multiple states X_i on Y at t is determined using a *combination function* $c_Y(..)$: **aggimpact$_Y$(t) = c$_Y$(impact$_{X1,Y}$(t), ..., impact$_{Xk,Y}$(t))**

$$= c_Y\big(\omega_{X1,Y} X_1(t), \ldots, \omega_{Xk,X} X_k(t)\big)$$

where X_i are the states with connections to state Y

The effect of **aggimpact$_Y$(t)** on Y is exerted over time gradually, depending on speed factor η_Y:

$$Y(t + \Delta t) = Y(t) + \eta_Y\big[\textbf{aggimpact}_\textbf{Y}(\textbf{t}) - Y(t)\big]\Delta t$$
$$\text{or:} \quad \textbf{d}Y(\text{t})/\textbf{dt} = \eta_\text{Y}[\textbf{aggimpactY}(\textbf{t}) - Y(t)]$$

Thus, the following *difference* and *differential equation for Y* are obtained:

$$Y(t + \Delta t) = Y(t) + \eta_Y\big[c_Y\big(\omega_{X_1,Y} X_1(t), \ldots, \omega_{X_k,X} X_k(t)\big) - Y(t)\big]\Delta t$$
$$\textbf{d}Y(t)/\textbf{d}t = \eta_Y\big[c_Y\big(\omega_{X_1,Y} X_1(t), \ldots, \omega_{X_k,Y} X_k(t)\big) - Y(t)\big]$$

Examples of numerical representation

A few example difference equations for the states SS_c, LC and DA will be given, in accordance with the model in Fig. 1.

$$SS_c(t + \Delta t) = SS_c(t) + \eta_{SS_c}[\omega_{WSc,SSc} WS_c(t) - SS_c(t)]\Delta t$$

$$LC(t + \Delta t) = LC(t) + \eta_{LC}[ssum_{LC}((\omega_{PFC,LC}PFC(t), \omega_{VTA,LC}VTA(t),$$
$$\omega_{esMPH,LC}es_{MPH}(t)) - LC(t)]\Delta t$$
$$DA(t + \Delta t) = DA(t) + \eta_{DA}[ssum_{DA}(\omega_{NA,DA}NA(t), \omega_{VTA,DA}VTA(t) - DA(t))]\Delta t$$

4 Example Simulation

Figure 2 shows an example simulation of the influence of MPH on anxiety levels. The mechanisms and connections between states, brain areas and hormones are based on the literature, as can be read in [34, 35]. Table 2 shows the values of connection weights used in the model. With the exception of ω_{27} and ω_{40}, which had an increasing connection weight due to Hebbian learning, all connections were stable. $\Delta t = 1$ for time steps was used. Table 3 displays the used scaling factors (λ_i) for each state with multiple incoming connections. The states that had one incoming connection, such as X_2, used the identity function those are not displayed in Table 3. However, X_3 (ws$_c$) and X_{10} (goal) both used the adaptive advanced logistic adalogistic function. The steepness was, respectively, 100 and 120 and the threshold factor 0.7 and 0.4. The simulation is implemented in MATLAB [48, 49].

First, the contextual stimulus (ws$_c$) of an exam is induced. The constant contextual stimulus will eventually affect brain areas and hormones, via the other states via sensory representation and the preparation state for fear (ps$_f$). This reaction leads to anxiety and concentration problems and is sensed by the individual. This induces the

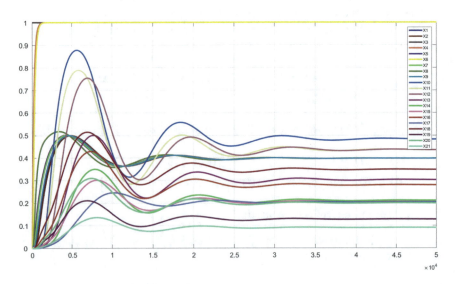

Fig. 2 Example scenario in which MPH is taken

Table 2 The implemented connection weights and their values

Connection weight	ω_1	ω_2	ω_3	ω_4	ω_5	ω_6	ω_7	ω_8
Value	1	1	1	1	1	1	1	1
Connection weight	ω_9	ω_{10}	ω_{11}	ω_{12}	ω_{13}	ω_{14}	ω_{15}	ω_{16}
Value	1	1	0.2	0.9	1	1	0.7	−0.9
Connection weight	ω_{17}	ω_{18}	ω_{19}	ω_{20}	ω_{21}	ω_{22}	ω_{23}	ω_{24}
Value	1	1	1	1	1	0.7	1	−0.5
Connection weight	ω_{25}	ω_{26}	ω_{27}	ω_{28}	ω_{29}	ω_{30}	ω_{31}	ω_{32}
Value	−0.8	−0.8	0.3	0.8	0.8	1	1	0.8
Connection weight	ω_{33}	ω_{34}	ω_{35}	ω_{36}	ω_{37}	ω_{38}	ω_{39}	ω_{40}
Value	1	1	1	0.6	0.8	−0.9	−0.2	0.3
Connection weight	ω_{41}	ω_{42}	ω_{43}					
Value	1	1	1					

increasing response of wanting to improve concentration and other learning abilities. The goal is activated by a bodily response of taking in MPH just after $t = 0.2 * 10^4$.

The intake of MPH triggers multiple brain areas. First of all, it activates the LC and the VTA. These two areas cause a cascade of activations, eventually leading to the suppression of the anxiety at $t = 0.45 * 10^4$. The activity of the brain areas is also reduced around $0.8 * 10^4$; however, this is mostly visible after anxiety levels already have been decreasing. As the anxiety decreases, the goal of wanting to increase learning abilities and the execution of taking MPH decreases as well (from $0.6 * 10^4$ until $1.2 * 10^4$).

The principle of state-connection modulation between state NA and (DH, VH) is presented in [42–50].

Non-Intake of MPH

Figure 3 shows the network when the goal is (artificially) blocked, and thus, the preparation state and execution state have not been activated. No decrease in anxiety is seen here, and all states reach an equilibrium.

5 Mathematical Analysis

To verify the temporal cognitive model of the effect of MPH on test anxiety, a mathematical analysis was executed with the following equations:

Table 3 Connection weights and scaling factors for the example simulation

State	X_5	X_8	X_{13}	X_{15}	X_{16}	X_{17}	X_{18}	X_{19}	X_{20}	X_{21}
λ_i	2	2.8	1.7	4.1	2	2.8	3	1.4	4	2
State	X_1	X_2	X_3	X_4	X_6	X_7	X_9	X_{10}	X_{11}	X_{12}
Function	id	id	alog	id	id	id	id	alog	id	id
Steepness threshold			100 0.7					120 0.4		

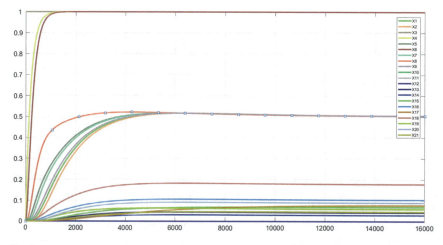

Fig. 3 Example scenario in which there is no goal, and thus, MPH is not taken

$x_3 = 1$

$x_4 = x_3$

$x_6 = x_4$

$2.8 * x_8 = x_6 + x_7 - 0.9 * x_{15} - 0.9 * x_{17} + 0.8 * x_{19} - 0.2 * x_{21}$

$x_9 = x_8$

$4.1 * x_{15} = x_{14} + x_{16} + 0.8 * x_{17} + x_{19} + 0.20 * x_{20}$

$2.8 * x_{17} = 0.8 * x_{15} + x_{16} + x_{19}$

$1.4 * x_{19} = -0.8 * x_{15} + 0.8 * x_{16} + 0.6 * x_{20}$

$0.7 * x_{14} = 0.7 * x_{13}$

$2 * x_{16} = x_{14} + x_{18}$

$1.7 * x_{13} = x_{12} - 0.8 * x_{15} + 0.7 * x_{18}$

$3 * x_{20} = x_{16} + x_{18} + x_{19} + 0.0909 * x_{21} + 0.2057 * x_{15}$

$2 * x_{21} = -0.5 * x_{14} + x_{16} + 0.20 * x_{20}$

$3 * x_{18} = x_9 + x_{12} + x_{14}$

$x_7 = x_5$

$2 * x_5 = x_2 + x_8$

$x_2 = x_1$

$x_1 = x_9$

$x_{12} = x_{11}$

$x_{11} = x_{10}$

$x_{10} = 0.4717$

These equations are based on our states and connections used in our model. For the alog functions for state X_3 and X_{10}, end values were used to implement its alogistic function in the equation solver. In this case, the world state of x_3 was 1, and the end value for x_{10} was 0.4717 (Table 4).

Table 4 Comparing analysis and simulation

States	x_1	x_2	x_5	x_7	x_{12}	x_{13}	x_{14}
Observed	0.3971	0.3971	0.3971	0.3971	0.4345	0.3022	0.2116
Solver	0.3552	0.3552	0.3552	0.3552	0.4717	0.3063	0.3063
Difference	0.0419	0.0419	0.0419	0.0419	0.0369	0.0041	0.0947
States	x_{15}	x_{16}	x_{17}	x_{18}	x_{19}	x_{20}	x_{21}
Observed	0.2057	0.2797	0.2047	0.3477	0.1280	0.2000	0.0909
Solver	0.2691	0.3421	0.2632	0.3778	0.1798	0.3221	0.1267
Difference	0.0634	0.0624	0.0585	0.0301	0.0518	0.1221	0.0358

Based on the unique solutions of the online solver in Fig. 3, a comparison is made between the equilibria in our model and the equilibria based on the equations. Optimal differences of 0.06 and below are seen as accurate. Based on this value, we conclude that our values of the used model are close to the equations that are integrated in our model and thus are seen as accurate.

6 Discussion

A theoretical, cognitive model has been constructed based on biological aspects of the human cognitive anxiety pathway in combination with the intake of methylphenidate. Methylphenidate is used to enhance cognitive tasks and concentration for patients suffering from ADHD and ADD. Based on previous literature, methylphenidate was seen as an altering factor for perceived anxiety and conditioned fear. In this model, the effect of methylphenidate on anxiety and conditioned fear was modulated and showed a decrease in perceived anxiety and conditioned fear which are in accordance with previous literature described in Chap. 2 of this report. To strengthen these findings, another model was made which excluded the intake of methylphenidate to show the effects on anxiety. Anxiety and conditioned fear were increased which also were in accordance with previous literature.

Future research can utilize this model to research the effects of methylphenidate on other brain tissues, long-term anxiety effects but also the distinction between ADHD and non-ADHD persons. The distinction between these target groups can be seen as important as it has different effects on these groups [2]. Until now, research is based on enhancement of concentration but future research on anxiety and the distinction between these groups can be helpful for fear and anxiety treatments.

References

1. Abraham, A.D., Neve, K.A., Lattal, K.M.: Dopamine and extinction: a convergence of theory with fear and reward circuitry. Neurobiol. Learn. Mem. **108**, 65–77 (2014). https://doi.org/10.1016/j.nlm.2013.11.007
2. Agay, N., Yechiam, E., Carmel, Z., Levkovitz, Y.: Non-specific effects of methylphenidate (Ritalin) on cognitive ability and decision-making of ADHD and healthy adults. Psychopharmacology **210**(4), 511–519 (2010). https://doi.org/10.1007/s00213-010-1853-4
3. Arnsten, A.: Fundamentals of attention-deficit/hyperactivity disorder: circuits and pathways. J. Clin. Psychiatry **67**, 7–12 (2006)
4. Bagot, R.C., Parise, E.M., Peña, C.J., Zhang, H.-X., Maze, I., Chaudhury, D., Nestler, E.J.: Ventral hippocampal afferents to the nucleus accumbens regulate susceptibility to depression. Nat. Commun. **6**, 7062 (2015). https://doi.org/10.1038/ncomms8062
5. Barendregt, B.: Predictors for the illicit use of methylphenidate among Dutch college and university students, aged ≥18 years (2017)
6. Bast, T., Zhang, W.-N., Feldon, J.: The ventral hippocampus and fear conditioning in rats. Exp. Brain Res. **139**(1), 39–52 (2001)

7. Blaha, C.D., Allen, L.F., Das, S., Inglis, W.L., Latimer, M.P., Vincent, S.R., Winn, P.: Modulation of dopamine efflux in the nucleus accumbens after cholinergic stimulation of the ventral tegmental area in intact, pedunculopontine tegmental nucleus-lesioned, and laterodorsal tegmental nucleus-lesioned rats. J. Neurosci. **16**(2), 714–722 (1996)
8. Butler, G., Mathews, A.: Anticipatory anxiety and risk perception. Cogn. Therapy Res. **11**(5), 551–565 (1987)
9. Bymaster, F.P., Katner, J.S., Nelson, D.L., Hemrick-Luecke, S.K., Threlkeld, P.G., Heiligenstein, J.H., Perry, K.W.: Atomoxetine increases extracellular levels of norepinephrine and dopamine in prefrontal cortex of rat: a potential mechanism for efficacy in attention deficit/hyperactivity disorder. Neuropsychopharmacology **27**(5), 699–711 (2002). https://doi. org/10.1016/S0893-133X(02)00346-9
10. Chai, N., Liu, J.-F., Xue, Y.-X., Yang, C., Yan, W., Wang, H.-M., Lu, L.: Delayed noradrenergic activation in the dorsal hippocampus promotes the long-term persistence of extinguished fear. Neuropsychopharmacology **39**, 1933 (2014). https://doi.org/10.1083/nnp.2014.42
11. de Oliveira, A.R., Reimer, A.E., de Macedo, C.E.A., de Carvalho, M.C., de Souza Silva, M.A., Brandão, M.L.: Conditioned fear is modulated by D2 receptor pathway connecting the ventral tegmental area and basolateral amygdala. Neurobiol. Learn. Mem. **95**(1), 37–45 (2011)
12. Deutch, A.Y., Tam, S.-Y., Roth, R.H.: Footshock and conditioned stress increase 3,4-dihydroxyphenylacetic acid (DOPAC) in the ventral tegmental area but not substantia nigra. Brain Res. **333**(1), 143–146 (1985)
13. Furini, C.R.G., Behling, J.A.K., Zinn, C.G., Zanini, M.L., Assis Brasil, E., Pereira, L.D., de Carvalho Myskiw, J.: Extinction memory is facilitated by methylphenidate and regulated by dopamine and noradrenaline receptors. Behav. Brain Res. **326**, 303–306 (2017). https://doi. org/10.1016/j.bbr.2017.03.027
14. Greba, Q., Kokkinidis, L.: Peripheral and intraamygdalar administration of the dopamine D_1 receptor antagonist SCH 23390 blocks fear-potentiated startle but not shock reactivity or the shock sensitization of acoustic startle. Behav. Neurosci. **114**(2), 262 (2000)
15. Jackson, M.E., Frost, A.S., Moghaddam, B.: Stimulation of prefrontal cortex at physiologically relevant frequencies inhibits dopamine release in the nucleus accumbens. J. Neurochem. **78**(4), 920–923 (2001). https://doi.org/10.1046/j.1471-4159.2001.00499.x
16. Jin, J., Maren, S.: Fear renewal preferentially activates ventral hippocampal neurons projecting to both amygdala and prefrontal cortex in rats. Sci. Rep. **5**, 8388 (2015). https://doi.org/10. 1038/srep08388
17. Johnston, L.D., O'Malley, P.M., Bachman, J.G., Schulenberg, J.E.: Monitoring the Future National Survey Results on Drug Use, 1975–2010, vol. I. Secondary School Students (2011)
18. Kharas, N., Reyes-Vazquez, C., Dafny, N.: Locus coeruleus neuronal activity correlates with behavioral response to acute and chronic doses of methylphenidate (Ritalin) in adolescent rats. J. Neural Transm (Vienna) **124**(10), 1239–1250 (2017). https://doi.org/10.1007/s00702-017-1760-5
19. Kim, M.J., Loucks, R.A., Palmer, A.L., Brown, A.C., Solomon, K.M., Marchante, A.N., Whalen, P.J.: The structural and functional connectivity of the amygdala: From normal emotion to pathological anxiety. Behav. Brain Res. **223**(2), 403–410 (2011). https://doi.org/10.1016/j. bbr.2011.04.025
20. Legault, M., Rompré, P.-P., Wise, R.A.: Chemical stimulation of the ventral hippocampus elevates nucleus accumbens dopamine by activating dopaminergic neurons of the ventral tegmental area. J. Neurosci. **20**(4), 1635–1642 (2000). https://doi.org/10.1523/jneurosci.20-04-01635. 2000
21. Lisman, J.E., Grace, A.A.: The hippocampal-VTA loop: controlling the entry of information into long-term memory. Neuron **46**(5), 703–713 (2005)
22. Linssen, A.M.W., Sambeth, A., Vuurman, E.F.P.M., Riedel, W.J.: Cognitive effects of methylphenidate in healthy volunteers: a review of single dose studies. Int. J. Neuropsychopharmacol. **17**(6), 961–977 (2014)
23. Maren, S., Aharonov, G., Fanselow, M.S.: Neurotoxic lesions of the dorsal hippocampus and Pavlovian fear conditioning in rats. Behav. Brain Res. **88**(2), 261–274 (1997). https://doi.org/ 10.1016/S0166-4328(97)00088-0

24. Mattson, M.E.: Emergency department visits involving attention deficit/hyperactivity disorder stimulant medications (2013)
25. Phelps, E.A., Delgado, M.R., Nearing, K.I., LeDoux, J.E.: Extinction learning in humans: role of the amygdala and vmPFC. Neuron **43**(6), 897–905 (2004). https://doi.org/10.1016/j.neuron.2004.08.042
26. Ponnusamy, R., Nissim, H.A., Barad, M.: Systemic blockade of D2-like dopamine receptors facilitates extinction of conditioned fear in mice. Learn. Memory **12**(4), 399–406 (2005)
27. Roth, R.H., Tam, S.Y., Ida, Y., Yang, J.X., Deutch, A.Y.: Stress and the mesocorticolimbic dopamine systemsa. Ann. N. Y. Acad. Sci. **537**(1), 138–147 (1988)
28. Sara, S.J.: The locus coeruleus and noradrenergic modulation of cognition. Nat. Rev. Neurosci. **10**, 211 (2009). https://doi.org/10.1038/nrn2573
29. Sattler, S., Wiegel, C.: Cognitive test anxiety and cognitive enhancement: the influence of students' worries on their use of performance-enhancing drugs. Subst. Use Misuse **48**(3), 220–232 (2013)
30. Segev, A., Gvirts, H.Z., Strouse, K., Mayseless, N., Gelbard, H., Lewis, Y.D., Bloch, Y.: A possible effect of methylphenidate on state anxiety: a single dose, placebo controlled, crossover study in a control group. Psychiatry Res. **241**, 232–235 (2016). https://doi.org/10.1016/j.psychres.2016.05.009
31. Sotres-Bayon, F., Cain, C.K., LeDoux, J.E.: Brain mechanisms of fear extinction: historical perspectives on the contribution of prefrontal cortex. Biol. Psychiat. **60**(4), 329–336 (2006). https://doi.org/10.1016/j.biopsych.2005.10.012
32. Sturm, V., Lenartz, D., Koulousakis, A., Treuer, H., Herholz, K., Klein, J.C., Klosterkötter, J.: The nucleus accumbens: a target for Deep-brain stimulation in obsessive-compulsive and anxiety disorders. In: Paper presented at the Proceedings of the Medtronic Forum for Neuroscience and Neuro-Technology 2005 (2007)
33. Trifoni, A., Shahini, M.: How does exam anxiety affect the performance of university students. Mediterranean J. Soc. Sci. **2**(2), 93–100 (2011)
34. Treur, J.: Network-Oriented Modeling: Addressing Complexity of Cognitive, Affective and Social Interactions. Springer, Cham (2016)
35. Treur, J.: The Ins and Outs of network-oriented modeling: from biological networks and mental networks to social networks and beyond. Trans. Comput. Collect. Intell. Paper for keynote lecture at ICCCI'18 (2018)
36. Volkow, N.D., Wang, G.-J., Fowler, J.S., Logan, J., Gerasimov, M., Maynard, L., Ding, Y.S., Gatley, S.J., Gifford, A., Franceschi, D.: Therapeutic doses of oral methylphenidate significantly increase extracellular dopamine in the human brain. J. Neurosci. **21**(2), RC121–RC121 (2001)
37. Wager, T.D., Davidson, M.L., Hughes, B.L., Lindquist, M.A., Ochsner, K.N.: Prefrontal-subcortical pathways mediating successful emotion regulation. Neuron **59**(6), 1037–1050 (2008). https://doi.org/10.1016/j.neuron.2008.09.006
38. Wang, M.E., Fraize, N.P., Yin, L., Yuan, R.K., Petsagourakis, D., Wann, E.G., Muzzio, I.A.: Differential roles of the dorsal and ventral hippocampus in predator odor contextual fear conditioning. Hippocampus **23**(6), 451–466 (2013). https://doi.org/10.1002/hipo.22105
39. Whitaker Sena, J.D., Lowe, P.A., Lee, S.W.: Significant predictors of test anxiety among students with and without learning disabilities. J. Learn. Disabil. **40**(4), 360–376 (2007)
40. Wilens, T.E., Adler, L.A., Adams, J., Sgambati, S., Rotrosen, J., Sawtelle, R., Utzinger, L., Fusillo, S.: Misuse and diversion of stimulants prescribed for ADHD: a systematic review of the literature. J. Am. Academy of Child Adolescent Psychiatry **47**(1), 21–31 (2008)
41. Zheng, X., Liu, F., Wu, X., Li, B.: Infusion of methylphenidate into the basolateral nucleus of amygdala or anterior cingulate cortex enhances fear memory consolidation in rats. Sci. China Ser. C Life Sci. **51**(9), 808–813 (2008). https://doi.org/10.1007/s11427-008-0105-x
42. Ziabari, S.S.M., Treur, J.: Integrative biological, cognitive and affective modeling of a drug-therapy for a post-traumatic stress disorder. In: International Conference on Theory and Practice of Natural Computing, pp. 292–304. Springer, Cham (2018a)

43. Ziabari, S.S.M., Treur, J.: Cognitive modelling of mindfulness therapy by autogenic training. In: Proceedings of the 5th International Conference on Information System Design and Intelligent Applications, INDIA'18. Advances in Intelligent Systems and Computing. Springer, Berlin (2018b)
44. Ziabari, S.S.M., Treur, J.: An adaptive cognitive temporal-causal network model of a mindfulness therapy based on music. In: Proceedings of the 10th International Conference on Intelligent Human-Computer Interaction, IHCI'18. Springer, India (2018c)
45. Ziabari, S.S.M., Treur, J.: An adaptive temporal-causal network model for decision making under acute stress. Vietnam J. Comput. Sci. (2018d) (submitted)
46. Ziabari, S.S.M., Treur, J.: Integrative Biological, Cognitive and affective modeling of a drug-therapy for a post-traumatic stress disorder. In: Proceedings of the 7th International Conference on Theory and Practice of Natural Computing, TPNC'18, Springer, Berlin (2018e)
47. Ziabari, S.S.M., Treur, J.: Computational analysis of gender differences in coping with extreme stressful emotions. In: Proceedings of the 9th International Conference on Biologically Inspired Cognitive Architecture (BICA2018), Elsevier, Czech Republic (2018f)
48. Ziabari, S.S.M., Treur, J.: A modeling environment for dynamic and adaptive network models implemented in matlab. In: Proceedings of the 4th International Congress on Information and Communication Technology (ICICT2019), 25–26 Feb. Springer, London, UK (2019a)
49. Ziabari, S.S.M.: Integrative cognitive and affective modeling of deep Brain stimulation. In: Proceedings of the 32nd International Conference on Industrial, Engineering and Other Applications of Applied Intelligent Systems (IEA/AIE 2019) (2019b) (submitted)
50. Ziabari, S.S.M., Treur, J.: An adaptive cognitive temporal-causal network model of a mindfulness therapy based on humor. In: International Conference on Computational Science (ICCS 2019) (2019c) (submitted)

Stroke Diagnosis Algorithm Based on Similarity Analysis

Sung-Jong Eun, Eun-Young Jung, Hyun Ki Hong and Dong Kyun Park

Abstract This paper proposes algorithm processing calculation of ASPECT ratio automatically for diagnosis of stroke. In this paper, we propose a CT image-based automatic ASPECT ratio determination algorithm to diagnose stroke, one of the central nervous system diseases. ASPECT ratio to be judged is classified grades and increases according to severity of the disease. The proposed method compares the correlation information of the left and right regions based on the detected gray matter regions. We propose the method to judge difference of ASPECT score according to difference of correlation information to be compared. This paper proposes an effective method featuring accuracy and simplicity.

Keywords Stroke diagnosis · ASPECT ratio · Correlation · Statistic analysis

1 Introduction

In this paper, we propose an algorithm to calculate the ASPECT ratio for stroke diagnosis. Computer work done manually by clinicians is important. There have been many attempts to use IT-based image analysis techniques. In this paper, we propose a CT image-based automatic ASPECT ratio determination algorithm to diagnose stroke, one of the central nervous system diseases. The percentage to be judged is classified as 0–10 and increases with the severity of the disease. This paper

S.-J. Eun (✉) · E.-Y. Jung · H. K. Hong · D. K. Park
Health IT Research Center, Gachon University Gil Medical Center, Incheon, Korea
e-mail: asclephios@naver.com

E.-Y. Jung
e-mail: eyjung@gilhospital.com

H. K. Hong
e-mail: hyunki85@gmail.com

D. K. Park
e-mail: pdk66@gilhospital.com

© Springer Nature Singapore Pte Ltd. 2020
X.-S. Yang et al. (eds.), *Fourth International Congress on Information and Communication Technology*, Advances in Intelligent Systems and Computing 1027, https://doi.org/10.1007/978-981-32-9343-4_27

347

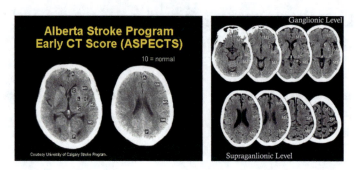

Fig. 1 The concept of ASPECTS ratio

proposes an effective method featuring accuracy and simplicity. The details of the process are described in the chapter.

2 Related Work

Mechanical thrombectomy (MT) in acute ischemic stroke (AIS) due to a large intracranial vessel occlusion in the anterior circulation has been proven to be an effective therapy. In 2000, Barber et al. [1] introduced the Alberta Stroke Program Early CT Score. ASPECTS divide the area of the middle cerebral artery into 10 predefined anatomical regions and grade the presence of an initial infarct signal by the parenchymal hypodensity of the NCCT. This gradually increasing trend has been a reliable predictor of clinical outcome in patients with AIS already treated with intravenous rtPA and/or MT [2, 3]. In particular, we propose an algorithmic computation of ratios automatically for the diagnosis rate of stroke (ASPECT ratio) (Fig. 1).

In addition, many studies have been conducted to confirm the degree of stroke based on image processing [4–6]. Various methods have been tried, including a method of dividing a region based on a local pixel [5] and a technique combining CT and MRI information processing [6]. Various methods have been applied to stroke diagnosis such as image processing methods and deep learning studies.

3 Proposed Stroke Diagnosis Method

In this paper, we propose a CT image-based automatic ASPECT ratio determination algorithm to diagnose stroke, one of the central nervous system diseases. The percentage to be judged is classified as 0–10 and increases from 0 (weak) to 10 (strong) depending on the severity of the disease. In this paper, the input CT images are used to determine the ASPECT score and the grade of the results, which are used as a secondary means for expert diagnosis. Figure 2 shows the proposed whole algorithm

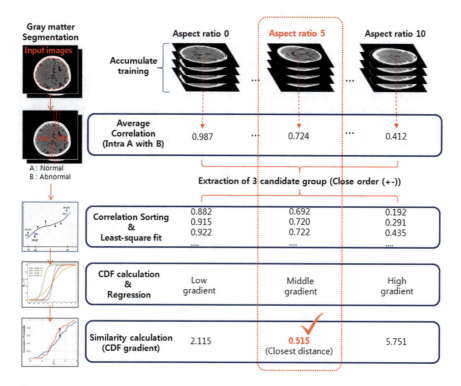

Fig. 2 Proposed stroke diagnosis method

concept chart.

3.1   *Extraction of Brain Gray Matter*

Pixel values (50 more) appearing first on upper and lower sides and right and left are recorded to detect boundary of outside area in CT image. Closing operation is performed that links the recorded outline to the closed object. Closed objects with internal information are created through fill operations, and the initial contours for detecting gray matter in the brain are generated by 3×3 mask-based erosion. When applying the Snake [7] method, the ACM model, the associated outline, is used as the initial outline to extract the final gray boundary.

3.2   Calculation of Correlation

Center of gravity of an object is calculated and the shortest axis is calculated based on central point to detect central boundary line in gray matter detected. Angle is adjusted by rotating calculated shortest axis so that it becomes 180° horizontally. A vertical crossing at 90° with the shortest axis horizontal line is created. When dividing created vertical into right and left by exploring it right and left by 10 pixels, a point which proportion of two areas become nearest to 1:1 is determined as optimum boundary line. Correlation of detected two areas is calculated and it is processed through the following Eq. (1) where hat A and B are mean values.

$$r = \frac{\sum_m \sum_n \left(A_{mn} - \bar{A}\right)\left(B_{mn} - \bar{B}\right)}{\sqrt{\left(\sum_m \sum_n \left(A_{mn} - \bar{A}\right)^2\right)\left(\sum_m \sum_n \left(B_{mn} - \bar{B}\right)^2\right)}} \tag{1}$$

3.3   Decision of ASPECT Candidate

Guideline was prepared by calculating correlation average of stroke patient group dataset (70 people) to establish a standard for grades of ASPECT ratio (0–10). Prior to the calculation of the mean, the correction is carried out with minimum sum of squares taking the impulse correlation into account. For the correlation value entered as a distance difference, a total of three candidates are determined, including the shortest rank, one rank higher than the relevant rank, and a rank lower than the relevant rank.

3.4   Final Decision Based on Cumulative Density Function

As in the creation of the ASPECT ratio guidelines, the cumulative density function (CDF) is created by sorting the correlation information calculated in each of the 0–10 groups in ascending order. The representative function is calculated by regression analysis of the cumulative value of the dataset for each grade. The calculated representative function for each grade is provided as a second guideline. The smallest difference between the three ratings is obtained by comparing the guidance function with the gradient function.

Table 1 Comparison of the proposed method with the other exist method

Confusion matrix (50 images)	Proposed method	Canny edge (contour based)	Region growing (region based)
True positive	44	21	40
False positive	6	29	10
True negative	46	41	41
False negative	4	9	9

4 Evaluation

In order to experiment, we use the confusion matrix. Using the matrix, the accuracy of the suggested algorithm and the typical algorithm were compared (Table 1).

Comparisons with the proposed method were confirmed by comparing the accuracy of the initially detected region information. As a result of comparison, it was confirmed that the proposed method is more effective than the existing methods. This is because it is difficult to compare objective methods in stroke judgment, and the accuracy of region information input to judgment is compared.

5 Conclusion

In this paper, we propose an algorithm to calculate the ASPECT ratio for stroke diagnosis. Getting and growing more than average data are required for ASPECT ratios, which have many lessons. We propose an effective method that features accuracy and simplicity. Compared with the conventional image processing method and machine learning method, the proposed method has higher accuracy and processing speed than the existing image processing method and machine learning method. We recommend that you add a weighting factor at each step to improve the accuracy of the proposed method.

Acknowledgements This research was supported by the Bio and Medical Technology Development Program of the National Research Foundation (NRF) funded by the Ministry of Science and ICT (2017M3A9E2072689).

This work was supported by Institute for Information and Communications Technology Promotion (IITP) grant funded by the Korea government (MSIT) (2018-2-00861, Intelligent SW Technology Development for Medical Data Analysis).

References

1. Barber, P.A., Demchuk, A.M., Zhang, J., et al.: Validity and reliability of a quantitative computed tomography score in predicting outcome of hyperacute stroke before thrombolytic therapy: ASPECTS Study Group—Alberta Stroke Programme Early CT Score. Lancet **355**, 1670–1674 (2000)
2. Menon, B.K., Puetz, V., Kochar, P., et al.: ASPECTS and other neuroimaging scores in the triage and prediction of outcome in acute stroke patients. Neuroimaging Clin N Am. **21**, 407–423 (2011). (xii)
3. Liebeskind, D.S., Jahan, R., Nogueira, R.G., et al.: SWIFT investigators. Serial Alberta stroke program early CT score from baseline to 24 hours in Solitaire flow restoration with the intention for thrombectomy study: a novel surrogate end point for revascularization in acute stroke. Stroke **45**, 723–727 (2014)
4. Ikeda, N., et al.: Automated segmental-IMT measurement in thin/thick plaque with bulb presence in carotid ultrasound from multiple scanners: stroke risk assessment. Comput. Methods Programs Biomed. **141**, 73–81 (2017)
5. Rajinikanth, V., et al.: Evaluation of ischemic stroke region from CT/MR images using hybrid image processing techniques. In: Intelligent Multidimensional Data and Image Processing, pp. 194–219. IGI Global (2018)
6. Rajinikanth, V., et al.: Shannon's entropy and watershed algorithm based technique to inspect ischemic stroke wound. In: Smart Intelligent Computing and Applications, pp. 23–31. Springer, Singapore (2019)
7. Williams, D., Shah, M.: A fast algorithm for active contours and curvature estimation. Comput. Vis. Graph. Image Process. Image Underst. **55**, 14–25 (1992)

A Multilevel Graph Approach for Predicting Bicycle Usage in London Area

Francesco Colace, Massimo De Santo, Marco Lombardi, Francesco Pascale, Domenico Santaniello and Allan Tucker

Abstract Our cities are significantly changing their structure and organization. These changes have given rise to the need to introduce new services able to rationalize the activities present in such a complex context. Within this scenario, one of the most important services is related to transport management. A proper transport system management can significantly improve the overall quality of life of citizens, in terms of improving air quality, reducing road traffic and ensuring the public transport schedule. In addition to the traditional services and urban traffic management approaches, a significant contribution may come from the adoption of all those IT systems, which are part of the so-called Internet of things. According to this paradigm, it is possible to design an added value and pervasive services in order to assist the users involved in the system. Although this approach could be considered interesting and promising, it is necessary to introduce methodologies able to manage data coming from several heterogeneous sensors in order to process and propose coherent information. In this paper, we propose the using of three graphic approaches able to process the information coming from various sources in order to manage urban transport systems. The three models of representation, on which to conduct inference processes are context dimension tree, ontology and Bayes network. These three

F. Colace · M. De Santo · M. Lombardi · F. Pascale · D. Santaniello (✉)
University of Salerno, Via Giovanni Paolo II, 132, 84084 Fisciano, SA, Italy
e-mail: dsantnaiello@unisa.it

F. Colace
e-mail: fcolace@unisa.it

M. De Santo
e-mail: desanto@unisa.it

M. Lombardi
e-mail: malombardi@unisa.it

F. Pascale
e-mail: fpascale@unisa.it

A. Tucker
Brunel University, Kingston Ln, Uxbridge, London UB8 3PH, UK
e-mail: allan.tucker@brunel.ac.uk

© Springer Nature Singapore Pte Ltd. 2020   353
X.-S. Yang et al. (eds.), *Fourth International Congress on Information and Communication Technology*, Advances in Intelligent Systems and Computing 1027, https://doi.org/10.1007/978-981-32-9343-4_28

approaches allow the creation of inference processes, which represent the basis of value-added services to be offered to several users. The aim of this paper is to present a service that through a multilevel approach, which takes advantage of three models of graphic representation, is able to analyse data from various sensors in an urban area in order to predict the bicycle-sharing public service usage in the city of London. Through the intersection and analysis of data from cameras, weather and transport sensors, it will be possible to establish in which condition there will be an increase or decrease of bicycle rental in order to manage the service. The results obtained on data collected in real scenarios are very satisfying.

Keywords Smart city · Knowledge management · Pervasive system

1 Introduction

The growing urbanization has created large urban areas, which include increasingly pervasive services that aim to increase the citizen livability. The changes made by these services are so deep that give us a chance to talk about smart city. We find ourselves in the smart city paradigm, where integrated technologies are exploited to optimize resources and improve sectors such as economy, mobility, productivity, environmental quality, etc. The integrated technologies used in the smart cities belong to the Internet of things (IoT) field, which represents a valid support to the added value services that a modern city should be provided [1].

According to Ashton, the IoT refers to a paradigm in which "objects" are considered "smart" considering their ability to communicate between them and with human beings [2]. This continuous exchange of information produces a considerable amount of data called Big Data. Many efforts in scientific literature have been done around this phenomena in order to provide solution to manage this amount of data, which represent a good starting point to increase quality of life in the smart city [3–5].

The transport sector was not an exception to these changes that within a modern urban environment includes a wide range of new services geared towards sustainable transport, which aims to reduce air and noise pollution, road congestion, accident rate, consumption of land due to infrastructure, etc. Taking advance of integrated technologies of a smart city, trying to manage the urban environment has become a fundamental field. This can be done through sustainable transport such as bicycle-sharing tsstem (BSS). A BSS is one of the main tools in the field of sustainable transport, which help public administrations whose objective is to increase the use of public transport. In general, this service is integrated with the classic means of transport (buses, trams and subways) covering areas difficult to reach by public transport. This system is also useful as an additional service to reach easily the near public transport station regardless of the final destination of the user. These services can be used through stations where bicycles can be requested through authentication, generally RFID services, which allow 24/7 usability. In addition to being useful for tourists to visit a city, sports lovers as a means of healthy travel and useful for environmental sustainability policies as a means of transport without pollution, BBS can be a crucial service for users of some urban areas that could use this service during their daily routine. As with all services, in this case too there may be some problems related to usability, such as any ordinary and extraordinary maintenance and disruptions that may affect the availability of the service. The possibility of having available data on the use of the service could be suitable in order to improve the quality and integration with other means of transport (i.e. know if at a specific BBS station there is bicycles availability). In addition to the data related to bicycle sharing, be given the chance to analyse other data coming from urban environment (such as road traffic, meteorological data, subways status, etc.) could be useful to improve the usability of services, besides a chance to have a general overview of the relationships between the city services.

At this stage there is reason to wonder if is possible to mine information among the data in order to support the management of services in a smart city. The aim of this article is to provide evidence of that using a methodology with forecasting ability and context adaptability. The proposed methodology consists of a system capable of combining three graph-based approaches such as ontology, context dimension tree (CDT) and Bayesian network (BN). Using that methodology, it is possible, taking advantage of a data set, to analyse, forecast and manage the smart services in real time. Ontology and CDT are valid elements of representation of context and of services available within it as they are able to provide a clear and interconnected framework of reality [6]. Due to several publications [7–12], it was possible provide a specific ontology and CDT views, which are fundamental elements in order to apply the proposed approach. Moreover, by exploiting the capabilities of the Bayesian networks, which from the experimental evidence and through the probabilistic approaches are able to identify the possible events occurrence, it has been possible to validate the methodology through an experimental campaign that has shown satisfactory results.

The paper is organized as follows: In the second paragraph, the proposed methodology is illustrated; the paragraph three is deal with the case of study, explanation of the available data set details and experimental results; the conclusions is presented in paragraphs five.

2 The Multilevel Graph Approach

In this section, we will introduce an approach based on three interconnected levels of representation, which is called multilevel graph approach (MuG) [13–16] in order to predict the use of a bicycle-sharing service (BSS). In particular, this approach is based on:

Representation of all the possible contexts that can be had in the application domain of interest through a tree model called context dimension tree (CDT) [10, 17];

Knowledge representation, able to model the reality of interest in concepts and in relations between concepts through the use of ontologies [18];
Graphical representation of a probabilistic model, which identifies the possible relationships that exist in a set of variables in uncertainty conditions using Bayesian networks [19, 20].

The architecture of the system is shown in Fig. 1 and includes three main blocks.

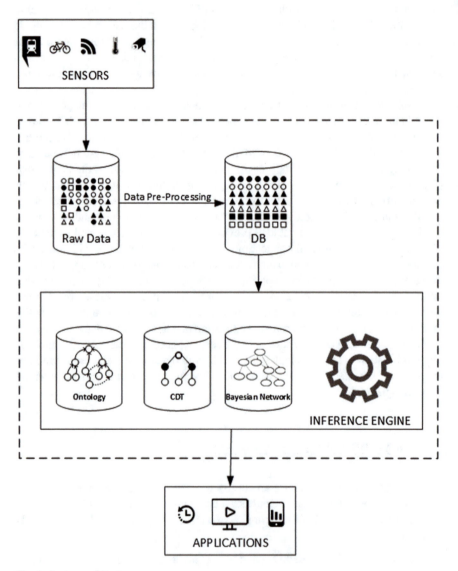

Fig. 1 System architecture

The last phase of this approach involves the definition and use of decision models that are continuously improved based on new data collected. For this purpose, the architecture deals with managing and organizing the knowledge base, providing a representation of the context and using inferential engines.

A formal and shared representation of the concepts and mutual relations that characterize a certain knowledge domain can be obtained through the use of ontologies and the construction of a knowledge base.

The context representation phase allows modelling the domain as a set of variables whose values can interest users and influence their actions. This phase assumes particular importance due to the need to reduce the informative noise generated by the large mass of data available by adapting the behaviour of the system during changing operational and/or environmental condition.

The main task of the inferential engine is to apply rules to data. This engine controls the entire process needed to obtain system outputs. The approach chosen is relative to the Bayesian networks, which allow the updating of all the used variables whenever new data is available.

3 Case of Study

This section aims to describe the case of study analysed, the methodology is applied in attempt to predict the traffic of bicycles in a specific London borough (Westminster), which the presented data set is referred. The CDT (Fig. 4) and ontological (Fig. 3) view have been designed, according to some scientific literature contributions [7–12]. The system, in accordance with the forecast target, makes a search of all the possible relationships present in the graph approaches defining a constraints list useful to Bayesian network structure design, which is achieved by an automatic-learning algorithm. The analysis is performed through R language and bnlearn package [21]. To validate the proposed methodology, the approach is applied comparing a standard-learning algorithm. The data set is divided into training data set and test data set, and the results of the analysis are provided in a classification matrix through which is possible to give results in terms of:

- Precision refers to the probability that a recovered instance (randomly selected) is relevant.

$$\gamma = \frac{TP}{TP + FP}$$

- Recall refers to the probability that a relevant instance (randomly selected) is retrieved in a search.

$$\rho = \frac{TP}{TP + FN}$$

Table 1 Data set

Data	Details
Day date	This data refers to the date of the day with the following format: yyyy-mm-dd and HH:MM:SS
Radiation	This data refers to the solar radiation expressed in W/m^2
Rain rate	This data refers to the instantaneous measure of precipitation, expressed in mm/h
Storm rain	This data refers to the amount of rain accumulated during a storm, expressed in mm
Temperature	This data refers to the outside temperature, expressed in Celsius degree
Wind speed	This data refers to the instantaneous speed of the wind, expressed in m/s
Start trips	This data refers to the number of bike rented in the borough
End trips	This data refers to the number of bike returned in the borough
Accidents. Slight	This data refers to the modest severity accidents that occurred in the borough
Accidents. Serious	This data refers to the high severity accidents that occurred in the borough

3.1 Data Set

In this section, we will describe the data set used in the experimental phase. Table 1 shows the complete data set details collected in one year of observation. It was possible to collect this data through several sensors scattered around London's boroughs, which provide heterogeneous data. There is data related to weather condition (temperature, wind speed, rain rate and radiation) and other related to "human behaviour" such as bicycles return and rent (end trips, and start trips) and road accident data (slight and serious). This data comes from open Transport for London API service.[1] These data has been aggregated in ranges, which allow us to perform better analysis. The BBS data has been aggregated in three ranges which refers to the low, medium or high demand and returns of bicycle, the accident data has been aggregated in a binary range: Yes or No.

3.2 Experimental Results

In the first phase, the Bayesian network structure is provided by an automatic-learning algorithm, which is hill-climbing learning algorithm [22]; subsequently, it is evaluated the reliability of the network through precision and recall parameters. Start trip and end trip variables measure the rents and return traffic of bicycles in the whole Westminster borough, which can be summarized into a three ranges: Low, medium and high.

---

[1] https://tfl.gov.uk/info-for/open-data-users/ (Link web visited on December 2018).

As shown in Fig. 2a, b, the network structures present many correlations, however, several of them are incorrect (i.e. radiation cannot influence the temporal interval). The used training data set seems to suggest some correlations, however, they are not enough in order to build a reliable Bayesian network. However, in Table 2 are reported results in terms of precision and recall according to those relationships, where may be notice the precision value in some cases reach over the 80% of accuracy. That would mean that the system has succeeded in correctly classifying events, even if the result could be improved in order to define the Bayesian network structure as reliable.

In Fig. 2c, d is provided the case in which is applied the proposed innovative methodology is applied. In this phase, the network structure learning is automatic and adopts the ML algorithm used in the previous phase but takes advance of a constraints list produced by CDT and ontological view data intersection. In this case, as can be seen in Table 2, the Bayesian network structure is not exhaustive, however, there is an improvement of results in terms of precision and recall, which allow us to appreciate the benefits of using the proposed approach.

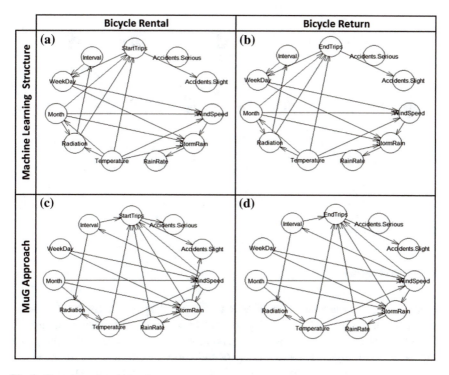

Fig. 2 Bayesian network structures

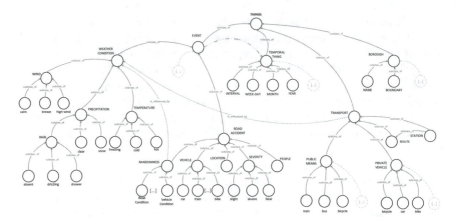

Fig. 3 Ontology view

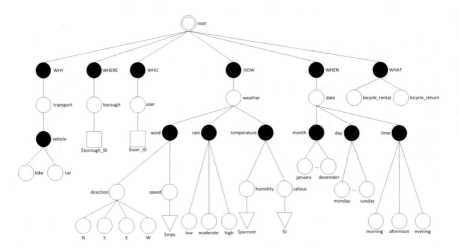

Fig. 4 Context dimension tree view

4 Conclusions

In this work is evaluated the application of a novel approach (multilevel graph approach) in order to predict a bicycle-sharing service usage in urban area with the aim to manage smart city services. The proposed system connects three graph approaches (ontologies, Bayesian network and the context dimension tree) and has given acceptable results in road accident risks field [13] (Figs. 3 and 4).

In regard to validate the proposed methodology to BBS predicting usage, we evaluate first results coming from BN structures learned by a learning score algorithm, subsequently the proposed methodology was applied in order to evaluate the improvement in terms of classification performance.

Table 2 Results

		Bicycle rental			Bicycle return		
		Low	Medium	High	Low	Medium	High
ML	Precision (%)	80.01	82.52	79.37	88.68	53.92	12.85
	Recall (%)	89.89	52.97	10.02	72.19	61.26	43.31
MuG	Precision (%)	76.92	85.80	83.85	80.53	84.44	76.24
	Recall (%)	93.63	73.33	65.95	92.43	74.22	63.77

Despite the structures are not complete in all its aspects, Table 2 shows positive results in terms of prediction and recall. The study of descriptive methodology, which represents the environment through graphs and a Bayesian network approach is able to bring us not only in forecast particular events, but in addition, allow us to understand the triggers behind the events leads us towards the emergency management. Furthermore, the described system architecture presents resources able to automatically adapt and update, increasing the performance according as much data as is available.

References

1. Zanella, A., Bui, N., Castellani, A., Vangelista, L., Zorzi, M.: Internet of Things for smart cities. IEEE Internet Things J. **1**(1), 22–32 (2014)
2. Ashton, K.: That 'Internet of Things' thing. RFID J. **22**(7), 97–114 (2009)
3. Abate, F., Carratù, M., Liguori, C., Ferro, M., Paciello, V.: Smart meter for the IoT. In: I2MTC 2018—2018 IEEE International Instrumentation and Measurement Technology Conference: Discovering New Horizons in Instrumentation and Measurement, Proceedings, pp. 1–6 (2018)
4. Castiglione, A., Dambrosio, C., De Santis, A., Castiglione, A., Palmieri, F.: On secure data management in health-care environment. In: Proceedings—7th International Conference on Innovative Mobile and Internet Services in Ubiquitous Computing. IMIS 2013 (2013)
5. Colace, F., Santaniello, D., Casillo, M., Clarizia, F.: BeCAMS: a behaviour context aware monitoring system. In: 2017 IEEE International Workshop on Measurement and Networking, M and N 2017—Proceedings (2017)
6. Colace, F., De Santo, M., Moscato, V., Picariello, A., Schreiber, F.A., Tanca, L. (eds.): Data Management in Pervasive Systems. Springer International Publishing, Cham (2015)
7. Junli, W., Zhijun, D., Changjun, J.: An ontology-based public transport query system. In Proceedings—First International Conference on Semantics, Knowledge and Grid. SKG 2005, pp. 62–62 (2006)
8. Zgaya, H., Hammadi, S.: Multi-agent information system using mobile agent negotiation based on a flexible transport ontology. In: Proceedings of the 6th International Joint Conference on Autonomous Agents and Multiagent Systems—AAMAS'07, p. 1 (2007)
9. Yang, W.-D., Wang, T.: The fusion model of intelligent transportation systems based on the urban traffic ontology. Phys. Proc. **25**, 917–923 (2012)
10. Panigati, E., Rauseo, A., Schreiber, F.A., Tanca, L.: Aspects of pervasive information management: an account of the green move system. In: Proceedings—15th IEEE International Conference on Computational Science and Engineering, CSE 2012 and 10th IEEE/IFIP International Conference on Embedded and Ubiquitous Computing. EUC 2012 (2012)

11. Annunziata, G., Colace, F., De Santo, M., Lemma, S., Lombardi, M.: ApPoggiomarino: a context Aware app for e-citizenship. In: ICEIS 2016—Proceedings 18th International Conference on Enterprise Information Systems, vol. 2, pp. 273–281 (2016)
12. Patel, A.S., Ojha, M., Rani, M., Khare, A., Vyas, O.P., Vyas, R.: Ontology-based multi-agent smart bike sharing system (SBSS). In: Proceedings—2018 IEEE International Conference on Smart Computing. SMARTCOMP 2018, pp. 417–422 (2018)
13. Colace, F., Lombardi, M., Pascale, F., Santaniello, D., Tucker, A., Villani, P.: MuG : A multilevel graph representation for big data interpretation. In: IEEE 20th International Conference on High Performance Computing and Communications; IEEE 16th International Conference on Smart City; IEEE 4th International Conference on Data Science and Systems, pp. 1410–1415 (2018)
14. Colace, F., Lombardi, M., Pascale, F., Santaniello, D.: A multi-level approach for forecasting critical events in smart cities. In: Proceedings—DMSVIVA 2018: 24th International DMS Conference on Visualization and Visual Languages (2018)
15. Clarizia, F., Colace, F., Lombardi, M., Pascale, F., Santaniello, D.: A multilevel graph approach for road accidents data interpretation. In: Lecture Notes in Computer Science. Lecture Notes in Artificial Intelligence and Lecture Notes in Bioinformatics, vol. 11161. LNCS, pp. 303–316 (2018)
16. Clarizia, F., Colace, F., De Santo, M., Lombardi, M., Pascale, F., Santaniello, D., Tuker, A.: A multilevel graph approach for rainfall forecasting: a preliminary study case on London area. Concurr. Comput. Pract. Exp., p. e5289, May 2019
17. Chen, H., Finin, T., Joshi, A.: An ontology for context-aware pervasive computing environments. Knowl. Eng. Rev. **18**(3), 197–207 (2003)
18. D'Aniello, G., Gaeta, A., Loia, V., Orciuoli, F.: Integrating GSO and SAW ontologies to enable Situation Awareness in Green Fleet Management. In: 2016 IEEE International Multidisciplinary Conference on Cognitive Methods in Situation Awareness and Decision Support, CogSIMA 2016 (2016)
19. Weber, P., Medina-Oliva, G., Simon, C., Iung, B.: Overview on Bayesian networks applications for dependability, risk analysis and maintenance areas. Eng. Appl. Artif. Intell. 25(4), 671–82 (2012)
20. Colace, F., Lombardi, M., Pascale, F., Santaniello, D.: A multilevel graph representation for big data interpretation in real scenarios. In: Proceedings—2018 3rd International Conference on System Reliability and Safety (ICSRS) 2018 (2019)
21. Scutari, M.: Learning Bayesian Networks with the bnlearn R Package, J. Stat. Softw. 35(i03) 2010
22. Cooper, G.F., Herskovits, E.: A Bayesian method for the induction of probabilistic networks from data. Mach. Learn. **9**(4), 309–347 (1992)

Machine Learning and Digital Heritage: The CEPROQHA Project Perspective

Abdelhak Belhi, Houssem Gasmi, Abdelaziz Bouras, Taha Alfaqheri, Akuha Solomon Aondoakaa, Abdul H. Sadka and Sebti Foufou

Abstract Through this paper, we aim at investigating the impact of artificial intelligence technologies on cultural heritage promotion and long-term preservation in terms of digitization effectiveness, attractiveness of the assets, and value empowering. Digital tools have been validated to yield sustainable and yet effective preservation for multiple types of content. For cultural data, however, there are multiple challenges in order to achieve sustainable preservation using these digital tools due to the specificities and the high-quality requirements imposed by cultural institutions. With the rise of machine learning and data science technologies, many researchers and heritage organizations are nowadays searching for techniques and methods to value and increase the reliability of cultural heritage digitization through machine learning. The present study investigates some of these initiatives highlighting their added value and potential future improvements. We mostly cover the aspects related

A. Belhi (✉) · H. Gasmi · A. Bouras
CSE, Qatar University, Doha, Qatar
e-mail: abdelhak.belhi@qu.edu.qa

H. Gasmi
e-mail: houssem.gasmi@qu.edu.qa

A. Bouras
e-mail: abdelaziz.bouras@qu.edu.qa

A. Belhi · H. Gasmi
DISP Laboratory, University Lumière Lyon 2, Lyon, France

T. Alfaqheri · A. S. Aondoakaa · A. H. Sadka
Brunel University, London, UK
e-mail: taha.alfaqheri@brunel.ac.uk

A. S. Aondoakaa
e-mail: akuha.aondoakaa@brunel.ac.uk

A. H. Sadka
e-mail: abdulsadka@brunel.ac.uk

S. Foufou
Le2i Lab, University of Burgundy, Dijon, France
e-mail: sfoufou@u-bourgogne.fr

© Springer Nature Singapore Pte Ltd. 2020
X.-S. Yang et al. (eds.), *Fourth International Congress on Information and Communication Technology*, Advances in Intelligent Systems and Computing 1027, https://doi.org/10.1007/978-981-32-9343-4_29

to our context which is the long-term cost-effective digital preservation of the Qatari cultural heritage through the CEPROQHA project.

Keywords Cultural heritage · Machine learning · Artificial intelligence · CEPROQHA project · 3D-holoscopic imaging

1 Introduction

From generation to generation, cultural heritage tends to represent the best way to express and transfer historical information. Cultural heritage items or artifacts have an important moral value and are mostly considered priceless. However, the risks that these artifacts face nowadays are more serious than ever before due to natural disasters, wars, degradation, etc. Many efforts are spent either by governments or by NGOs in order to prevent such risks. But through these efforts, it was rather clear that the physical preservation of cultural artifacts is a costly and labor-intensive process that not only is ineffective for the long term but may also lead to different kinds of issues. Digital preservation, which was first proposed by product lifecycle management (PLM) tools, was suitable in order to keep track and preserve cultural assets using digital tools, especially with the increasingly approved reliability of IT systems and the extreme high-quality digitization tools (photography, 3D, etc.) [1]. Applying these technologies in the cultural context is not straight forward, unfortunately. Several challenges related to the specificities of cultural content that are directly related to the effectiveness of these digitization approaches have arisen. For example, in the cultural context, the metadata is as important as the asset itself. If an asset history is lost, its value is heavily degraded. Much research was thus undertaken in order to find methods and techniques that can effectively label and annotate heritage assets based on their partially available information. This information can be their visual capture (visual features), partially annotated metadata or a combination of both. Through time, the accuracy of such systems kept improving until reaching very high-precision levels. Thanks to the advancements in computer vision and machine learning, and with the maturity of visual acquisition technologies, multiple heritage institutions began using their 2D and 3D collections in information retrieval and data mining tasks [2].

Overall, the main focus either for researchers or heritage specialists remains to how effectively take advantage of the recent innovations in the field of data science and machine learning in order to empower, increase the value, preserve, and promote cultural heritage. In our context and through the CEPROQHA project, our team aims at providing a framework intended for cultural data preservation through the use of advanced and effective digitization techniques such as the 3D-holoscopic imaging framework developed by our Brunel university team. We also aim at integrating cutting edge machine learning technologies in our digitization process mostly to increase the quality of the digitization and also to enrich the cultural assets to make their preservation more effective and sustainable for the long term.

The remainder of this paper is organized as follows. In section two, we present digital heritage where we explain how heritage organizations are leveraging information technology to add value to their collection and the potential applications using the collected data. In section three, we present some of the techniques and tools developed by our team and highlight some on progress and future work related to applications of machine learning in the context of cultural heritage. In section four, we give our conclusion with some perspectives for forthcoming work.

2 Digital Heritage

Digital heritage is the term used for digital information that represents a real physical or moral cultural heritage asset. Nowadays, cultural institutions such as museums rely heavily on digital technologies not only to manage their inventory of assets but also to make their collections more attractive and visible through digital media. Techniques such as 2D and 3D capturing, along with their respective visualization tools, have introduced new methods of content consumption and broadcasting. Digital heritage is widely used not only for entertainment and historical transfer but also for long-term digital preservation and data analytics [3]. Figure 1 outlines the cultural data lifecycle where the input is usually a visual capture of an asset and the output is a digitally preserved copy.

Digital heritage is nowadays widespread across heritage institutions mainly due to the increasing reliability and falling costs of IT systems. As a result, cultural assets are now more accessible for larger audiences than ever before. However, one of the limiting factors is the quality and effectiveness of cultural asset digitization. An asset with unavailable or incomplete metadata is automatically devalued. Consequently, researchers are looking for methods related to recent machine learning techniques to promote and increase the value of underlooked cultural assets.

Fig. 1 Cultural data lifecycle

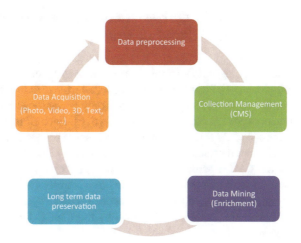

Fig. 2 Sony α7 II camera body fitted with the H3D lens array

3 The CEPROQHA Project Context

In this section, we present the most notable contributions and techniques studied and introduced by the CEPROQHA project for the promotion and the enrichment of digital heritage. These contributions and techniques mostly cover three topics related to cultural data digitization, enrichment, management, and long-term preservation with a focus on machine learning-related techniques. The first topic is related to the improvement at the post-processing stages of the 3D-holoscopic imaging framework. These improvements mainly utilize machine learning techniques such as image super-resolution and video motion interpolation to ensure the cost-effectiveness of the H3D framework while maintaining high-quality standards. In the collection management and enrichment context, our team is working on numerous approaches to complete and curate cultural collections through machine learning-based approaches mainly to save costs and time. Traditionally, heritage institutions have to refer to long-time experts to complete these tasks. Instead, we leverage the power of machine learning in order to annotate, classify, and visually complete missing cultural data [3]. In the following, we present some of the approaches that we designed to tackle the previously mentioned challenges.

3.1 The 3D-Holoscopic Imaging Framework Adapted for the Cultural Content

The 3D-holoscopic technology is not recent. Its principle was proposed in 1908 by Lippmann [4]. The technology is often referred to as lightfield imaging. The principle is inspired by Fly's eyes using an evenly spaced macrolens array fitted to a normal camera (either DSLR or mirrorless) [5, 6]. Each of these lenses captures the scene from a slightly shifted angle in comparison with neighboring lenses in the array. The fundamental principle of H3D is described by Fig. 3. The lightfield data is recorded by the CMOS sensor and stored as a 2D capture. At the display stage, the same process used for capture is reversed. A Macrolens array (MLA) is placed in front of the screen and the object can be reconstructed in space [5, 6]. Figure 2 represents the H3D camera prototype developed by the CMCR Laboratory at Brunel University, London.

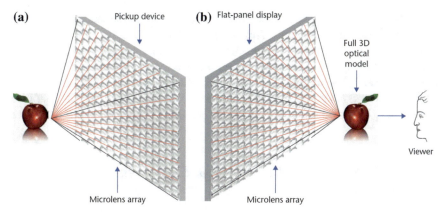

Fig. 3 3D-holoscopic capture and display principle

The principle of the H3D technology and its ability to preserve the depth information with a single capture made it one of the best alternatives to 3D scanning and photogrammetry for the acquisition and display of cultural data due to its cost-effectiveness and ease of use. However, there are still several challenges regarding the requirements of heritage organizations such as the preservation of output quality, etc. The CEPROQHA team focused on adapting the H3D acquisition framework to comply with these requirements, and thus introduced some novelties to the capture; post-processing and display stages of the framework (see Fig. 4) [7].

One of the limitations found in 3D-holoscopic cameras is that the output resolution is smaller in comparison with normal 2D captures. Increasing this resolution through software is thus a must to preserve the main selling point of the H3D technique being its cost-effectiveness. Our team designed and implemented a content adaptation framework to mitigate the low spatial pixel density of H3D captures through using machine learning techniques such as super-resolution to upscale the images.

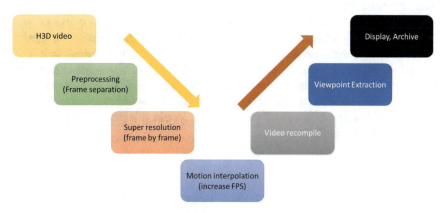

Fig. 4 H3D post-processing framework design

The 360° video scenario was also studied. The video mode of the camera we used has 4 K as the maximum resolution. This is a 1/5 of the 41 megapixels that our CMOS sensor has. We thus developed an adapted 360° video framework that takes 72 still captures shifted by 5° of the assets using the maximum resolution of the camera (41 megapixels). These captures are first upscaled using super-resolution. Then, we apply video motion interpolation to compute the in-between frames. The result is a smooth very high-resolution 360° video that preserves the fine details of the asset.

3.2 Data Analytics for Cultural Data Enrichment and Curation

Cultural data annotation

The CEPROQHA team designed several annotation and classification approaches for multiple scenarios. These approaches mainly focus on either the full annotation of the metadata or its partial annotation. These approaches rely mostly on the visual features and characteristics of cultural assets to predict the desired labels using a combination of deep learning-based approaches such as convolutional neural networks (CNN) and transfer learning. The frameworks we designed achieved excellent performance and were validated on several datasets of paintings collected from multiple museums and institutions such as the Museum of Islamic Art in Doha, Qatar, Wikiart, the Rijksmuseum and the Metropolitan Museum of New York [8].

- Multi-task Hierarchical classification

The main intuition behind this approach came up after analyzing the collected datasets and reviewing the related works on cultural heritage classification. We mainly found out that it is inefficient to deal with different types of assets with the same classification model and tools. Indeed, each type of objects has some meta-data properties that can easily be predicted when a specific classifier is concisely implemented, but this classifier cannot be generalized to other types of assets using the same approach. For example, the genre and style fields can be found for a painting, but can never be found for a sword even if both are hosted in the same museum or collection. We thus designed a multi-task hierarchical classification that has mainly two stages. In the first stage, a general type classification CNN takes as input the asset image and predicts its type. In the second stage, the asset image is forwarded to the assigned multi-task classifier for that specific type in order to recover the missing metadata (see Fig. 5) [8, 9].

Fig. 5 Hierarchical multi-task classification

- Multimodal classification

Traditional visual classification approaches rely on a single visual capture of a cultural asset to perform prediction on the desired labels. However, in reality, the visual capture of the misannotated asset is generally not the unique information that we have. Often, some labels can be found along this capture and can be used to enrich the input data. This, in fact, is a more realistic scenario which was validated with a handful of heritage institutions. Our team designed and implemented a multimodal classifier for paintings based on convolutional neural networks and transfer learning that takes as input the visual capture of a painting along with the available information at hand. The task of the designed model is to use both visual and textual features to predict the missing labels. We compared this approach with a two-task multi-task classifier with the same output labels and the result was that the multimodal classifier was more effective and yielded a higher validation accuracy across the dataset we used for validation [3] (see Fig. 6).

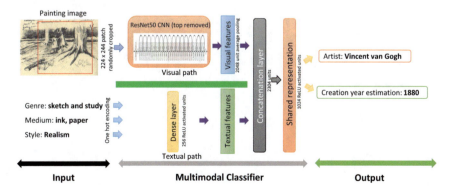

Fig. 6 Our multimodal classifier architecture

Information Extraction for Cultural Ontology Learning

There is a growing need for tools that facilitate the transformation of cultural data to a common form that is shareable and accessible by the domain community as Linked Open Data. Unfortunately, the information manipulated by different organizations is not commonly structured. Each organization has its own data scheme that makes sharing this information with other institutions and organizations difficult. There are several standards for sharing cultural information such as the CIDOC CRM scheme [10]. The manual transformation of this information into a standardized scheme is impractical as it is both time-consuming and labor-intensive. Natural language processing (NLP) techniques are automatic text processing techniques and can help with the metadata scheme transformation. NLP techniques can also help in tackling one of the main limitations that prevent institutions in adopting ontologies as a replacement to standard databases as it is impractical to manually populate ontologies [11]. These activities are usually performed by a domain expert and are labor-intensive and time-consuming. The manual population of ontologies is generally unfeasible except for very small domains. To be practical, the system needs to automate or semi-automate the definition of item metadata from resources such as item descriptions that are available in Web sites, blogs, etc. As these resources are generally available in a free text format, effective methods need to be developed to be able to extract the entities and their relations which are the building block of any ontology. This area of research is called *Ontology Learning* [12].

The two NLP research areas that are today active and mostly relevant to ontology learning are *named-entity recognition* (NER) and *relation extraction* (RE). *Named-entity recognition* (NER) focuses on extracting domain entities from unstructured text such as dates, places, and people names. The definitions of entities are either taken from knowledge sources such as Wikipedia and other sources that are easily and freely accessible or discovered through NLP techniques [13, 14] (see Fig. 7). *Relation Extraction* (RE) on the other hand is concerned with extracting the occurrence of relations in the text which would facilitate the discovery of the relation between the domain entities mentioned in the text [12].

Figure 8 shows an example of the entities extraction and the relation between them from a description of a pot from the Web site of The Metropolitan Museum of Art in New York. Our research concentrates on the transformation of the information available online from museums and cultural heritage institutions. This information is mainly stored as a free text into a more formal structure as an ontology. By representing them as an ontology, this opens the door for many applications such as semantic search, browsing, and visualization, etc.

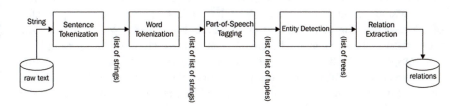

Fig. 7 Typical natural language processing pipeline

Stews and soups were staples of the colonial diet, so cooking pots such as this example would have seen daily use. With no tripod base to stand on, this pot had to be hung over the fire from a trammel. Similar pots were made early in the colonial era at the Saugus Iron Works outside Boston (in operation from 1644 to 1670).

Fig. 8 Named-entity recognition and relation extraction

4 Conclusion

Through this paper, we presented the different challenges and the most important uses of machine learning and data science in cultural heritage context. We presented digital heritage, which represents the set of tools and techniques used to digitize and transfer cultural heritage from the physical to the digital world. This transformation is nowadays necessary as it opens the way to a large spectrum of applications that positively impact cultural heritage. We presented some of the work that was performed by our team in the course of the CEPROQHA project which targets the application of machine learning and data science in the cultural domain. We mostly covered topics such as the automated labeling of misannotated assets, the visual completion of incomplete or damaged assets, and the 3D-holoscopic content adaptation framework, which was improved with super-resolution and motion interpolation. We have also highlighted the importance of using natural language processing to facilitate the adoption of ontologies in cultural heritage context. Some of these techniques still require some work to achieve better maturity and to be ready for real-world use. As future work, we plan to integrate all these tools as well as a tailored digital preservation process in a customized collection management system that links artificial intelligence to cultural heritage.

Acknowledgements This publication was made possible by NPRP grant 9-181-1-036 from the Qatar National Research Fund (a member of Qatar Foundation). The statements made herein are solely the responsibility of the authors (www.ceproqha.qa).
The authors would also like to thank Mr. Marc Pelletreau, the MIA Multimedia team, the Art Curators and the management staff of the Museum of Islamic art, Doha Qatar for their help and contribution in the data acquisition.

References

1. Belhi, A., Foufou, S., Bouras, A., Sadka, A.H.: Digitization and preservation of cultural heritage products. In: IFIP International Conference on Product Lifecycle Management, pp. 241–253. Springer (2017)
2. Van Noord, N., Hendriks, E., Postma, E.: Towards discovery of the artist's style: learning to recognize artists by their artworks. IEEE Sig. Process. Mag. **32**, 46–54 (2015)
3. Belhi, A., Bouras, A., Foufou, S.: Leveraging known data for missing label prediction in cultural heritage context. Appl. Sci. **8**, 1768 (2018)
4. Lippmann, G.: Epreuves reversibles donnant la sensation du relief. J. Phys. Theor. Appl. **7**, 821–825 (1908)
5. Swash, M.: Holoscopic 3D Imaging and Display Technology: Camera/Processing/Display. Brunel University, London (2013)
6. Aggoun, A., Tsekleves, E., Swash, M.R., Zarpalas, D., Dimou, A., Daras, P., Nunes, P., Soares, L.D.: Immersive 3D holoscopic video system. IEEE Multimedia **20**, 28–37 (2013)
7. Alfaqheri, T., Nasri, S.A.E.M., Sadka, A.H.: 3D Holoscopic Imaging for Cultural Heritage Digitalisation. arXiv preprint arXiv:1810.03916 (2018)

8. Belhi, A., Bouras, A., Foufou, S.: Towards a hierarchical multitask classification framework for cultural heritage. In: 2018 IEEE/ACS 15th International Conference on Computer Systems and Applications (AICCSA), pp. 1–7. IEEE 2018
9. Belhi, A., Bouras, A.: Towards a multimodal classification of cultural heritage. In: Qatar Foundation Annual Research Conference Proceedings, pp. ICTPD1010. HBKU Press, Qatar (2018)
10. Doerr, M.: The CIDOC conceptual reference module: an ontological approach to semantic interoperability of metadata. AI Mag. **24**, 75 (2003)
11. Gasmi, H., Bouras, A., Laval, J.: LSTM recurrent neural networks for cybersecurity named entity recognition. ICSEA **2018**, 11 (2018)
12. Buitelaar, P., Cimiano, P., Magnini, B.: Ontology learning from text: an overview. Ontology learning from text: methods, evaluation and applications. Front. Artif. Intel. Appl. Ser. **123** (2005)

13. Belhi, A., Bouras, A., Foufou, S.: Digitization and preservation of cultural heritage: the CEPROQHA approach. In: Software, Knowledge, Information Management and Applications (SKIMA), 2017 11th International Conference on, pp. 1–7. IEEE 2017
14. Nadeau, D., Sekine, S.: A survey of named entity recognition and classification. Lingvisticae Invest. **30**, 3–26 (2007)

Developing a Business Model for a Smart Pedestrian Network Application

George Papageorgiou, Eudokia Balamou and Athanasios Maimaris

Abstract This paper examines the field of business models for successfully developing a smartphone application in order to promote walkability in urban environments. Carrying out a review of the literature on the theory of business models, it was revealed that successful business development today should be accompanied by a holistic view of the organization and its relationships with multiple stakeholders at all sides: supply, demand and partners sides. As a result, we propose a suitable framework for creating an effective business model—system architecture for a Smart Pedestrian Network (SPN) application. Due to the complexity and dynamic nature of the SPN domain, it is necessary to adopt an open innovation (OI) approach, where end users–citizens, businesses, application programming interface (API) gateways, system integrators (SI), external service providers, external service developers, well as municipalities–city authorities work together in a co-creation atmosphere.

Keywords Business models · Smartphone applications · Smart cities · Municipal organizations · Strategic management

1 Introduction

With the recent rapid advancements of technology, new opportunities arise for creating business models in order to efficiently and effectively serve people's needs. In this way, organizations can implement strategies to develop sustainable comparative advantages. Associated with such strategies is the field of business models, which

G. Papageorgiou · E. Balamou · A. Maimaris (✉)
E.U.C. Research Center, 6 Diogenous Street, 2404 Engomi, Cyprus
e-mail: a.maimaris@cycollege.ac.cy

G. Papageorgiou
European University Cyprus, 6 Diogenous Street, 2404 Engomi, Cyprus
e-mail: g.papageorgiou@euc.ac.cy

A. Maimaris
Cyprus College, 6 Diogenous Street, 2404 Engomi, Cyprus

© Springer Nature Singapore Pte Ltd. 2020
X.-S. Yang et al. (eds.), *Fourth International Congress on Information
and Communication Technology*, Advances in Intelligent Systems
and Computing 1027, https://doi.org/10.1007/978-981-32-9343-4_30

375

has developed over the past three decades with the phenomenal exponential growth of the Internet. Several studies have contributed to the field of business models, such as the work of Timmers [9] on business models for electronic markets, Mahadevan [4] on developing models based on e-commerce, Afuah and Tucci [1] with their work on strategies for Internet business models, Bouwman and van den Ham [2] on developing eMetrics for business models, Osterwalder and Pigneur [6] on an ontology for modeling e-business, Weill and Vitale [11] on migrating from place to electronic space markets as well as Chen and Nath [3] on developing a framework for mobile business applications. The aforementioned studies and other work are further examined in the literature review section of this paper.

The field of business models is currently under heavy development and transforms the field of strategic management. It also has a significant effect on the process of innovation, and especially on the concept of co-creation and open innovation (OI) [7]. In a critical review on business models research, Massa et al. [5] interpret business models as attributes of a real firm, as cognitive linguistic schemas, as well as how a business actually functions. Clearly, this finding adds high complexity as to what constitutes an effective business/organization. This further calls for thorough analysis of the many elements that are related to the business/organization. Particularly, for municipalities/city government organizations, for success in business model development, a holistic approach is necessary, which involves both the demand side and supply sides of an organization as well as the partners' side. Thus, in order to create real value, we need to take a multi-stakeholder approach in our business model. Such an approach will be especially useful, due to the dynamic nature of the municipal environment and the many stakeholders involved.

The next section presents the concept of the Smart Pedestrian Network and its importance for smart cities. This is followed by an analysis of the topic of business model development As a result, we develop and present a system architecture for the proposed SPN business model. Finally, conclusions are drawn, and a statement is made for future work.

2 The Concept of a Smart Pedestrian Network (SPN)

The idea of a Smart Pedestrian Network (SPN) [8] arises from the tragic situation that today people drive more and walk less. This situation has negative effects on health aspects, environmental conditions but also on the socioeconomic state of a city, as well as on people's quality of life. For the development of smart cities, it seems that local governments have placed a lot of emphasis on efficiently moving motorized traffic and ignored to some degree active transport modes. As a result, people are attracted to driving even more, with a corresponding increase in noise and air pollution.

The SPN concept is an effort to provide relevant information, so that people are attracted toward walking more and driving less. Such information could come from a variety of sources such as electronic sensors, social media analytics and Geographical

Information Systems (GIS). This information, though, needs to be assessed and consolidated so that it is valid and useful in order to promote walkability. Further, the SPN system could serve as an evaluation tool for municipalities in order to identify areas of improvement of their cycling and pedestrian infrastructure, as well as invest more in their public transport system. With the implementation of the SPN system, active transportation could be enhanced, which leads to a healthier lifestyle and more energized and productive workers. In this way, we can create real smart cities, with sustainable urban environments.

SPN uses multiple criteria to analyze and evaluate the pedestrian conditions. The criteria are based on external aspects, such as the morphology of the built environment, land use, accessibility, connectivity, safety and so on. Also, SPN considers human factors on the decision-making process for walking, as well as the satisfaction arising from walking. As a result, information can be given to citizens about the best route to take personally for their own walk.

Moreover, the SPN system will be useful to municipal organizations in order to formulate and implement their urban transport plans for improving walkability. Such plans could include educational seminars, investment on the pedestrian infrastructure, as well as the creation of attractive open public spaces. The main idea is that the provision of relevant information could bring about a behavioral change, provided that the pedestrian conditions are attractive, safe, comfortable, with high public transport connectivity.

Implementing the SPN concept will involve a high initial investment cost for municipalities. Therefore, it is necessary to consider a sustainable business model for the implementation and further development/upgrading of the SPN system. The business model should be holistic and have a socioeconomic nature, so that it takes into consideration monetary benefits, health benefits, environmental benefits, transportation savings, as well as appreciation of property value, as a result of a more walkable urban environment.

SPN takes the form of a smartphone application, which is maintained by the city/municipality authorities, with the support of the local business community. The next section provides a literature review of business models, with a discussion on their relevance to the SPN concept.

3   Development of an Architecture for the SPN System

Considering the various options that exist in the literature for the choice of business models, this section presents the development of an architecture for the SPN business model. As shown in Fig. 1, this architecture is comprised of six different entities. These are the end users part, the application programming interface (API) gateway to services, the system integrator data center, the city/municipality, external service providers and potential external service developers.

In the end user entity, we see five possible business models to be used by the SPN smartphone app. End users could have access via outright purchase, on a subscription

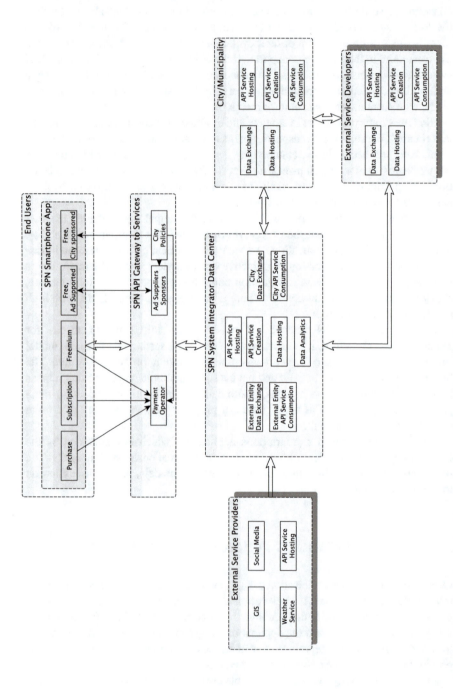

Fig. 1 Business model system architecture for a Smart Pedestrian Network Application

basis, a freemium model with limited choice of features, a free ad-supported model and a free city-sponsored model. The API gateway to services involves a payment operator, ad suppliers/sponsors and specific city policies. Through the API gateway, the SPN developers can have control on these entities. The system integrator data center is comprised API service creation, hosting and consumption as well as data hosting and analytics. The data center also has data exchange with the city and other external entities through appropriate API service consumption. The external service providers could be GIS, social networks, weather service and many others. Potentially, the proposed architecture includes external service developers, which interact with both the city/municipality and the system integrator data center. Finally, the city/municipality entity, which interacts with the system integrator data center, is comprised also with API service creation, hosting and consumption as well as data hosting and exchange.

The city needs to ascertain whether it has the infrastructure and know-how to handle the SPN data storage, processing and end user services provision in-house that is own data center, personnel and Internet bandwidth or to procure such a service that System Integrator Data Center as a public–private partnership (PPP).

The city has a wealth of information at its disposal, such as traffic information from traffic light controllers, CCTV, location and speed of municipal fleet, ongoing and future roadworks, GIS database of road and pedestrian network and so on. This information needs to be available to the data center in a secure manner and without affecting the privacy rights of individuals.

The data center needs to collect information from the city and also from numerous external service providers, such as GIS databases for local information e.g. Open-StreetMap and Google Maps for shops, interesting places, and amenities and social media e.g., Facebook, Twitter, Instagram and so on. This is necessary in order to derive information about local issues such as the quality of sidewalks and the level of security of particular areas. Also, it is important to have access to information such as weather conditions and particulate levels in the atmosphere. All of this wealth of information can then be accessed by the SPN service application in order to handle the end user queries on the best route to take, depending on pedestrian's concerns and priorities.

The information that the city and data center hold can be further processed and used by external service providers, in a responsible, secure and privacy-respecting manner, to provide novel services to people for the sustainable development of the urban environment.

End users have a choice of multiple smartphone app purchase models. Depending on their needs and their willingness to use the application, they could go for a direct purchase without ads, a subscription scheme, freemium with limited features, a free version with ads and a free city-sponsored option. When the city is rich enough, or there are sponsors willing to support it, the SPN app can be used for free in that particular city. There is a case that the sponsors may want the city to allow advertisements in the app as a condition of their sponsorship. This is a possible way for a city to recover expenses, but care and willingness are needed to vet the adverts for content and how invasive are to the end user.

In freemium SPN app, where the end user can download and use a limited set of features for free, the end user can see if the SPN app is useful and may decide to pay a small amount of money to unlock further features on demand. Also, an end user may elect to pay a small subscription fee monthly or yearly to receive further enhancements and new versions of the SPN app as they are made available. In the case of direct purchase, the SPN app has all features available from day one, and in addition all newer versions as they are made available.

Access to the connection between the SPN smartphone app and the SPN data center is through an API gateway. Depending on the city's choice of business model (free, ad-supported, freemium, subscription or purchase), the API gateway would determine what features the end user can access. Also, this gateway would handle payments by the end users. Through the API gateway, SPN could be promoted. Especially if the intermediary is Apple App Store or Google Play Store, the application will be available worldwide. Such a development will be very beneficial regardless of any commission amount given.

4 Conclusion

This paper examines the concept of business models and their application in the case of municipal organizations. Based on the current developments on business model theories, a business model system architecture for a Smart Pedestrian Network Application is developed and presented. The proposed architecture aims at promoting walkability, satisfying the needs of citizens, while supporting the operations and decision processes of municipalities. To ensure the implementation, viability and sustainability of the envisioned Smart Pedestrian Network Application, the proposed framework should be considered, as it involves all the main stakeholders.

The presented framework could serve as a common platform for communication purposes and open innovation, where end users, API gateways, system integrators, external service providers and the city/municipalities could contribute in successfully implementing SPN. This will be especially useful to municipalities and local authorities, but also to software developers and businesses who wish to innovate and build services for smart cities. The next step is to test and demonstrate an implementation of the proposed framework in specific municipalities in Europe.

Acknowledgements The research presented in this paper is co-funded by the Republic of Cyprus and the European Regional Development Fund as part of ERA–NET Cofund Smart Urban Futures (ENSUF) Joint Programming Initiative (JPI) Urban Europe, through the Research Promotion Foundation, protocol no. ΚΟΙΝΑ/ΠΚΠ URBAN EUROPE/1215/11. This framework is supported by the European Commission and funded under the HORIZON 2020 ERA–NET Cofund scheme.

References

1. Afuah, A., Tucci, C.L.: Internet Business Models and Strategies: Text and Cases. McGraw-Hill Higher Education, New York (2000)
2. Bouwman, H., van den Ham, E.: Designing Metrics for Business Models Concerning Mobile Services Delivered by Networked Organisations. In: 16th Bled Electronic Commerce Conference eTransformation, pp. 1–20 (2003)
3. Chen, L.d., Nath, R.: A framework for mobile business applications. International Journal of Mobile Communications **2**(4), 368–381 (2004). 10.1504/ijmc.2004.005857
4. Mahadevan, B.: Business Models for Internet-Based E-Commerce: An Anatomy. California Management Review **42**(4), 55–69 (2000). https://doi.org/10.2307/41166053
5. Massa, L., Tucci, C.L., Afuah, A.: A Critical Assessment of Business Model Research. Academy of Management Annals **11**(1), 73–104 (2017). https://doi.org/10.5465/annals.2014.0072
6. Osterwalder, A., Pigneur, Y.: An e-Business Model Ontology for Modeling e-Business. In: 15th Bled Electronic Commerce Conference (2002)
7. Papageorgiou, G., Efstathiades, A., Milikouri, G.: Innovation, Co-creation and the New Product Development Process in Small to Medium-Sized Enterprises (SMEs). In: 2016 European Conference on Innovation and Entrepreneurship, pp. 488–495. Academic Conferences International Limited, Reading (2017)
8. Papageorgiou, G., Maimaris, A.: Towards the development of Intelligent Pedestrian Mobility Systems (IPMS). In: 2017 International Conference on Electrical Engineering and Informatics (ICELTICs), pp. 251–256 (2017). 10.1109/ICELTICS.2017.8253267
9. Timmers, P.: Business models for electronic markets. Electronic markets **8**(2), 3–8 (1998). https://doi.org/10.1080/10196789800000016
10. Walravens, N.: Validating a Business Model Framework for Smart City Services: The Case of FixMyStreet. In: 2013 27th International Conference on Advanced Information Networking and Applications Workshops (WAINA), pp. 1355–1360. IEEE (2013). 10.1109/waina.2013.11
11. Weill, P., Vitale, M.R.: Place to Space: Migrating to eBusiness Models. Harvard Business School Press (2001)

The Role of Supervised Climate Data Models and Dairy IoT Edge Devices in Democratizing Artificial Intelligence to Small Scale Dairy Farmers Worldwide

Santosh Kedari, Jaya Shankar Vuppalapati, Anitha Ilapakurti, Sharat Kedari, Rajasekar Vuppalapati and Chandrasekar Vuppalapati

Abstract Climate change is impacting milk production worldwide. For instance, increased heat stress in cows is causing average-sized dairy farms losing thousands of milk gallons each year; drastic climate change, especially in developing countries, pushing small farmers, farmers with less than 10–25 cattle, below the poverty line and is triggering suicides due to economic stress and social stigma. It is profoundly clear that current dairy agriculture practices are falling short to counter the impacts of climate change. What we need are innovative and intelligent dairy farming techniques that employ best of traditional practices with data-infused insights to counter negative effects of climate change. We strongly believe that "climate" is a data problem and the democratization of artificial intelligence-based dairy IoT devices to farmers is not only empowers farmers to understand the patterns and signatures of climate change but also provides the ability to forecast the impending climate change adverse events and recommends data-driven insights to counter the negative effects of climate change. With the availability of new data tools, farmers can not only improve their standard of life but, importantly, conquer perennial "climate change-related suicide" issue. It's our staunch believe that the *gold standard* for the success of the democratization of artificial intelligence is no farmer life loss due to *negative effects of climate*

S. Kedari · J. S. Vuppalapati
Hanumayamma Innovations and Technologies Private Limited, HIG-II, Block-2/Flat-7, Baghlingumpally, Hyderabad 500044, India
e-mail: skedari@hanuinnotech.com

J. S. Vuppalapati
e-mail: jaya.vuppalapati@hanuinnotech.com

A. Ilapakurti · S. Kedari · R. Vuppalapati · C. Vuppalapati (✉)
Hanumayamma Innovations and Technologies, Inc., 628 Crescent Terrace, Fremont, CA, USA
e-mail: cvuppalapati@hanuinnotech.com; chandrasekar.vuppalapati@sjsu.edu

A. Ilapakurti
e-mail: ailapakurti@hanuinnotech.com

S. Kedari
e-mail: Sharath@hanuinnotech.com

R. Vuppalapati
e-mail: raja@hanuinnotech.com

© Springer Nature Singapore Pte Ltd. 2020
X.-S. Yang et al. (eds.), *Fourth International Congress on Information and Communication Technology*, Advances in Intelligent Systems and Computing 1027, https://doi.org/10.1007/978-981-32-9343-4_31

change. In this paper, we propose an innovative machine learning edge approach that considers the impact of climate change and develops artificial intelligent (AI) models that is validated globally but enables localized solution to thwart impacts of climate change. The paper presents prototyping dairy IoT sensor solution design as well as its application and certain experimental results.

Keywords Internet of things (IoT) · Machine learning · Climate change · Climate change-related suicides · Decision tree · Regression analysis · Embedded device · Edge analytics · Humidity sensors · Dairy IoT sensor · Hanumayamma dairy IoT sensor · Climate models · Farmer suicides

1 Introduction

The Climate change is real and "climate is now a data problem" [1]. Increases of atmospheric carbon dioxide (CO_2), rising temperatures, and inconsistent precipitation patterns are impacting milk production worldwide. For instance, increased heat stress in cows is causing average-sized dairy farms losing thousands of milk gallons each year; drastic climate change, especially in the developing and the developed countries, pushing small farmers below the poverty line and is triggering suicides due to economic stress and social stigma.

The study [2] from the University of California, Berkeley, has clearly corroborated the link between the rising temperatures and the resultant stress on India's agricultural sector and its impact on increase in suicides over the past 30 years (see Fig. 1); the

Fig. 1 Farmer protest [3]

Fig. 2 Climate models [6]

study has categorically quantified suicides of nearly 60,000 Indian farmers [3, 4] are related to climate change.

The problem not only confined to the developing countries, ironically, in the Unites States of America, a recent report by the Guardian Newspaper [5] establishes the facts that "the Suicide rate for farmers is more than double that of veterans." Similarly, Forbes magazine's article "dairy Farmers across the United States have been committing suicide at an alarming rate because their respective industry no longer supports them in the same way that it has for generation." The National Public Radio (NPR) [6], importantly, ran the program that underscores [7] "milk prices decline to the worries about dairy farmers suicides rise" (see Fig. 2). Finally, a recent New York Times article summarizes the plight of farmers in Australia [8]: "being a breadbasket to the world and a globalization success story, so why are Australian farmers killing themselves?" [8].

The farmer's suicide is sad and heart-breaking. Many of the farmers are small-scale farmers with no financial, social, or political support to fight against the impacts of the climate change. They have no adequate tools at their disposal to apply to the change their practices to accommodate rising temperatures. And the paths many of these farmers are traversing are leading to the unthinkable perennial issue.

It is profoundly clear that current dairy agriculture practices are falling short to counter the impacts of climate change. The boomerang effect of current state, in terms of what future holds, is dire as multiplicative stress vectors of population growth with projected depletion of fertile lands for dairy and agriculture due to urbanization would exert unprecedented scarcity of milk and milk-based products, leaving small farmers helpless and resulting millions of malnourished children worldwide, especially in developing countries. Without the application of data-intense practices, the farmers of our generations seldom to exist as climate change is our problem and not of our future generations.

What we need is innovative and intelligent dairy farming techniques that employ best of traditional practices with data-infused insights and continuous inclusion of adaptive data models that factor-in climate changes to counter the negative effects. For instance, a new study conducted by Columbia Engineering [9] has indicated that machine learning to better represent clouds in climate models, and thus better predict the global and regional climate's responses to rising greenhouse gas concentrations.

We strongly believe that by creating network effect of reinforcement learning from the statistical supervisor models on the data collected from the real-time IoT dairy sensors that are distributed across different thermal and geolocation regions with each sensor capturing cattle's productivity, health, behavioral, and clinical data with underpinning climate markers; the collective intelligence of such data can be used to provide *recommendations* that exploit network benefits and tacit knowledge of farmers across the world. Thus, developed models can be perpetually improved as the new data elements are collected and reenergizes and creates anti-climate-change data fusion; the data insights, albeit the light, enable the best defense for our current and future generations. It is our ardent believe that the ***data is our best defense*** and savior against the negative effects of climate change. The sooner we embark on democratization of AI, the better we leave our progeny a wonderful life on the earth, i.e., better than what we have inherited.

As part of the paper, we have proposed our approach to address climate change. We have developed mathematical, electrical, software, and data science solution! We have developed dairy sensors to empower small farmers to counter the climate change. As part of our presentation, we would like to demonstrate the workings and positive impact of our sensors to small farmers and present our sensor deployments in South Asia. The structure of this paper is presented as follows: Sect. 2 discusses the basic concepts and methods about mathematical models, data attributes, partial derivatives, and data science algorithms. Section 3 presents our system architecture. Section 4 discusses its related design and implementation decisions, and Sect. 5 shows a case study. The conclusion and future work is included in Sect. 6.

2   Understanding Climate Change and Dairy Sensor—Software, Hardware and Analytics

2.1   Internet of Things (IoT) Sensor

The IoT is characterized by data collection and processing at various stages (see Fig. 3): the edge layer is very close to the source as it ensembles a sensor or an actuator. The data at this level is called data in motion and need to process with very less latency. The next layer is typically called a fog layer. This is the central network hub that assembles the data from the edge devices. The fog layer inspects or performs a minimal data operation. And the cloud layer, the data center, stores the data for historical and other data reporting or operations purposes.

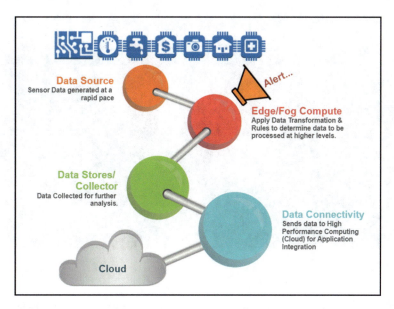

Fig. 3 IoT stack

2.2 Data Definitions

- **Cold start**: Developing dairy analytics with climate models imbed into it is a classical "cold start" problem. There is less or no available data.
- **Holdout issue**: Obtaining dairy cattle data is often expensive and time consuming. On top of it, obtaining dairy data with climate change and weather infused is practically non-existing; this is our findings.
- **Stratification issue**: Initial data models that we have developed suffered from data stratification issues.

2.3 NOAA

The National Centers for Environmental Information (NCEI[1]), hosted by NOAA, delivers public access to one of the most important records for environmental data on Earth (see Fig. 4). The center provides over 25 petabytes of weather and climate data.

The climate and weather forecast models can be used to model agriculture data.

---

[1]NOAA—https://www.ncei.noaa.gov/.

Fig. 4 NOAA

For instance, wind chill warning[2] can be used to take proactive steps to mitigate the side effects (see Fig. 5). Additionally, the data points could be baselined to

Fig. 5 National weather services

[2]National Weather Services—https://www.weather.gov/safety/cold-wind-chill-warning.

Fig. 6 AQI

functioning of the dairy sensors.

2.4 Air Quality Index

Having air quality index (AQI[3]) data (see Fig. 6) also helps to prepare climate models.

2.5 Hanumayamma Dairy Sensor

Generally, Class 10 classification[4] is reserved to medical apparatus such as: Surgical, medical, dental and veterinary apparatus and instruments, artificial limbs, eyes and teeth; orthopedic articles; suture materials [10–13].

Specifically, for dairy sensor (see Fig. 7), also known as cow necklace, we have been assigned "Class 10: Wearable veterinary sensor for use in capturing a cow's vital signs, providing data to the farmer to monitor the cow's milk productivity, and improving its overall health."

We have based above sensor design based on reference architecture provided by Hanumayamma Innovations and Technologies, Inc., IoT dairy sensor.[5]

---

[3] AQI—https://airnow.gov/index.cfm?action=airnow.local_city&zipcode=94536&submit=Go.

[4] Trademark administration, general provisions, and definitions—https://www.sec.state.ma.us/cor/corpdf/trademark_regs_950_cmr_62.pdf.

[5] Hanumayamma Innovations and Technologies, Inc., http://hanuinnotech.com/dairyanalytics.html.

Fig. 7 Dairy IoT sensor

2.6 Machine Learning

(1) *ANOVA*: Analysis of variance, a statistical method in which the variation in a set of observations is divided into distinct components. In our case, we have analyzed the data samples collected by sensors for different thermo-regions to milk and health of the cattle.

(2) *Regression techniques*: We have used statistical linear regression techniques to influence the parameter the of temperature change to the number of input parameters.

We have developed two edge analytics models: (1) Crowdedness to temperature and (2) Temperature to humidity correlation. Both models have used case level applicability to the climate model analytics.

We have downloaded Kaggle dataset (see figure Kaggle Dataset) to get crowdedness to temperature machine learning model.

Dataset: https://www.kaggle.com/nsrose7224/crowdedness-at-the-campus-gym.

The dataset consists of 26,000 people counts (about every 10 min) over the last year. In addition, extra information was gathered, including weather and semester-specific information that might affect how crowded it is. The label is the number of people, which is predicted, given some subset of the features.

Label:

- Number of people

  Features:

- date (string; datetime of data)
- timestamp (int; number of seconds since beginning of day)
- day_of_week (int; 0 [monday] - 6 [sunday])
- is_weekend (int; 0 or 1) [boolean, if 1, it's either saturday or sunday, otherwise 0]

- is_holiday (int; 0 or 1) [boolean, if 1 it's a federal holiday, 0 otherwise]
- temperature (float; degrees fahrenheit)
- is_start_of_semester (int; 0 or 1) [boolean, if 1 it's the beginning of a school semester, 0 otherwise]
- month (int; 1 [jan] - 12 [dec])

```
Temp_Model
= pd.read_csv(r'C..MachineLearning\
Crowdedness_To_Temperature_20170403.csv')
df = DataFrame(Stock_Market,columns = ['number_people','day_of_week',
'is_weekend','is_holiday','month','hour','temperature'])
```

```
X = df[['number_people','day_of_week','is_weekend','is_holiday','month',
'hour']] # here we have 5 variables for multiple regression.
If you just want to use one variable for simple linear regression,
then use X = df['Interest_Rate'] for example.Alternatively,
you may add additional variables within the brackets
Y = df['temperature']
```

```
# with sklearn
regr = linear_model.LinearRegression()
regr.fit(X, Y)
```

```
# prediction with sklearn
New_number_people = 48
New_day_of_week = 5
New_is_weekend = 0
New_is_holiday = 0
New_month = 9
New_hour = 20
print ('Predicted Temperature: \n',
regr.predict([[New_number_people ,New_day_of_week,
New_is_weekend,New_is_holiday,New_month,New_hour]]))
```

```
# with statsmodels
X = sm.add_constant(X) # adding a constant
```

```
model = sm.OLS(Y, X).fit()
predictions = model.predict(X)
```

```
print_model = model.summary()
print(print_model)
```

For the model output, see Fig. 8.

(3) *Naive Bayesian*: The conditional probability makes very easier to model the changing values of one parameter with respect to a given sample set of data. The Naive Bayesian model helped to parameterize the model.

(4) *Clustering techniques*: The clustering is the process of arranging items that are similar in nature. The clustering technique functions as a backbone for identifying and creation of recommendations.

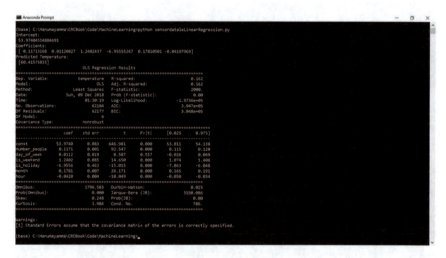

Fig. 8 Console output

2.7 Climate and Dairy Sensor Parametric Model

A parametric model is a collection of probability distributions such that each member of this collection, P_θ is defined by finite-dimensional parameters. The set of all allowable values for the parameter is denoted by $\odot \leq R^k$ where R^k defines all possible values.

Climate events: To derive climate change to dairy sensor parametric model, let us start with following climate events (see Table 1): (other extreme climate events can be logically deduced through the established process):

Chicago's El trains move along snow-covered tracks Monday, January 28, 2019, in Chicago. The plunging temperatures expected (see Fig. 9) later this week that have forecasters especially concerned. Wind chills could dip to negative 55 degrees in northern Illinois, which the National Weather Service calls "possibly life threatening" (AP Photo/Kiichiro Sato).

Source: https://phys.org/news/2019-01-record-breaking-cold-midwest-snowstorm.html#jCp

Table 1 Climate events

Climate event	Symbol
Rise of tide[a]	$\tilde{U}$
Rise of temperature	$\mathring{U}$
Rise of humidity	$\breve{U}$
Flooding[b]	$\acute{U}$
Rise of noise	ȸ
Depression	$\underline{U}$
Wind chill warning	$\hat{W}$
Snow storm[c]	$\ddot{Y}$

[a]Water Quality for Live Stock—https://www.agric.wa.gov.au/livestock-biosecurity/water-quality-livestock
[b]Flooding—Contaminated   farm   dams—https://www.agric.wa.gov.au/water-management/contaminated-farm-dams?page=0%2C1
[c]Mid-West Snow Storm—https://phys.org/news/2019-01-record-breaking-cold-midwest-snowstorm.html

Fig. 9 Record-breaking cold coming to mid-west after snowstorm

Table 2 Climate events to dairy sensor

Climate event	$A(x, y, z)$	$G(x, y, z)$	T	H	Physiological	Emotional
$\tilde{U}$	↑	↑	↑	↑	↑	↑
$\mathring{U}$	↓[a]	↓	↑	↓	↑	→
$\breve{U}$	↑	↑	↓	↑	→	→
$\tilde{U}$	→	→	→	→	↑	↑
♊	↑	↑	↑	↑	↑	↑

[a]Heat Stress in Dairy Cows—http://article.sapub.org/pdf/10.5923.j.zoology.20120204.03.pdf

Let us consider the following climate events to dairy sensor parametric model table (see Table 2).

Please note:

↓ means decreases
↑ Increases
→ no change or do not care

2.8 Factor-in Climate Conditions

To factor-in the effects of climate events on dairy cattle, through dairy sensors, first, consider the mathematical models of dairy sensor on the cattle activities and, next, bring in the effects of climate events. Third, apply partial derivatives that zoom in the attribute in-focus leaving all other constant. Finally, once influencing coeffects are calculated, apply network effects and cohort cluster techniques to apply to forecast models.

(1) *Mathematical modeling*:

For instance, consider activity function of dairy sensor, of course, based on the cattle's activity:

Let F_a (x, y, z): F_a is accelerometer function with varying values based on the time where $\frac{dx}{dt}$, $\frac{dy}{dt}$ and $\frac{dz}{dt}$ are change of x, y, and z values with respect to time.

$$F_a(t) = F\left(\frac{dx}{dt}, \frac{dy}{dt}, \frac{dz}{dt}\right) \tag{1}$$

Normal climate conditions: Under normal climatic condition, the F_a has the influence of climatic factor U as part of dairy sensor values. Equation 1 can be rewritten as:

$$F_a(t) = F\left(\frac{dx}{dt}, \frac{dy}{dt}, \frac{dz}{dt}\right)$$
$$= F_{aU}\left(U\left(\frac{dx}{dt}\right), U\left(\frac{dy}{dt}\right), U\left(\frac{dz}{dt}\right)\right) \tag{2}$$

That is, by default, Eqs. 1 and 2 are same.

(2) *Factor-in Climate Events*:

For example, consider extreme (see Fig. 10) flood event has happened ($\acute{U}$):
The corresponding impact of $\acute{U}$ on dairy sensor accelerometer values be:

$$F_{a\acute{U}}\left(\acute{U}\left(\frac{dx}{dt}\right), \acute{U}\left(\frac{dy}{dt}\right), \acute{U}\left(\frac{dz}{dt}\right)\right) \tag{3}$$

(3) *The partial factor*:

Partial derivatives are defined as derivatives of a function of a multiple variable when all, but the variable of interest is held fixed during the differentiation. Partial derivatives are very useful to see the effects of the climate on cohort clusters that are geofenced or spread across different geolocations. The partial derivatives factors calculate the climate change effect or delta on the factor.

$$\text{The Climate Factor} = \frac{F_{a\acute{U}}}{F_{aU}}$$

$$= F_{a\acute{U}}\left(\acute{U}\left(\frac{dx}{dt}\right), \acute{U}\left(\frac{dy}{dt}\right), \acute{U}\left(\frac{dz}{dt}\right)\right) \Big/ F_{aU} \tag{4}$$

$$\left(U\left(\frac{dx}{dt}\right), U\left(\frac{dy}{dt}\right), U\left(\frac{dz}{dt}\right)\right)$$

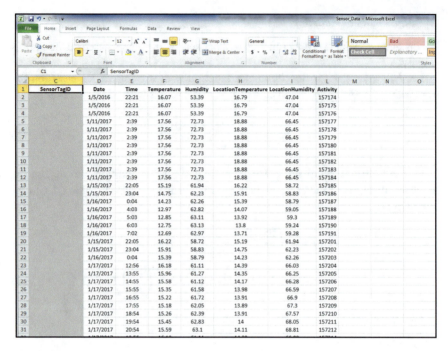

Fig. 10 Data capture—sensor in India

Since F_{aU} and $F_{a\breve{U}}$ have three different values (x, y, z), the partial values for each vector can be deduced:

Climate partial for X variance:

$$\left(\frac{\partial \breve{U}_x}{\partial F_{a\breve{U}}}\right) \bigg/ \frac{\partial U_x}{\partial F_{aU}}$$

Simplifies to $\qquad\qquad\qquad\qquad\qquad\qquad\qquad\qquad\qquad$ (5)

$$\frac{\partial \breve{U}_x}{\partial U_x} \text{ given } \partial F_{aU} \equiv \partial F_{a\breve{U}}$$

Therefore, partial accelerometer x variance due to climate fluid event is $\dfrac{\partial \breve{U}x}{\partial Ux}$.

Similar approach will calculate partial accelerometer y variance due to climate fluid and partial accelerometer z variance due to climate fluid.

(4) *Persisting values for network effect calculations*:

This would be the most valuable final step that stores the values of climate event to the partial derivative values of each attribute for the cohort of similar attributes.

For instance, dairies of similar cattle with can store the partial climate factors in the database to utilize by the recommendation systems when a forecastable climate event is predicated.

(5) *Meta data for the database*:

The database that persists climate partials is of huge value for future data analysis purposes and climate model rendition purposes.
    The structure of the data would include the following categories:

- Dairy cattle breed details
- Medication details
- Age/height/weight
- Geolocation
- Seasonal values
- Temperature
- Humidity's
- Climate event
- Partial derivatives of climate events
- Date and timestamp
- Sensor version.

2.9 Data as a Climate Modeler (DAACM)

To climate model, we need to have data. The toughest challenge to develop data-infused techniques was non-availability of dairy cattle heat stress-to-climate data. To overcome, we have developed around 150 low-cost (around $20 per cattle) [14, 15] small farmer-friendly dairy sensors and deployed in Punjab and Telangana states in India and collected cattle activity, body temperature, and humidity values at every 5-min interval from September 18, 2015 to November 22, 2017 (see Fig. 9). The application of data science technique known as decision tree (ID3 Iterative Dichotomiser) on collected data (around 168 million records) with cattle medication, geospatial data (ambient temperature/humidity) could help us to develop cattle heat stress-to-climate decision models with predictive accuracy of around 73%. Similarly, the use of linear regression and ANOVA techniques on the collected data could help us to develop water consumption and milk productivity forecasting models with accuracy of 65%. The deployment of these data science algorithms into sensors, making it precision sensor, enabled to deliver early warning heat-stress notification to dairy farmers. This has led small farmer to take proactive actions such as cooling dairy cattle by watering or improving dairy farm ventilation air-flow systems. The water consumption and milk productivity forecasting models help to improve overall (Fig. 11).

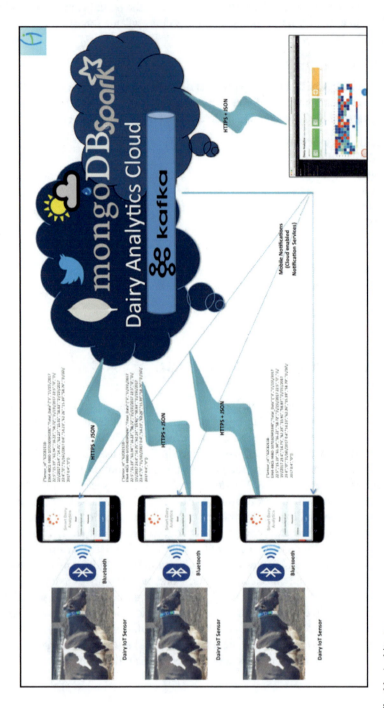

Fig. 11 Architecture

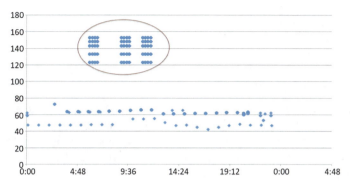

Fig. 12 Time–humidity values

2.10 Geolocation Data Markers

We have collected humidity and temperature data by installing dairy IoT sensors in state of Punjab, India, for the period of October 30, 2016 to January 18, 2017. We have observed the reporting of humidity values greater than 100% during the five months as shown in Fig. 12.

We also collected the data in South India, *Vizag—a costal of Andhra Pradesh*, and observed humidity values more than 140%. The discrepancies could be (a) sensor corruption or (b) could be due to battery current or voltage corruption.

Sensor hysteresis behavior could have played a role in creating data, geolocation temperature and humidity,[6] and corruption scenarios (see Fig. 12).

3 System Overview

The system architecture consists of collecting real-time data values from dairy cattle using Hanumayamma dairy IoT sensors (see Fig. 11). The collected values are locally stored on the sensors until a host is connected using mobile phone.

Upon host establishment, the data is taken from the farmer's phone and uploads to dairy analytics IoT cloud. The IoT cloud stores the data based on tenant specification.

3.1 System Architecture

The system has the following major components:

- Dairy IoT sensor

---

[6]Temperature and humidity data: https://www.wunderground.com/history/wmo/42101/2016/12/1/DailyHistory.html?MR=1.

- Mobile application
- Middleware
- Dairy IoT cloud
- Dairy analytics engine, and
- Dairy sensor values install base.

At high-level, the dairy IoT sensor is an intelligent edge device that has minimal data processing ML engine and collects data 24 × 7 and 365 days.

The embedded system consists of following components: Battery power supply, microcontroller, EPROM, terminal blocks, timer sensor, accelerometer, Bluetooth Low Energy (BLE), and temperature and humidity sensors.

Tenant Specification: The storage values are de-identified by dairy company or small farmer unique tenant ID. Each sensor used by farmer will have unique UUID (universally unique ID) that is linked to Bluetooth peripheral identification purposes. The cattle details will contain real-time sensor collected values on the cloud.

At any point, the system ensures, no two tenants are allocated with the same ID. And, also the manufacturing process ensures the dairy IoT sensors are produced with universally unique ID for the sensors.

For detailed designs into mobile application, Middleware, install base structure, please refer [10–15].

3.2 Dairy IoT Cloud

The dairy IoT cloud performs two vital operations: one, it ingests data from sensors that are distributed globally. The resilience, elasticity, and availability are achieved using Cloud compute.

The next major function that the dairy IoT cloud performs is computation of climate models and storage of climate partials derivatives into the database for lateral use purposes (see Fig. 13).

3.3 Climate and Dairy Sensor Parametric Model

(1) *Dairy IoT sensor values ingest*:
    The embedded system [12–15] was deployed in fields to collect cattle activity, the temperature, and humidity data. The sensors are deployed globally and are periodically uploads data on day-to-day basis.
(2) *Geolocation details and climate data*:
    We have integrated Weather API on SaaS level to get the geolocation details. The weather API provides ambient temperature and humidity values that are stored per sensor location basis.

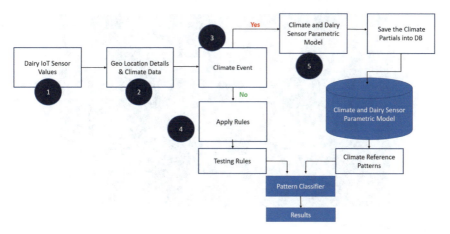

Fig. 13 Parametric model

The Climate data is retrieved from geolocation air quality data.

4 System Design and Implementation

4.1 Parametric Model

The purpose of parametric model is to evaluate any climate event and calculate partial derivatives that could have impacted by the event. Additionally, to store the partial derivatives for future climate forecast—Eqs. 1–5.

4.2 Climate Reference Pattern

To ensemble the impact of a forecasted climate event, the saved derivative values are combined with model equation to simulate the negative effects of climate change. To show in action, let us consider Eq. 1:

$$F_{aU}\left(U\left(\frac{dx}{dt}\right), U\left(\frac{dy}{dt}\right), U\left(\frac{dz}{dt}\right)\right) \tag{6}$$

Let us say National Weather Services has predicted climate event—Wind chill warning.

To calculate the impact of wind chill warning on Eq. 7, retrieve all partial derivatives from the database that have influence on the model equation. Of course, take into consideration the similarities of cohort cluster need to be validated for to apply:

Fig. 14 Dairy sensor

$$F_{a\widehat{W}}\left(\hat{W}\left(\tfrac{\mathrm{d}x}{\mathrm{d}t}\right), \hat{W}\left(\tfrac{\mathrm{d}y}{\mathrm{d}t}\right), \hat{W}\left(\tfrac{\mathrm{d}z}{\mathrm{d}t}\right)\right) = \int FaU\,\mathrm{d}\hat{W}$$

$$F_{a\widehat{W}}\left(\hat{W}\left(\tfrac{\mathrm{d}x}{\mathrm{d}t}\right), \hat{W}\left(\tfrac{\mathrm{d}y}{\mathrm{d}t}\right), \hat{W}\left(\tfrac{\mathrm{d}z}{\mathrm{d}t}\right)\right) = \int FaU\left(U\left(\tfrac{\mathrm{d}x}{\mathrm{d}t}\right) * \hat{W}\left(\tfrac{\mathrm{d}x}{\mathrm{d}t}\right), U\left(\tfrac{\mathrm{d}y}{\mathrm{d}t}\right)\right.$$
$$\left. * \hat{W}\left(\tfrac{\mathrm{d}y}{\mathrm{d}t}\right), U\left(\tfrac{\mathrm{d}z}{\mathrm{d}t}\right) * \hat{W}\left(\tfrac{\mathrm{d}z}{\mathrm{d}t}\right)\right) \tag{7}$$

The forecasted climate change event dairy sensor data is equal to the raw sensor model plus any partial derivatives that have influence on the model. Through this computation, the system provides recommendation to the farmer the data insights that are derived from similar cohort of dairies plus superimposed to the readings of the dairy sensor.

5   A Case Study

As part of adaptive edge analytics, the temperature and humidity embedded system that we have developed used in several agricultural and dairy settings (see Fig. 14). We have captured real-time customer-related data for creating an adaptive edge analytics paper. Finally, the product is used and tested by precision dairy agriculture company.[7]

6   Conclusion and Future Work

The Climate change is real and "climate is now a data problem." The data-driven insights to counter the negative issues of climate change are possible through sheer democratization of artificial intelligence to the bottom of the pyramid. By empowering every farmer, the point of view of a data scientist, farmer citizen AI, we can

---

[7]Hanumayamma Innovations and Technologies, Inc.—Dairy IoT sensor.

conquer the climate change and, importantly, overcome perennial "climate change-related suicide" issues. It is our staunch believe that the gold standard for the success of the democratization of artificial intelligence is no farmer life loss due to negative effects of climate change period.

References

1. Jones, N.: How machine learning could help to improve climate forecasts. Nature magazine. The Scientific American. https://www.scientificamerican.com/article/how-machine-learning-could-help-to-improve-climate-forecasts/. 23 Aug 2017
2. Umar, B.: India's shocking farmer suicide epidemic. https://www.aljazeera.com/indepth/features/2015/05/india-shocking-farmer-suicide-epidemic-150513121717412.html. 18 May 2015
3. Safi, M.: Suicides of nearly 60,000 Indian farmers linked to climate change, study claims. https://www.theguardian.com/environment/2017/jul/31/suicides-of-nearly-60000-indian-farmers-linked-to-climate-change-study-claims. 31 Jul 2017
4. Doshi, V.: 59,000 farmer suicides in India over 30 years may be linked to climate change, study says. https://www.washingtonpost.com/news/worldviews/wp/2017/08/01/59000-farmer-suicides-in-india-over-three-decades-may-be-linked-to-climate-change-study-says/?noredirect=on&utm_term=.38c8866059de. 1 Aug 2017
5. Weingarten, D.: The suicide rate for farmers is more than double that of veterans. A former farmer gives an insider's perspective on farm life—and how to help. https://www.theguardian.com/us-news/2017/dec/06/why-are-americas-farmers-killing-themselves-in-record-numbers. Dec 2017
6. Adams, M.: Suicidal dairy farmers should consider marijuana industry. https://www.forbes.com/sites/mikeadams/2018/03/16/suicidal-dairy-farmers-should-consider-marijuana-industry/#30a8ae8c1d1e. 16 Mar 2018
7. Smith, T.: As milk prices decline, worries about dairy farmer suicides rise. https://www.npr.org/2018/02/27/586586267/as-milk-prices-decline-worries-about-dairy-farmer-suicides-rise. 27 Feb 2018
8. Williams, J.: A booming economy with a tragic price. https://www.nytimes.com/2018/05/20/world/australia/rural-suicides-farmers-globalization.html. 20 May 2018
9. Blogger, G.: Machine learning may be a game-changer for climate prediction. https://blogs.ei.columbia.edu/2018/06/21/machine-learning-may-game-changer-climate-prediction/. 21 June 2018
10. Karwowski, W., Ahram, T. (eds.): Proceedings of the 1st International Conference on Intelligent Human Systems Integration (IHSI 2018): Integrating People and Intelligent Systems. Dubai, United Arab Emirates, 7–9 Jan 2018. https://www.springer.com/us/book/9783319738871
11. Kedari, S., Vuppalapati, J.S., Ilapakurti, A., Vuppalapati, C., Sharat, K., Vuppalapati, R.: Chapter 35 precision dairy edge, albeit analytics driven. In: A Framework to Incorporate Prognostics and Auto Correction Capabilities for Dairy IoT Sensors. Springer Nature, Berlin (2019)
12. Ilapakurti, A., Vuppalapati, J.S., Kedari, S., Kedari, S., Vuppalapati, R., Vuppalapati, C.: Adaptive edge analytics for creating memorable customer experience andvenue brand engagement, a scented case for Smart Cities. In: 2017 IEEE SmartWorld, Ubiquitous Intelligence & Computing, Advanced & Trusted Computed, Scalable Computing & Communications, Cloud & Big Data Computing, Internet of People and Smart City Innovation (SmartWorld/SCALCOM/UIC/ATC/CBDCom/IOP/SCI) (2017)
13. Vuppalapati, J.S., Kedari, S., Ilapakurti, A., Vuppalapati, C. et al.: Cognitive secure shield—a machine learning enabled threat shield for resource constrained IoT Devices. In: 2018 17th IEEE International Conference on Machine Learning and Applications (ICMLA) (2018)

14. Ilapakurti, A., Vuppalapati, J.S., Kedari, S., Kedari, S., Chauhan, C., Vuppalapati, C.: iDispenser—big data enabled intelligent dispenser. In: 2017 IEEE Third International Conference on Big Data Computing Service and Applications (BigDataService) (2017)
15. Kedari, S., Vuppalapati, J.S., Ialapakurti, A., Kedari, S., Vuppalapati, R., Vuppalapati, C.: Adaptive edge analytics. In: A Framework to Improve Performance and Prognostics Capabilities for Dairy IoT Sensor. Springer Nature, Berlin (2018)
16. Han, J., Kamber, M., Pei, J.: Data Mining: Concepts and Techniques, 3rd edn. Morgan Kaufmann, Burlington (2011)
17. Rajaraman, A., Ullman, J.D.: Mining of Massive Datasets. Cambridge University Press, Cambridge (2011)

Security Benchmarks for Wearable Medical Things: Stakeholders-Centric Approach

Swapnika Reddy Putta, Abdullah Abuhussein, Faisal Alsubaei, Sajjan Shiva and Saleh Atiewi

Abstract Internet of Medical Things (IoMT) is a fast-emerging technology in healthcare with a lot of scope for security vulnerabilities. Like any other Internet-connected device, IoMT is not immune to breaches. These breaches can not only affect the functionality of the device but also impact the security and privacy (S&P) of the data. The impact of these breaches can be life-threatening. The proposed methodology used a stakeholder-centric approach to improve the security of IoMT wearables. The proposed methodology relies on a set of S&P attributes for IoMT wearables that are identified to quantify S&P in these devices. This work aimed to (1) Guide hesitant users when choosing a secure IoMT wearable device, (2) Encourage healthier competition among manufacturers of IoMT wearables, and therefore, (3) Improve the S&P of IoMT wearables.

Keywords IoT (Internet of Things) · IoMT (Internet of Medical Things) · Wearable devices · Sensors · Security · Privacy · Healthcare

S. R. Putta · A. Abuhussein (✉)
St. Cloud State University, St. Cloud, MN, USA
e-mail: aabuhussein@stcloudstate.edu

S. R. Putta
e-mail: sputta@stcloudstate.edu

F. Alsubaei · S. Shiva
The University of Memphis, Memphis, TN, USA
e-mail: flsubaei@memphis.edu

S. Shiva
e-mail: sshiva@memphis.edu

S. Atiewi
Al Hussein Bin Talal University, Ma'an, Jordan
e-mail: saleh@ahu.edu.jo

© Springer Nature Singapore Pte Ltd. 2020
X.-S. Yang et al. (eds.), *Fourth International Congress on Information and Communication Technology*, Advances in Intelligent Systems and Computing 1027, https://doi.org/10.1007/978-981-32-9343-4_32

1 Introduction

Internet of Medical Things (IoMT) wearables are Internet-connected electronic devices that can be worn on the body to improve patient's quality of medical treatments. These devices are available from head to toe in many forms such as smart wristbands, watches, eyeglasses, belts, necklaces, patches, etc. They can track physical activity, temperature, glucose, sleep, heart rate, and much more. These devices can monitor the health signs of the patients/users and send them wirelessly to the physicians to cut down personal visits. The use of IoMT wearables is increasing rapidly, and the global wearable medical device market is anticipated to reach an estimated $9.4 billion by 2022 [1].

Although this radical change is much appreciated, we need to take a step back to review the security and privacy (S&P) of such devices. Ensuring their S&P is very important because the consequences of insecure medical devices are very dire as many patient's lives depend on them. Due to the rush to embrace IoMT technologies, the S&P of these devices are often overlooked by manufacturers. While shopping for the IoMT wearables, customers also often focus on the design, price, and performance of these devices. This is because customers are unable to choose or rank these devices in terms of S&P. Also, different stakeholders have different objectives and tolerance for risks. Hence, this work aims to assist hesitant users to evaluate and select IoMT wearables based on their ability to protect customers from potential S&P issues. This work also encourages healthier competition among manufacturers of IoMT wearables and therefore helps to improve the security of IoMT wearables.

2 Related Work

Many researchers and manufactures of IoMT devices are concentrating on the S&P of these medical devices. Also, many regulatory authorities have recognized the importance of this problem and started serious steps toward ensuring the protection of patient health information and the compliance of medical devices. However, the main gaps in these efforts can be summarized as follows: (1) Considering S&P attributes that are specific to a set of IoMT scenarios (e.g., patient monitoring) [2, 3]. (2) Providing generalized S&P recommendations that target manufacturers and considering whole IoMT ecosystems and all IoMT device types [4–8]. (3) Lacking an evaluation method that helps adopters to quantify and compare the S&P of potential IoMT wearables [9–11]. (4) Focusing only on assessing existing IoMT devices by utilizing post-adoption parameters such as configurations and current users' feedback, which requires technical knowledge that often most IoMT stakeholders lack [2, 7, 12].

Even though these works are considered necessary, these works lack methods of measuring security and do not integrate easily into an effective evaluation method for IoMT. Complementing the previous works, this paper presents a method for measuring S&P in IoMT wearables. The contribution of this paper is a list of S&P evaluation attributes for the IoMT wearables as well as an easy-to-use evaluation method that measures the S&P in IoMT wearables. Our presented solution is designed to help users compare candidate IoMT wearables in terms of their S&P levels in order to make well-informed decisions such as, choosing, or replacing current, IoMT wearables.

3   Wearable IoMT Security Evaluation

Evaluating the S&P in IoMT wearables is *multiple criteria decision-making* problem, which considers multiple conflicting criteria for decision-making. In order to solve this problem, key attributes that are critical for the S&P of IoMT wearables are identified in this work. Each of these attributes is represented by its definition (i.e., what is it?), its rationale (i.e., why is it important?), and its S&P functionality (i.e., how is it important?). Furthermore, each attribute is represented by a set of considerations. These considerations are a set of polar questions (i.e., yes/no questions). The attributes and their considerations are explained in the next chapter.

These attributes along with their considerations are integrated in a two-step methodology to assist the stakeholders when choosing of IoMT wearables. This steps of methodology are as follows:

- **Step 1**: The S&P of IoMT wearables are evaluated by answering the attribute questions. Device S&P specifications can be used by stakeholders to answer these questions. These specifications are publicly available on manufacturers' websites.
- **Step 2**: The score of each attribute is computed using its considerations. The scores for all the attributes are normalized to a score of 10.

Every stakeholder's interaction with the device is different. Hence, not all the attributes are necessary for all the stakeholders. This stakeholder-centric approach helps to satisfy the requirements of the stakeholders who have different needs, goals, and tolerance to risks. The identified stakeholders for these devices include the patient, doctor, hospital, nurse, manufacturer, security researcher, and regulatory authorities, and insurance.

4   IoMT Wearable S&P Attributes

This section discusses our identified S&P attributes. We investigated the attributes that define the S&P of the wearable IoMT devices using devices specification, FAQs and best practices in medical IoMT from research organizations, government agencies, and industry associations. Our attributes are discussed in the following subsections.

4.1 Authentication and Identity Management

This measures the device ability to verify the user identity. Identity is associated with a user with a unique username or unique ID. Authentication verifies the identity of the user with a password or a key. This attribute is important because it defines the effectiveness in protecting device and user data from unauthorized access. The following questions are used to determine the strength of the authentication and identity management of a wearable device:

1. Does the device allow MFA?—Multi-factor authentication (MFA) is a combination of two or more types of authentication. It is always harder to bypass multi-layer security than single-layer security.
2. Is the minimum size of password eight characters?—Recommendations for a minimum length of the password is eight.
3. Does the password require each of uppercase/lowercase/number/special characters?—A strong password is a combination of all the different types of characters.
4. Does the device password expire?—Users should change their passwords regularly at least for every 90 days and they should be notified to do so before they expire.
5. Does the device have a password recovery option?—It is always important to have a password recovery option and to identify the user before resetting the password.
6. Does the device have a password history option?—A password history stores the previous passwords and prevents the same password to be re-used.
7. Does the device have biometric authentication?—Biometric methods use a physical characteristic such as fingerprint, retina, iris, voice recognition or facial recognition.
8. Does the device allow to choose the same password as your username?—setting the same password as your username can be easily guessed by any hacker. The device should display an error message if the user picks a password similar to username.

4.2 Access Control and Profiling

This measures the device ability to grant access and privileges to the resources for the users. These resources can be data, applications, or the device. This access is defined by the permissions assigned based on the authorization to the data and the device. It also measures the ability to define and customize profiles of the users. This helps device owners to limit the access to the device and the privileges that each user has. Only the owner of the device (e.g., patient) should have the highest privilege. The strength of access control and profiling can be determined by the following questions:

1. Does the wearable device have role-based access control?—Role-based access control uses roles to grant/deny permissions. The stakeholders can be categorized by their roles so that a stakeholder has all the necessary privileges for the role.
2. Does the device have rule-based access control?—Rule-based access control uses rules. These rules typically remain static until changed by device owner.
3. Does the device have discretionary access control?—In discretionary access control, the device owner decides and grants access to the other stakeholders.
4. Does the device have mandatory access control?—mandatory access control uses labels (e.g., data sensitivity or security labels) to determine access.
5. Does the device have attribute-based access control?—Attribute-based access control evaluates attributes and grants access based on the value of these attributes.

4.3 Storage Location

This attribute measures the device ability to store data in a secure location(s). Data storage locations include cloud storage, mobile storage, and device storage. This helps the user to recognize the locations where the data is stored so that the user can limit storage locations. Storing the data in multiple locations helps to backup data in redundant locations. However, it also expands the attack surface. A good storage location(s) attribute can be defined by the following questions:

1. Does the device allow the user to store the data in the device itself?—Data stored in the device is easily accessible and can be tracked easily since the device is always worn on the user's body.
2. Does the device allow the user to manage, and control the device data from the device itself?—Since the device is a wearable device and is worn on the body, it is more secure as the device, and the data can be controlled from the device itself.
3. Can users store, access, manage and control the device using a smartphone?— Wearable medical device that can be controlled and managed by the smartphone, can also be accessed by anyone and/or other applications that have access to it.
4. Can users store, access, manage and control the device from cloud/third party apps?—Most of the cloud applications are in the control of a third party. If these applications are hacked and if the device can be managed from these applications, the hacker can easily control the device on the patient's body.
5. Does the device allow the user to select the locations where the data can be stored?—Storing the data in cloud/third parties is always at risk. Users should be able to decide the storage location depending on the requirements and security.

4.4 Encryption

This measures the device ability to make the data unreadable at various levels like data at rest, in transit, and in use. Data can only be read by the user who has the encryption key which converts data to clear text. This attribute guarantees the confidentiality of the patient's data. If data is stored in plain text, it can be read by intruders during the data transmission or while the data is being processed. A good encryption attribute can be defined using the following questions:

1. Does the device allow users to select what data to encrypt and where?—Although encryption guarantees data confidentiality, it consumes time and space. Hence, the customer should be given an option to encrypt only sensitive data to reduce the time and space.
2. Does the device encrypt and hash passwords?—Passwords are usually stored in plaintext in these devices. If the passwords are stored in plaintext, anyone who has access to the device can easily read the password.
3. Does the device encrypt the data at rest?—Data at rest is when the data is not being used but is stored physically on the device, smartphone, and or cloud. If data is not encrypted when it is at rest, the data can be easily viewed if the device is lost/stolen.
4. Does the device encrypt data in transit?—Data in transit is when the data is being transmitted from one location to the other. If data is being communicated in clear text between the two devices, it can be read by eavesdropper.
5. Does the device encrypt data in use?—Data should be encrypted even when it is being used by the device or applications because it helps to protect sensitive data.
6. Does the device follow any of the standard encryption techniques?—There are encryption standards that are proven to be immune to attacks or very hard to be broken.
7. If yes, Does the device comply with regulations in the country where it is used?—Every country has its own laws and regulations to be followed. It is also necessary for the user to check if the device complies with the regulations in the country where the device is being used.
8. Does the device allow the user to choose an encryption technique?—There are different encryption technologies available depending on time and reliability.

4.5 Compliance

Compliance attribute measures the device ability to follow the guidelines set by the regulatory authorities which this increases the trustworthiness of the device. A good compliance attribute can be determined using the following questions:

1. Is the device FDA compliant?—Food and Drug Administration (FDA) is a federal agency of the United States Department of Health and Human Services which is responsible for protecting and promoting public health.
2. Is the device HIPAA compliant?—"Health Insurance Portability and Account-ability Act of 1996 (HIPAA) is United States legislation that provides data privacy and security provisions for safeguarding medical information."
3. Is the device ISO/IEC 80001 compliant?—International Organization for Stan-dardization (ISO)/IEC 80001 is application of risk management for IT-networks incorporating medical devices. The key properties are risk management of IT-networks incorporating medical devices to address safety, effectiveness and sys-tem security.
4. Is the device ISO 14971 compliant?—ISO 14971:2007 specifies a process for a manufacturer to identify the hazards associated with medical devices, including in vitro diagnostic medical devices, to estimate and evaluate the associated risks, to control these risks, and to monitor the effectiveness of the controls.
5. Is the device compliant with any other medical device regulatory?—There are different regulations and authorities for medical data and security.
6. Is the device compliant with any regulatory authority where the device is manu-factured?—Different authorities have different guidelines based on the location and rules. The device should be compliant to the regulations of manufacturer location.
7. Is the device compliant with any regulatory authority where the device is used?—Different bodies have different guidelines based on the location and rules. The device should be compliant to the regulatory where it is being used.

4.6 Connectivity

This attribute measures the device ability to connect to other devices through a different medium. It defines how the device can be connected to the other devices. Each method of connectivity has its own security challenges. Secure connectivity is determined by the following questions:

1. Does the device allow to select the type of connectivity?—A user might feel comfortable using connectivity methods. Hence, the device should allow the user to choose how to connect to the Internet or the other devices.
2. Does the device allow to connect with other devices via the Internet?—Internet helps to connect and view data in different devices globally. It helps the user to view, edit and share data at any time through the private network. Connecting to the Internet in public places increases hackers' chances to gain unauthorized access to the devices.
3. Does the device have the ability of mutual authentication when connecting with other devices?—Mutual authentication is a two-way authentication that helps both the devices to authenticate before they connect to each other.

4. Can the device anonymously connect to other devices?—Being anonymous can make the user feel safer even when device is breached.

4.7 Data Shredding

This measures the device ability to ensure that all patient identifiable data is securely and correctly deleted from the equipment prior to disposal or reuse. It ensures that the device does not retain previous user's data or malicious code. Data shredding permanently wipes data so that it cannot be recovered. A good data shredding attribute can be defined by the following questions:

1. Is the device under MDISS agreement?—Medical Device Innovation, Safety and Security (MDISS) consortium checks if the medical device is ready to use, no previous data is present in the device, and helps the device to fix any vulnerabilities.
2. Does the device use any data shredding mechanisms?—Data shredding mechanisms help the medical devices to clear all data that was previously stored by other users.
3. Is the device capable of installing any data shredding tools?—There are many open-source data shredding tools that help to wipe all the previously stored data.

4.8 Classification of Data

This attribute helps to determine the type of data that is stored in the device and helps to categorize the data depending on its sensitivity. The more sensitive data on the device, the more risk it introduces if device hacked. This can be verified by the following questions:

1. Does the device allow to catalog, categorize or classify data based on their sensitivity?—Categorizing the data helps the user to know which data can be encrypted.
2. Does the device allow the owner to select the validity for the data stored in the device?—A user can delete the less important data after a period of time which helps the device to increase the storage space and helps to improve the processing time.

4.9 Simultaneous Data Accessed

This attribute measures the device ability to access the data by different users/programs at the same time. It is vital because data accessed at the same time by different users/programs increases the attack surface. This is measured by the following questions:

1. Can the device restrict multiple users from connecting to the device at the same time?—Each device can have different users. If all the users connect at the same time, the functionality of the device decreases, and it also becomes hard to track attacks.
2. Does the device have an option to limit the number of users accessing the device at the same time?—If the user can restrict the connections to the device, it helps to track the user or the program if the device is compromised.

4.10 Number of Stakeholders

This measures the device ability to identify how many users can have access to the data and who they are. As the number of stakeholders increases the attack surface also increases. This is measured by the following questions:

1. Does the device allow the owner to select the users?—The owner of the device should have the privilege to choose the users.

4.11 Device Bandwidth

This measures the device ability to control communication bandwidth. Bandwidth is measured in bits per second. If the traffic exceeds the allowed bandwidth threshold that may possibly mean that the device under a denial of service attack. To verify this attribute, the following questions can be used:

1. Does the device allow the owner/admin to limit the bandwidth?—Bandwidth should be limited based on the data that is being transmitted.
2. Does the device allow the owner to limit the bandwidth for different users?— There are different bandwidth recommendations for healthcare users such as the federal communication commission recommendations [13].
3. Does the device allow the user to limit the bandwidth based on the number of users?—There might be situations where the number of users accessing the device at the same time increases, at that time the device should allow the owner/admin to increase the bandwidth depending on the number of stakeholders.

4. Does the device allow owner/admin to limit bandwidth based on the user locations?—There might be unwanted traffic in public networks, so depending on the location, the bandwidth should be limited.

4.12 Tested Device

This attribute measures if the device is tested and all its vulnerabilities are addressed. This helps to determine the current level of security of the device, the device vulnerabilities and if they were patched? This can be defined by the following questions:

1. Is the device tested in terms of S&P?—According to a survey by Synopsys, 36% of the medical device makers, and 45% of the healthcare delivery organizations do not test their medical devices in terms of S&P [14].
2. Is the device known vulnerabilities addressed?—According to a survey by Synopsys, 35% of medical device makers, and 26% of healthcare organizations say that their medical devices contain significant vulnerabilities. 18.3% of device makers and 13% of healthcare organizations say their tested medical device contains malware [14].

4.13 Log Management

This measures the device ability to monitor and analyze all events. For example, this helps the owner of the device to know the users who logged in, to check the data and can find if any unauthorized user or program is able to see or alter the data. A good log management is measured by the following questions:

1. Does the device have a log management system?—Log management system helps the user to know the users who accessed the device and the data.
2. Is the log management system in the device trustworthy?—There are some basic log management systems available which are not reliable.
3. Is the device capable of installing a log management system?—There are different log management systems available which can be easily installed based on the requirement such as Logsign, Splunk, and Log packer.

4.14 Compatibility

This measures the device compatibility with other devices. If the device is compatible with other devices and applications, this means that it can automatically shares the data with those devices and applications. This is measured by the following questions:

1. Is the device compatible with iOS and notifies the user before sharing the data?
2. Is device compatible with macOS and notifies the user before sharing the data?—macOS is an operating system developed and marketed by Apple Inc. It is the primary operating system for Apple's Mac family of computers.
3. Is the device compatible with Android but notifies the user before sharing the data?—Android is a mobile operating system developed by Google, based on a modified version of the Linux kernel and other open-source software.
4. Is the device compatible with Windows but notifies the user before sharing the data?—Microsoft Windows is a group of graphical operating system families, all of which are developed, marketed, and sold by Microsoft.

5 Stakeholder–Centric Approach

The stakeholders for the IoMT wearables include patients, doctor, hospital, nurse, manufacturer, security researcher, and regulatory authorities, insurance. Not all these stakeholders require all the previously defined attributes. Hence, Table 1 depicts the 14 S&P attributes discussed in Sect. 4 as applied to the stakeholder's requirements.

This clearly shows that patients and manufacturers need to consider all the attributes. This is because patients are the main users and their personal data and health data will be stored and communicated from and to the device. Manufacturers also need all attributes to measure S&P in their products and their competitor products.

6 Case Study

In this section, we evaluate the S&P of two IoMT wearables (i.e., Dexcom g5 [15] and MiniMed 530G [16]) using our method. In step 1, we used devices specifications from manufacturer websites to answer the considerations questions for both devices as shown in Table 2. Step 2 of the methodology, the following equation is used to compute the score for each attribute using its considerations. The scores for all the attributes are also normalized to score out of 10.

$$\text{Attribute score} = \sum_{i=1}^{N} \text{Consideration}_i \times \frac{10}{N} \tag{1}$$

All the attribute scores were plotted in a graph for better visualization. Figure 1 shows all the attributes' scores for Dexcom g5 and MiniMed 530G.

Table 1 Wearable IoMT S&P attributes: stakeholder-centric approach

Stakeholders	1	2	3	4	5	6	7	8	9	10	11	12	13	14
Patient	X	X	X	X	X	X	X	X	X	X	X	X	X	X
Hospital		X				X			X					X
Doctor			X	X	X	X	X		X			X		X
Nurse				X		X			X					X
Manufacturer	X	X	X	X	X	X	X	X	X	X	X	X	X	X
Regulatory authorities		X		X							X	X		X
Insurance			X	X	X					X		X		
Security researchers		X	X	X	X	X	X	X			X			X

Table 2 The result of considerations for Dexcom g5 and MiniMed 530G

Attributes	Dexcom g5	MiniMed 530G
1. Authentication and identity management	2.3	5.5
2. Access control and profiling	2.2	2.2
3. Storage location	4.0	4.0
4. Encryption	0.0	0.0
5. Compliance	4.1	3.0
6. Connectivity	2.0	2.2
7. Data shredding	0.0	0.0
8. Classification of data	0.0	6.0
9. Simultaneous data accessed	10.0	10.0
10. Number of stakeholders	10.0	10.0
11. Device bandwidth	0.0	0.0
12. Tested device	4.9	0.0
13. Log management	0.0	0.0
14. Compatibility	0.0	0.0
Total	39.5	42.9

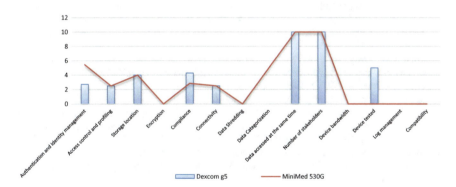

Fig. 1 Comparison of two IoMT wearable devices using the proposed methodology

7 Conclusion and Future Work

Stakeholders of IoMT wearables (i.e., healthcare practitioners and patients) often focus more on the functionality and performance of the device, but overlook the S&P issues associated with these devices. In most cases, the reason to overlook the S&P is lack of proper awareness. Hence, this work presented a methodology to assist IoMT stakeholders to rank IoMT wearables in terms of their protection and deterrence. The novelty of this work lies in that it defines security according to every stakeholder's interaction with the IoMT wearables. This method aims to assists IoMT stakeholders with different requirements, goals, and tolerance to risks, to be aware of the S&P issues in IoMT and to manage them. It can also help in dealing with stakeholders' conflicts of interests broadly and thoroughly in decision-making.

Our future works include developing a tool with this methodology to easily evaluate the values of any IoMT wearable device. This tool can also be developed to store the values of previously evaluated devices and help the customers to retrieve these values. Finally, for each attribute, weightage can be added as pertinent to the stakeholder.

References

1. Markets, R.: Global wearable medical device market opportunities to 2022. https://www.prnewswire.com/news-releases/global-wearable-medical-device-market-opportunities-to-2022-300542952.html
2. Abie, H., Balasingham, I.: Risk-based adaptive security for smart IoT in eHealth. In: Proceedings of the 7th International Conference on Body Area Networks. ICST (Institute for Computer Sciences, Social-Informatics and Telecommunications Engineering), pp. 269–275 (2012)
3. Savola, R.M., Savolainen, P., Evesti, A., Abie, H., Sihvonen, M.: Risk-driven security metrics development for an e-health IoT application. In: Information Security for South Africa (ISSA), pp. 1–6. IEEE (2015)
4. Alsubaei, F., Abuhussein, A., Shiva, S.: A framework for ranking IoMT solutions based on measuring security and privacy. In: Arai, K., Bhatia, R., Kapoor, S. (eds.) Proceedings of the Future Technologies Conference (FTC) 2018, pp. 205–224. Springer International Publishing, Cham (2019)
5. Food and Drug Administration: Postmarket management of cybersecurity in medical devices. https://www.fda.gov/downloads/MedicalDevices/DeviceRegulationandGuidance/GuidanceDocuments/UCM482022.pdf (2016)
6. MDRAP | Home Page. https://mdrap.mdiss.org/
7. McMahon, E., Williams, R., El, M., Samtani, S., Patton, M., Chen, H.: Assessing medical device vulnerabilities on the Internet of Things. In: 2017 IEEE International Conference on Intelligence and Security Informatics (ISI), pp. 176–178. IEEE (2017)
8. Medical equipment in general. https://www.iso.org/ics/11.040.01/x/
9. Laplante, P.A., Kassab, M., Laplante, N.L., Voas, J.M.: Building caring healthcare systems in the Internet of Things. IEEE Syst. J. 1–8 (2017)
10. Islam, S.M.R., Kwak, D., Kabir, M.H., Hossain, M., Kwak, K.S.: The Internet of Things for health care: a comprehensive survey. IEEE Access. 3, 678–708 (2015)
11. Williams, P.A., Woodward, A.J.: Cybersecurity vulnerabilities in medical devices: a complex environment and multifaceted problem. Med. Devices Auckl. NZ. 8, 305–316 (2015)
12. Leister, W., Hamdi, M., Abie, H., Poslad, S.: An evaluation framework for adaptive security for the IoT in ehealth. Int. J. Adv. Secur. 7(3&4), 93–109 (2014)
13. Recommended bandwidth for health care providers. https://www.greatsys.com/recommended-bandwidth-for-health-care-providers/
14. Medical-device-security-ponemon-synopsys.pdf. www.synopsys.com/content/dam/synopsys/sig-assets/reports/medical-device-security-ponemon-synopsys.pdf
15. zak.huber: Dexcom G5 Mobile CGM System | Glucose on your phone. https://www.dexcom.com/g5-mobile-cgm
16. MiniMed 530G Insulin Pump | Diabetes Pump System With SmartGuard Technology. https://www.medtronicdiabetes.com/products/minimed-530g-diabetes-system-with-enlite

Teacher Perception of OLabs Pedagogy

**Pantina Chandrashekhar, Malini Prabhakaran, Georg Gutjahr,
Raghu Raman and Prema Nedungadi**

Abstract Online Labs (OLabs) is a major Digital India initiative with over 135
online experiments mapped to high school curriculum. For each experiment, OLabs
provides background on the theory, animations, simulations, videos, viva voce questions, and links to additional resources. OLabs has been translated to multiple Indian
languages. As part of scaling OLabs to the nation, over 16,000 teachers in all Indian
states have been trained across India. The current manuscript presents a survey of
112 teachers who attended OLabs workshops and uses OLabs in the classroom. The
study's purpose is to understand the effective implementation of teacher training, to
understand how OLabs is used in school laboratory experiments, and to understand
how OLabs can supplement or replace real laboratories. A majority of teachers agree
that repetition of OLabs experiments helps improve understanding of the concepts.
There exists a strong correlation between the teachers' perception of the quality of
videos and animations and the teachers' attitude on the usefulness of virtual laboratory software. Whether or not teachers feel that virtual laboratory software is useful
to students is strongly associated with whether or not teachers feel that software is
sufficiently fast and responsive. With regard to the workshops, teachers place high
emphasis on the importance of establishing a clear agenda during the workshops.

P. Chandrashekhar (✉) · M. Prabhakaran · G. Gutjahr · R. Raman · P. Nedungadi
Center for Research in Analytics & Technologies for Education (CREATE), Amrita Vishwa
Vidyapeetham, Amritapuri, India
e-mail: pantinac@am.amrita.edu

M. Prabhakaran
e-mail: malinip@am.amrita.edu

G. Gutjahr
e-mail: georgcg@am.amrita.edu

R. Raman
e-mail: raghu@amrita.edu

P. Nedungadi
e-mail: prema@amrita.edu

R. Raman
School of Business, Amrita Vishwa Vidyapeetham, Amritapuri, India

© Springer Nature Singapore Pte Ltd. 2020
X.-S. Yang et al. (eds.), *Fourth International Congress on Information
and Communication Technology*, Advances in Intelligent Systems
and Computing 1027, https://doi.org/10.1007/978-981-32-9343-4_33

Finally, almost all teachers agree that OLabs can be an effective supplement to real laboratories.

Keywords Science experiments · Teacher training · E-learning · Online labs · Virtual labs · STEM education

1 Introduction

A virtual laboratory is a computer-based activity where students or teachers interact with the computers to get the laboratory work done. Its main purpose is to understand the learning objectives of a laboratory through interaction. It is useful for both teachers and students, especially in areas where there is a lack of real laboratories. Simulations and animations make it interesting and easy to learn the difficult topics.

Virtual labs have advantages over traditional laboratories [1]. Students can use it anytime and anywhere. It can be used during regular class hours thereby enabling teachers to make the understanding of concepts easy. Students can do an experiment multiple times so it avoids the risk of doing the laboratory only one time and thereby giving a chance to understand the concept thoroughly. Virtual labs software can be deployed as a website; so, the maintenance cost for schools is low. Virtual labs are available for free or at a low cost [2].

Hands-on sessions are one of the major learning methods in STEM education. By performing laboratory experiments, teaching initiatives such as learning by doing, converting theories into practice, manipulating the physical environments and understanding the errors and limitations can be attained [3, 4].

Virtual laboratory may include simulations, remote online labs and Virtual Reality Laboratories (VRL) categories. Simulations are self-contained software applications that emulate a real process. Remote laboratories are experiments where users can remotely control equipment. VRL are highly interactive and realistic in an artificial three- dimensional environment [5].

The present study deals with a virtual laboratory system called OLabs (online laboratories) [6, 7]. Online Labs provides students with the opportunity to perform simulations of science experiments. It consists of virtual practical experiments of schools' laboratories including physics, chemistry and biology experiments. There are 135 science OLabs aligned to CBSE/NCERT syllabus for classes 9 through 12. The online laboratories system contains procedure, theory, simulator, animation, viva voce and video and links for additional resources. Figure 1 shows a screenshot of the user interface.

This paper describes a survey conducted with teachers who attended OLabs teacher training and uses OLabs in the classroom. The aim of the survey was to understand the potentials and shortcomings of virtual laboratory software and also to understand the effectiveness of OLabs as perceived by the teachers.

The outline of the paper is given as follows. Section 2 tells about literature review. Section 3 tells about the teacher survey. Section 4 contains a discussion.

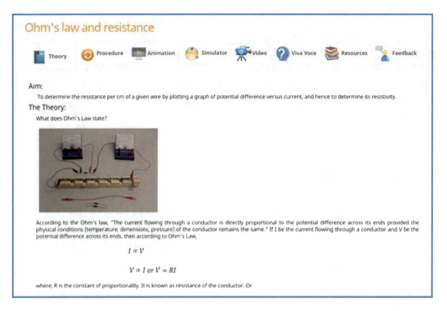

Fig. 1 Contents of OLab experiment

2 Literature Review

Information technologies have become an important aspect of education [8]. While laboratories are highly effective in teaching students concepts in science, the time and costs make it difficult to create a real laboratory environment in many institutions. Adoption of virtual laboratories is found to be effective in resolving this problem [9].

Although there is a big lacuna in understanding mathematics, scientific studies show that virtual laboratory has lots of positive contribution in improving the level of student concepts and understanding their behaviour towards the phenomena [10]. There are added benefits to the use of simulations in computers such as keeping away from dangerous laboratory experiments.

Remote laboratories [12] and simulations [13] are innovative learning tools for laboratory practical experiments. The laboratory devices, accessories, and substances of a virtual laboratory emulate a real laboratory. It consists of different workspaces for these users. The hands-on experience of the simulation aids the users to learn the functioning principle and experimental methodology [14].

Remote laboratories are widely used in mixed and distance learning. The main advantages of remote laboratories are that they are accessible at any time, they are accessible remotely, they allow sharing of resources at different schools and they can reduce the costs of laboratory ownership as the number of students increases [15].

Ease of use and trial ability of the virtual experiments significantly influence the students to adopt virtual lab into their studies [16]. Virtual labs have advantages over physical laboratories as they can support learning tools such as concept maps and inquiry learning [17, 18]. Medical case-based virtual patient simulations allow medical students to diagnose and treat the virtual patient [19].

With virtual labs, students have the opportunity to perform experiments that are structures in a pedagogical way [20]. Faculties can add using authoring platforms to develop laboratories by incorporating new content and reusing existing multimedia content [21, 22].

3 Teacher Survey

The aim of this present study is to understand how much effective OLabs is in the daily laboratory practice of students. Another aim is to find whether OLabs as a teaching aid is beneficial for students and also teachers to teach in a digital format.

A. Research Questions

- According to teachers, can OLabs be an effective supplement to real laboratories?
- According to teachers, does OLabs present concepts in a clear way?
- According to teachers, is OLabs effective for all kinds of students ranging from below average to above average?

B. Student Participants

The study included 112 responses of teachers who attended an OLabs workshop. Teachers teach ninth, tenth, eleventh and twelfth standard. Teachers' subject includes mathematics, physics, English, chemistry and biology.

C. Methodology

OLabs is the tools used in this study. Teacher accessed OLabs from the website www.olabs.edu.in. A brief introduction about OLabs was provided to teachers. Teachers were trained in the workshop by getting them to know two experiments from each subject. Physics included Ohm's law and resistance (class 12) and bell jar experiment (class 9). Chemistry included determination of concentration of $KMnO_4$ solution (class 11) and qualitative analysis of oils and fats (class 12). Biology experiments included study of pollen germination (class 12) and detection of starch in food samples (class 9). Mathematics included area of circle and Pythagoras theorem. English included singular to plural conversion. In total, nine experiments were shown to the teachers during the training.

For physics, chemistry and biology experiments, teachers got an idea about the objectives of the experiment and the concept behind the experiment from the 'Theory' section. Teachers go through the procedure sections, the animations and laboratory videos to correctly understand its methodology behind the experiments. Next, the teachers performed the simulation. After completing the simulations, teachers tested the viva voce questionnaire. Some contents of OLabs experiments (animation and simulator) are shown in Fig. 2.

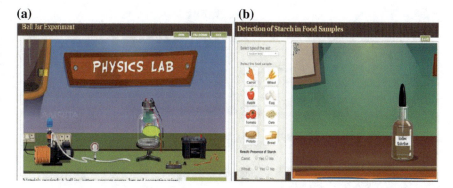

Fig. 2 **a** Animation of 'bell jar experiment', **b** simulator of 'detection of starch in food samples'

After the training workshop, teachers filled a questionnaire. The questionnaire includes 26 questions. For each question, the teacher answers on a Likert scale, with the scale being strongly agree, agree, neutral, disagree and strongly disagree.

4 Results and Discussions

D. Statistical Analysis

We used chi-squared tests to compare the numbers of teacher who agree or disagree with the research questions. We used bar charts to visualize the pattern of answers directly related to these research questions. We quantified the strength of association between various questions using Goodman's Gamma, Spearman's rank correlation rho and p-values from a rank correlation test.

E. Results

Figure 3 shows the positive responses of teachers towards OLabs experiment. Almost all the teachers agreed that the animations of all experiments have good quality and real laboratory videos of the experiments help students to gain better knowledge. Teachers agreed that the option to repeat the experiments in a number of times will help the students to improve their learning ability. Teachers also agreed that the online performances of our website were good and the resources given in the experiment are very helpful for students to understand the concept thoroughly.

Table 1 shows associations between some key questions. The table shows about those teachers who agreed that the online performance of the website is good also agreed that OLabs resources are a perfect way to understand the experiments thoroughly (gamma 0.39). Teachers who agreed that OLabs resources are better way to understand the experiments thoroughly also agreed that real laboratory videos in the experiments helps students to understand the experiment and that the equipment used for the experiment (gamma 0.34). Teachers who agreed

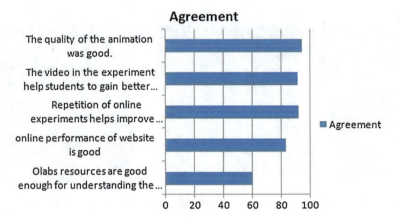

Fig. 3 Response of teachers towards online lab experiments

Table 1 Strength of association between questions using Goodman's gamma, spearman's rank correlation rho and *p*-values from a rank correlation test

	Gamma	Rho	p
OLabs helps understand the concepts vs. online performance of the website	0.39	0.64	0.01
OLabs helps understand the concepts vs. videos in the experiment	0.34	0.58	0.01
OLabs helps understand the concepts vs. quality of animation	0.26	0.46	0.01
Repetition of online experiments improves understanding vs. online performance of the website	0.32	0.66	0.01
Agenda of the workshop is clear vs. comparison of real and online laboratories	0.30	0.47	0.03

with the good quality of animation also agreed that OLabs resources are better way to understand the experiment clearly. The teachers who agreed with the online performance of website being better also agreed that repetition of online experiments helps improve learning (gamma 0.32). Teachers who felt that the agenda of the workshop was unclear also feel that real laboratories are a proper way to understand the concepts when compared to online labs (gamma 0.30).

5   Conclusion

In this paper, we examined the effectiveness of OLabs in daily laboratory experiments of students and the benefit of OLabs as a teaching aid to students and also teachers to teach in a digital format. A total of 112 teachers from different states of India actively participated in the workshop. The results from the case study suggested that features like better online performance of website, the good quality animation, well-documented theory and procedure, interesting laboratory videos and resources, etc. enable teachers to put OLabs into the student curriculum. Also, having covered laboratory experiments from CBSE, it becomes easy to implement the OLabs effectively in schools. Ultimately, the students benefit from the teachers who are trained in OLabs. After conducting the workshops, we have come to understand that more observation tables can be included especially in physics and also in chemistry.

The overall analysis results from the survey found that online laboratories are one of the effective methodologies for teachers to teach their students and has benefits like repetition of experiments n number of times, and animation plus simulation are an added advantage and it is equivalent or rather better than real laboratories.

In summary, the OLabs workshop is an effective way to train teachers as it clears the concepts in a better way. From the result of the survey analysis, we come to the conclusion that OLabs has features which are very useful and informative because of which it comes in handy for laboratory practical experiments.

The limitations of the study include potential selection bias and response bias inherent in questionnaire surveys. Also, since the study was concerned only with one particular virtual laboratory software, the extent to which these results can be generalized to other software is unclear. Furthermore, the specific details of the schools and the different levels of experience of the teachers have not been taken into account.

In the future, we want to conduct a similar survey with students to find out the effectiveness of student training workshops. In particular, we want to understand how OLabs benefits the students in their practical experiments and whether it is a learning aid for students in their day-to-day concepts in their subjects.

Acknowledgements The inspiration for writing this paper is Sri Mata Amritanandamayi Devi, Chancellor Amrita Vishwa Vidyapeetham. The project is funded by Ministry of Electronics and Information Technology (MeitY), Government of India and implemented jointly by Amrita and CDAC, Mumbai. We thank the teachers of classes 9–12 from different schools across India both CBSE and State Board for their survey responses. The given study has been approved by Ethical Research committee of our institution. The requisite consent was taken from the participants for the study.

References

1. Soni, S., Katkar, M.D.: Survey paper on virtual lab for E-learners. International J. Appl. Innov. Eng. Manag. (IJAIEM) **3**(1) (2014)
2. Chu, K.C.: What are the benefits of a virtual laboratory for student learning? In: HERDSA Annual International Conference Melbourne, 12–15 July 1999
3. Lynch, T., Ghergulescu, I.: NEWTON virtual labs: introduction and teacher perspective. In: IEEE 17th International Conference on Advanced Learning Technologies (2017)
4. Nedungadi, P., Raman, R., McGregor, M.: Enhanced STEM learning with Online Labs: empirical study comparing physical labs, tablets and desktops. In: Frontiers in Education Conference. IEEE (2013)

5. Achuthan, K., Murali, S.S.: Virtual lab: An adequate multi-modality learning channel for enhancing students' perception in chemistry. Adv. Intell. Syst. Comput. **574**, 419–433 (2017)
6. Nedungadi, P., Malini, P., Raman, R.: Inquiry based learning pedagogy for chemistry practical experiments using OLabs. Adv. Intell. Syst. Comput. **320**, 633–642 (2015)
7. Raman, R., Haridas, M., Nedungadi, P.: Blending concept maps with online labs for STEM learning. Adv. Intell. Syst. Comput. **320**, 133–141 (2015)
8. Garfield, J.B., Burrill, G.: Research on the role of technology in teaching and learning statistics. In: Proceedings of the 1996 IASE Round Table Conference University of Granada. Spain, 23–27 July 1996
9. Alexiou, A., Bouras, C., Giannaka, E.: Virtual laboratories in education: a cheap way for schools to obtain laboratories for all courses by using the computer laboratop. In: IFIP TC3 Technology Enhanced Learning Workshop (Tel'04), World Computer Congress, 22–27 August 2004
10. Durey, A., Beaufils, D.: L'ordinateurdansl'enseignement des sciences physiques: questions de didactique. In: Actes des 8ème Journées Nationales Informatique et pédagogie des Sciences Physiques, Udp et INRP, pp. 63–74 (1998)
11. Milot, M.C.: Place des nouvelles technologies dansl'enseignement de la physique-chimie (1996)
12. Frerich, S., Kruse, D., Petermann, M., Kilzer, A.: Virtual labs and remote labs: practical experience for everyone. In: IEEE Global Engineering Education Conference (EDUCON) (2014)
13. Ko, C.C., Chen, B.M., Hu, S., Ramakrishnan, V., Cheng, C.D., Zhuang, Y., Chen, J.: A web-based virtual laboratory on a frequency modulation experiment. In: IEEE Trans. Syst. Man Cybern. B Cybern. Part C: Appl. Rev. **31** (2001)
14. Ansgar, S.: Software development process model and methodology for virtual laboratories. Appl. Inf. Proc. **2**, 47–52 (2002)
15. Dobrzański, L.A., Honysz, R.: Building methodology of virtual laboratory posts for materials science virtual laboratory purposes. Arch. Mater. Sci. Eng. **28**, 695–700 (2007)
16. Herrera, O.A., Alves, G.R., Fuller, D., Aldunate, R.G.: Remote lab experiments: opening possibilities for distance learning in engineering fields, education for the 21st century—impact of ICT and digital resources. In: IFIP International Federation for Information Processing, vol. 210. Springer, Boston, MA (2006)
17. Raman, R., Achuthan, K., Nedungadi, P., Diwakar, S., Bose, R.: The VLAB OER experience: modeling potential-adopter student acceptance. IEEE Trans. Educ. **57**(4), 235–241 (2014)
18. Raman, R., Haridas, M., Nedungadi, P.: Blending concept maps with online labs for STEM learning. In: Advances in Intelligent Informatics, pp. 133–141. Springer, Cham (2015)
19. Zacharias, Z.C., Constantinos, C.P.: Comparing the influence of physical and virtual manipulatives in the context of the physics by inquiry curriculum: the case of undergraduate students' conceptual understanding of heat and temperature. Am. J. Phys. **76**(4&5), 425–430 (2008)
20. Nedungadi, P., Raman, R.: The medical virtual patient simulator (MedVPS) platform. In: Intelligent Systems Technologies and Applications, pp. 59–67. Springer, Cham (2016)
21. De Jong, T., Sotiriou, S., Gillet, D.: Innovations in STEM education: the Go-Lab federation of online labs. Smart Learn. Environ. **1**(1), 3 (2014)
22. Nedungadi, P., Ramesh, M.V., Pradeep, P., Raman, R.: Pedagogical support for collaborative development of virtual and remote labs: Amrita VLCAP. In: Auer, M.E., Azad, A.K.M., Edwards, A., de Jong, T. (eds.) Cyber-physical Laboratories in Engineering and Science Education. Springer, Cham (2018)

Comparing English and Malayalam Spelling Errors of Children using a Bilingual Screening Tool

Mithun Haridas, Nirmala Vasudevan, Georg Gutjahr, Raghu Raman and Prema Nedungadi

Abstract Despite the high prevalence of reading disabilities among Indian children, many school teachers are not adept at identifying and assessing these difficulties. Screening tools for reading disabilities are available in English but are unavailable in many Indian languages. Reading disabilities manifest differently depending on the characteristics of the language being studied. This paper compares reading difficulties that arise when studying English and Malayalam. In a previous study, we designed a bilingual screening test in English and Malayalam and tested it with 135 school children in Kerala. In the current study, the screening test was modified in light of the findings from our previous study. We administered our updated bilingual screening test to 25 second grade children, ages 7–8, who were studying at two other schools in Kerala. Student errors were classified into multiple categories. Similarities and differences between errors in English and Malayalam were identified, and the errors that were specific to Malayalam were analyzed in further detail.

Keywords Reading disabilities · Reading difficulties · Dyslexia · Screening tests · Malayalam · English · Bilingual

M. Haridas (✉) · N. Vasudevan · G. Gutjahr · P. Nedungadi
Center for Research in Analytics and Technologies for Education (CREATE),
Amritapuri, Kollam, India
e-mail: mithunh@am.amrita.edu

N. Vasudevan
e-mail: nirmalav@am.amrita.edu

G. Gutjahr
e-mail: georgcg@am.amrita.edu

P. Nedungadi
e-mail: prema@amrita.edu

N. Vasudevan
Department of Physics, Amritapuri, Kollam, India

R. Raman
School of Business, Amrita Vishwa Vidyapeetham, Coimbatore, India
e-mail: raghu@amrita.edu

© Springer Nature Singapore Pte Ltd. 2020
X.-S. Yang et al. (eds.), *Fourth International Congress on Information and Communication Technology*, Advances in Intelligent Systems and Computing 1027, https://doi.org/10.1007/978-981-32-9343-4_34

1   Introduction

Human beings are born with a "universal grammar" wired in their brains, and a child exposed to spoken language will easily and naturally learn to speak [1]. But while spoken language evolved 50,000–200,000 years ago, the writing system was invented only some 5,400 years ago [2]. And so the brain is not prewired to read and write. A child needs to be taught reading and writing. Most children are able to learn reading and writing without much difficulty. But there are children, sometimes from well-educated families with sufficient exposure to reading material, who still struggle to learn reading and writing [3]. Such children find school very difficult, since academic achievement depends largely on the reading and writing ability.

Dyslexia is a reading disability where the individuals have difficulties with accurate word recognition and spelling. Around 80% of dyslexics have phonological deficits and around 20% exhibit visual processing deficits. Some dyslexics have both phonological and visual deficits [4, 5].

The worldwide prevalence of dyslexia is estimated to be 5–10% [6]. Studies in India have shown that approximately 10% of the Indian population, or 28 million people, have some kind of reading disability [7]. But many Indian teachers have limited knowledge and awareness about these disabilities and their assessment and remediation. Since early intervention is critical for children with learning disabilities [8], this lack of knowledge and awareness has profound implications for the child and his/her family.

A short screening test, requiring around half an hour for its administration, could help in identifying children with reading disabilities. Ideally, these tests should be conducted in the local language spoken by the people of the region. Currently, no screening tests are available for many of the Indian languages. We developed a screening test in Malayalam, a language spoken by 35 million people in the south Indian state of Kerala. Like many other Indic scripts, the Malayalam script is an abugida (alphasyllabary) [9].

We first administered our paper-based screening tests to 135 primary school students in Kerala [10]. The results of this initial study were used to improve the tests. The modified screening tests were next administered to 25 children, ages 7–8, studying in two other Government schools in Kerala. This paper reports the results of the newer study that used the modified screening tests. Understanding the various ways that reading disabilities manifest is important not only for screening tests but also for the design of classroom intervention [11] and effective language learning software such as intelligent tutoring systems [12–15] and m-learning systems [16–19].

This paper is organized as follows: Sect. 2 describes the screening test and the study procedure. Section 3 discusses the results and Sect. 4 summarizes the findings of this study.

2 Materials and Methods

2.1 The Screening Tests

The children were tested only on material that was taught at school. All tests were carried out in both English and Malayalam. The screening test consisted of:

1. Visual discrimination tasks: Visual perception plays an important role in reading. Children with reading disabilities have problems in differentiating between the letters of the alphabet. In the visual discrimination task, the students were asked to identify a specific letter or word from a group of letters or words.
2. Naming tasks: The children were shown pictures of six objects and were asked to write the names of those objects.
3. Spelling tests: Students with reading disabilities sometimes experience greater difficulties in spelling than in reading. The process of spelling involves the integration of several cognitive abilities. Dyslexic children rely heavily on orthographic skills, and poor phonological representation and an incomplete knowledge of the alphabetic principle are largely responsible for their spelling errors in English. We dictated ten words in each of the spelling tests.
4. Vocabulary tests (free writing): The children were given 2.5 min to write as many words as they could think of.
5. Copying a passage: The children were asked to copy a passage in 2.5 min.
6. Reading comprehension: The children were asked to read a passage and write answers to questions based on the passage.

2.2 Procedure

The participants in this study consisted of 25 children (10 boys and 15 girls) studying in Grade 2 in two Government schools of Kollam District, Kerala State. Both the schools conducted classes in Malayalam, while English was taught as a second language. All the children spoke Malayalam at home.

The tests were of 30 min duration. Printed question papers were handed to the students at the start of the test. Specific instructions were provided orally before each task. Scores were obtained for each of the six tasks described in the preceding section. Furthermore, all mistakes in the English visual discrimination, spelling, and vocabulary tests were classified [4]. Mistakes in the Malayalam tests were similarly classified. The data were analyzed using Weka and SPSS Statistics.

The study was performed in full accordance with the ethical standards established by the ethical research committee at our institution. Informed consent was obtained from the participants in our study.

3   Results

The maximum possible score on the screening test was 120. The students obtained scores ranging from 24 to 115.6, with 25% of the students obtaining scores less than or equal to 60.6, as shown in Table 1.

Cluster analysis was performed using the k-means algorithm in Weka. Two clusters were obtained:

Cluster 1: 8 students (32%) who performed poorly on the screening test (students with reading difficulty). (The school teachers later confirmed that these students exhibited reading difficulties in the classroom.)

Cluster 2: 17 students (68%) who performed within the normal range for their age (students reading at grade level).

Table 1 Overall scores in screening tests

Mean	Standard deviation	Percentiles				
		0%	25%	50%	75%	100%
68.51	22.67	24	60.6	74.8	80	115.6

Table 2 Comparison of test scores of children with a reading difficulty and those reading at grade level

Tasks	Cluster 1 (Reading difficulty)	Cluster 2 (Reading at grade level)	t-statistic	p-value
Visual discrimination	7.86	9.39	−1.642	0.14
Naming task	4.57	5.89	−2.119	0.07
Spelling test—English	3.14	8.72	−6.387	0.00
Spelling test—Malayalam	4.29	8.44	−4.723	0.00
Vocabulary test—Malayalam	3.61	11.22	−3.282	0.00
Vocabulary test—English	3.29	14.67	−5.181	0.00
Copying a passage—English	5.71	9.67	−3.626	0.00
Copying a passage—Malayalam	3.71	5.67	−5.352	0.00
Reading comprehension—English	2.43	3.72	−2.489	0.02
Reading comprehension—Malayalam	0.57	2.78	−4.025	0.00

We compared the performance of students in Cluster 1 and Cluster 2 in each of the tasks (Table 2). The results show that their performance is somewhat comparable in visual discrimination and naming tasks (p-value > 0.05), but significantly different in all other tasks. We also performed a detailed analysis of all incorrect answers, separately studying the mistakes in English (Table 3) and Malayalam (Table 4).

Table 3 Mistakes in English tasks

Task	Mistake		Cluster 1 (Reading difficulty)	Cluster 2 (Reading at grade level)
	Description	Examples		
Visual discrimination	Letter reversal—for upper case alphabets	–	0	0
	Letter reversal—for lower case alphabets	A confusion between b and d, p and q, n and its mirror image, k and its mirror image	43	33
	Mistakes in the positions of letters in the word	"gril" instead of "girl"	29	22
Spelling test	Visual errors	Writing mirror image of s	57	22
	Omission of letters/syllables	"ca" instead of "car"	29	22
	Addition of letters/syllables	"care" instead of "car"	43	11
	Substitution of letters/syllables	"cet" instead of "cat"	71	39
	Not all sounds represented, and sounds represented in a wrong order	"had" instead of "hand" "elantph" instead of "elephant"	57	44
	All sounds represented but poor grapheme-phoneme correspondence	"sed" instead of "said"	14	6
	Miscellaneous	–	0	6
Vocabulary test	Visual errors	Writing mirror image of s	71	11

(continued)

Table 3 (continued)

Task	Mistake		Cluster 1 (Reading difficulty)	Cluster 2 (Reading at grade level)
	Description	Examples		
	Omission of letters/syllables	"ca" instead of "car"	29	22
	Addition of letters/syllables	"care" instead of "car"	14	6
	Substitution of letters/syllables	"cet" instead of "cat"	43	17
	Not all sounds represented, and sounds represented in a wrong order	"had" instead of "hand" "elantph" instead of "elephant"	43	22
	All sounds represented but poor grapheme-phoneme correspondence	"sed" instead of "said"	14	6
	Miscellaneous	–	29	17

Table 4 Mistakes in Malayalam tasks

Task	Mistake		Cluster 1 (Reading difficulty)	Cluster 2 (Reading at grade level)
	Description	Examples		
Visual discrimination	Letter reversal	Mirror image of the letter	14	0
Spelling test	Visual errors	Mistakes in the shapes of the letters	43	6
	Omission of syllables	Omission of syllables like ക(ka), കി(ki), കു(ku), etc.	57	0
	Omission of diacritics indicating a short vowel sound	Omission of the diacritic ി in കി(ki), etc.	14	0
	Omission of diacritics indicating a long vowel sound	Omission of the diacritic ീ in കീ(kee), etc.	86	6

(continued)

Table 4 (continued)

Task	Mistake		Cluster 1 (Reading difficulty)	Cluster 2 (Reading at grade level)
	Description	Examples		
	Substitution of a long vowel diacritic with the corresponding short vowel diacritic	Writing കി(ki) instead of കീ(kee), etc.	57	17
	Substitution of a vowel diacritic with another (incorrect) diacritic	Writing കൈ(kai) instead of കോ(ko)	29	50
	Substitution with a similar sounding syllable	Writing കുദിര(kudhira) instead of കുതിര(kuthira, horse)	0	22
	Addition of letters/syllables	ആനക (ānaka) instead of ആന(āna, elephant)	14	33
	Miscellaneous	–	29	8
Vocabulary test	Visual errors	Mistakes in the shapes of the letters. Mistakes in the positions of the letters in the word	14	6
	Omission of syllables	Omission of syllables like ക(ka), കി(ki), കു(ku), etc.	14	11
	Omission of diacritics indicating a short vowel sound	Omission of the diacritic ി in കി(ki), etc.	14	0

(continued)

Table 4 (continued)

Task	Mistake		Cluster 1 (Reading difficulty)	Cluster 2 (Reading at grade level)
	Description	Examples		
	Omission of diacritics indicating a long vowel sound	Omission of the diacritic ീ in കീ(kee), etc.	86	44
	Substitution of a long vowel diacritic with the corresponding short vowel diacritic.	Writing കി(ki) instead of കീ(kee), etc.	43	22
	Addition of letters/syllables	ആനക(ānaka) instead of ആന(āna, elephant)	29	6
	Mistakes in ligatures	Writing ക(ka) instead of ക്ക(kka)	57	28
	Miscellaneous	–	14	0

Goodman gamma correlations [11] were calculated to determine the association between each variable in the passage copying tasks in English and Malayalam. There was a strong correlation between English and Malayalam handwriting ($G = 0.437$, $p = 0.06$), the heights of ascenders and descenders in English and Malayalam ($G = 0.538$, $p < 0.05$), and the spacing between words ($G = 0.502$, $p < 0.005$). On the other hand, errors of "omission/addition of letters" were not associated with each other ($G = 0.088$, $p = 0.769$), implying that students who omitted letters in English words did not necessarily omit letters or syllables while copying the passage in Malayalam.

4 Conclusion

Summarizing our findings:

1. Most poor performers omitted diacritics for long vowel sounds and also incorrectly substituted long vowel diacritics with short vowel diacritics. Thus, spelling tasks on screening tests should include questions involving the long vowel diacritic.
2. Another common mistake among poor performers was missing ligatures, e.g. writing ക(ka) instead of ക്ക(kka). These types of errors will not manifest in English.

3. Well-developed phonological awareness (e.g. the ability to distinguish ക(ka) from വ(kha)) is required for spelling proficiency in Malayalam.
4. Additionally, the script is visually complex (e.g. ി(i) and ീ(ee) are diacritics indicating different sounds), and a deficit in visual processing can also contribute to reading difficulties in Malayalam.
5. Since the poor performers obtained lower scores on all tasks including reading comprehension, all the tasks can be included in the future screening tests.
6. Comparing performance on English and Malayalam copying tasks, there was a significant association in handwriting, ability to correctly write heights of ascenders and descenders, and correctly include spacing between words.
7. No relevant association between errors of omission and addition of letters in English and Malayalam was found in the copying tasks.

These findings indicate that both similarities and differences exist in the manifestation of reading disabilities in English and Malayalam. In the future, we plan to further investigate the differences between specific forms of reading disabilities in English and Malayalam, such as surface dyslexia and phonological dyslexia.

Acknowledgements We are grateful to the Chancellor of our University, Sri Mata Amritanandamayi Devi, for Her guidance and motivation. The first author is supported by the Visvesvaraya Ph.D. Scheme. This work was partly supported by the Department of Science and Technology—Cognitive Sciences Research Initiative (DST-CSRI), Government of India (DST SR/CSI/120/2013 and DST SR/CSI/121/2013).

References

1. Hulin, R., Na, X.: A study of Chomsky's universal grammar in second language acquisition. Int. J. Stud. Engl. Lang. Lit. (IJSELL) **2**, 1–7 (2014)
2. Reid, G., Fawcett, A., Manis, F., Siegel, L.: The sage handbook of dyslexia. Sage, London (2008)
3. Shaywitz, S.E.: Overcoming dyslexia: a new and complete science-based program for reading problems at any level. Knopf, New York (2003)
4. Phillips, S., Kelly, K., Symes, L.: Assessment of learners with dyslexic-type difficulties. Sage, London (2013)
5. Kibby, M.Y., Dyer, S.M., Vadnais, S.A., Jagger, A.C., Casher, G.A., Stacy, M.: Visual processing in reading disorders and attention deficit/hyperactivity disorder and its contribution to basic reading ability. Front. Psychology. **6**, 1635 (2015)
6. Siegel, L.S.: Perspectives on dyslexia. Paediatr. Child Health **11**(9), 581–587 (2006)
7. Vasudevan, N., Iyer, A.: Developmental dyslexia—its prevalence and treatment in India. In: Proceedings SGEM Multidisciplinary Scientific Conference on Social Sciences and Arts (SGEM 2015), vol. 1, pp. 425–430. Book 1 (2015)
8. Poulsen, M., Nielsen, A.V., Juul, H., Elbro, C.: Early identification of reading difficulties: a screening strategy that adjusts the sensitivity to the level of prediction accuracy. Dyslexia **23**(3), 251–267 (2017)
9. Daniels, P.T.: Fundamentals of grammatology. J. Am. Orient. Soc. **119**(4), 727–731 (1990)
10. Haridas, M., Vasudevan, N., Iyer, A., Menon, R., Nedungadi, P.: Analyzing the responses of primary school children in dyslexia screening tests. In: Proceedings IEEE International Conference MOOCs, Innovation and Technology in Education (MITE 2017). IEEE Press (2017)

11. Menon, R., Nedungadi, P.: New methodology to differentiate instructional strategies for ESL learners in the Indian context. In: Frontiers in education conference (FIE), pp. 1–6. IEEE (2015)
12. Nedungadi, P., Raman, R.: A new approach to personalization: integrating e-learning and m-learning. Educ. Tech. Res. Dev. **60**(4), 659–678 (2012)
13. Karmeshu, R.R., Nedungadi, P.: Modelling diffusion of a personalized learning framework. Educ. Technol. Res. Dev. **60**(4), 585 (2012)
14. Nedungadi, P., Raman, R.: Effectiveness of adaptive learning with interactive animations and simulations. In: 3rd International Conference on Advanced Computer Theory and Engineering (ICACTE) vol. 6, pp. 6–40. IEEE (2010)
15. Raman, R., Nedungadi, P.: Adaptive learning methodologies to support reforms in continuous formative evaluation. In: International Conference on Educational and Information Technology (ICEIT), vol. 2, pp. 2–429. IEEE (2010)
16. Jayakumar, A., Babu, G.S., Raman, R., Nedungadi, P.: Integrating writing direction and handwriting letter recognition in touch-enabled devices. In: Proceedings of the Second International Conference on Computer and Communication Technologies, pp. 393–400. Springer, New Delhi (2016)
17. Raman, R., Vachhrajani, H., Shivdas, A., Nedungadi, P.: Low cost tablets as disruptive educational innovation: modeling its diffusion within Indian K12 system. In: Innovations in Technology Conference (InnoTek), pp. 1–5. IEEE (2014)
18. Nedungadi, P., Jayakumar, A., Raman, R.: Low cost tablet enhanced pedagogy for early grade reading: In: Indian context. Humanitarian technology Conference (R10-HTC). IEEE Reg. **10**, 35–39 (2014)
19. Sanjanaashree, P., Kumar, M.A., Soman, K.P.: Language learning for visual and auditory learners using scratch toolkit. In: Proceedings IEEE Conference Computer Communication and Informatics, pp. 1–5. IEEE Press (2007)

Curriculum Enrichment in Empowering "Corporate-Ready" Individuals

Ganeshayya Shidaganti, S. Prakash and K. G. Srinivasa

Abstract It is turned out to be obvious in current trend that the curriculums are not adequately covering up the skills that a fresh graduate should possess to become employable. In order to become "corporate-ready," the industry expectations have to be incorporated into the curriculum. Curriculum enrichment supports cognitive development of the student fraternity through well-planned lessons spread throughout the period of education. This study focuses on bridging the gap of academic-industrial interface with necessary skill sets between the seekers and the employees and molding individuals as "corporate-ready." Once the Board of Study (BOS) identifies the existing gap, they will be able to incorporate the skills sets that are required to make the fresh graduates "corporate-ready." The results were very optimistic showing 85% of the study group members welcoming the prerequisite in comparison with only 15% for the control group members opting for other dimensions of skill sets.

Keywords Corporate-ready · Curriculum enrichment · Skill set

1 Introduction

Today, the educational perspective has drastically evolved and has observed many changes, as it uses Information and Communication Technology (ICT) in the education system. The main aim of any affiliating university today is to mold and equip the students so as to make them shoulder the responsibilities and face the challenges in the future.

G. Shidaganti (✉)
Ramaiah Institute of Technology, Bangalore, India
e-mail: ganeshayyashidaganti@msrit.edu

S. Prakash
East Point College of Engineering And Technology, Bangalore, India
e-mail: prakash.hospet@gmail.com

K. G. Srinivasa
National Institute of Technical Teachers Training & Research, Chandigarh, India
e-mail: kgsrinivasa@gmail.com

© Springer Nature Singapore Pte Ltd. 2020 437
X.-S. Yang et al. (eds.), *Fourth International Congress on Information and Communication Technology*, Advances in Intelligent Systems and Computing 1027, https://doi.org/10.1007/978-981-32-9343-4_35

"Expectation" is the word that every young mind of an engineering graduate dreams about. This been the central dogma of every educational institution. The expectations of the engineering graduates start from What to expect? What does the corporate world expect? Am I equipped enough? Am I provided with expected skill sets? Is my graduation program relevant to the corporate expectation? Is the training program attained in the college is up to the mark of corporate expectations? And finally, am I a "Corporate-ready" graduate?

In achieving the said above, a curriculum should be designed in view to cater the needs of all aspiring engineering graduates from all forms of socio-economic intellectual background. Curriculum is nothing but an overall plan for learning. The plan encapsulates community needs and a structure to transform the values into learning experiences [1]. Curriculum in the educational system is nothing but an educational plan [2]. Hence curriculum designing becomes a key aspect in providing overall development of an individual in knowledge-based domain. But in most of the case, generalized curriculum design does not cater the need of individuals aspiring for different roles to play in corporate world. Thus, this calls for curriculum enrichment program that would cater to the needs of every engineering graduating individual.

It has been a need of the hour to provide high quality education to meet the international standards which is hastily shifting and challenging. It adds a number of pressures on educational institutions to mold employability skills in equipping the student community to the global market. Previously known comfortable concepts "jobs for life" and "planned career path" are replaced by life-long learning, constant self-development, initiative, and personal motivation to stay employable [3]. It is unfortunate that many Indian universities are not able to achieve this goal; the main reasons being institutions are driven by people who lack industry knowledge, reduced capital investments, and exposure to the research domain. In the Indian state of affairs, this has turned into a noteworthy concern being voiced by the ventures and industry affiliation [4]. Adaptability and skill set acquired by today's fresh engineering graduates decide their sustainability of employment in the competitive global market. Capabilities/competency development is firmly coupled with the growth of employability [5]. Table 1 maps the key capabilities to the employability skills which is necessary for fresh engineering graduate. With due respect to the demand, enough emphasis to be given to supply. A multi-faceted system of education encounter on the supply decides the output. In spite of exponential growth in the field of education for the past few decades, yet the Indian education system is struggling to find its feet in the global market. Owing to the fact from interest level of students opting for the course to the quality of the students' influences from the family, peer pressure, and lack of proper guidance from the high school. The supply side can escalate its standards only with an intervention of the curriculum enrichment module.

Curriculum enrichment describes formulation of a curricular plan by inviting subject expert from academy and industry. Such enrichment program is provided by colleges in order to extend the skill set of the student's knowledge-based domain beyond their mail curriculum of study. A commitment to provide opportunities in broadening student's technological skill set in widespread throughout the education sector. Though such curriculum enrichment programs enhance student's efficiency

Table 1 Key capabilities mapped to employability skills

Key capabilities	Employability skills
Convey thoughts and information	Communication
Commitment toward learning	Learning
Achievement through collaboration	Team building
Adjust as per workplace situation	Adaptability
Solid assurance toward work	Task perseverance
Get to bottom of problems	Problem solving
Utilize technologies	Techno knowledge
Liability headed for organization	Responsibility
Look for work, exclusive of being told	Pro-activeness
Sketch and systemize work	Work performance

during the course duration, increase the motivation, achievements, etc. Such programs in a way are found non-responsive enough. This could be due to the way colleges respond to the teaching community, system and procedures of the organization, investment aspects, provision toward resources, etc. It becomes their deemed role to provide valuable links within the internal community, set standards of the college to grab the potential students, and to faster sense achievability during the graduating period. Figure 1 portrays the role of society/stakeholder in designing the curriculum.

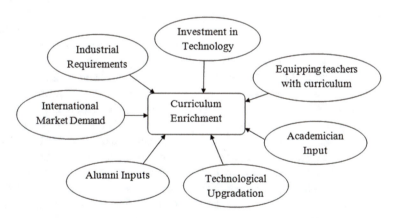

Fig. 1 Stakeholders of curriculum design

2 Literature Review

For the past two decades, there has been a continuous focus on studies analyzing the academy—industry relationships. There were many research contributions made toward improving academic standards and also toward the required contribution to be made by the industry. Thus, made contributions helped the student community to be responsive toward demographic style that has occurred in the system of education.

The advancement in educational system has been extremely slow in keeping pace with the global change. Montgomery and Porter [6] also reported that the academia is way behind the required standards of the industry. The curriculum of the management program emphasizes only on quantitative and analytical skill set, and oversees the human values-based education resulting in managers failing to meet the challenges of the global environment which is with diverse work force.

Debabrata et al. [7] found that in India, there were many methodologies being adopted in view with "Academic-Industry" interaction; his study also focused on the industry's contribution right from drafting of the frame work of the syllabi to absorbing the trained students. This initiative enables the industries to reduce the burden on orienting the fresh graduates. As a backup of this work, Sanjay [8] in this work concluded that fresh graduates require an induction program spanning between six months to two years to prove themselves. This is due to the gap between what is given to students and what is expected out of the graduating students. There are several studies conforming [9–12] that there is a gap in global market demand and graduates perception of employability. Hence the graduating period becomes very critical in equipping the students as corporate-ready individuals. Therefore, it requires coordination among the academia, industry, and government which is well documented by Udailal [13]. He also points that there should be a frequency in revision of curriculum incorporating the insight given by the industries, thus creating the professional with global mind set to face various cultural and social setups.

Patil and Popker [14] emphasized on a common platform which brings in the industry and academia/University/Colleges/Autonomous educational institutions to frame a value-based curriculum enrichment enabling in producing corporate-ready professionals. There has been a huge difference being what is provided to what is produced [15]. Though there are various drawbacks being projected in the past with this regards, like faculty-student ratio, lack of teacher quality, lack of funds in training the students with technological support, lack of government's participation, still there remains a huge vacuum created by the curriculum being provided to the student fraternity. This can be only addressed with an enriched curriculum. Various works with this regard were undertaken in the past. Work pertaining to (i) "Involvement of stakeholder in design, implementation and evaluation of the University curriculum" [16]. (ii) Employers involvement in designing the curriculum and explaining the clear positive impact on the ability of the graduates as "corporate-ready" [17]. (iii) Incorporation of practical methods and techniques to resolve the gap between theory and practice by bringing the change to the curriculum [18]. (iv) Introduction of curriculum that allows consistent assessment and revaluation in aim to achieve and cater the needs of the dynamic corporate world [19].

The curriculum structure necessarily starts from systematic theoretical framework of employment betterment and guidance [20]. This hypothesis was based primarily for career counseling and relies on the perception of ancient prognostic theories and latest constructivist counseling approaches [21–24]. This system functions as a connection between theory and practice, which is the center need behind graduate proficiency gap. In case of difficulty in changing the entire pattern of curriculum design, each educational institution at their ground level can frame an additional curriculum as enrichment program to have "Corporate-ready" graduates.

Kaur and Bhalla [25], in their study, concluded that educational institution is rated as best based on three factors, one being the teaching environment, two research environment (environment for innovation), and last but not the least the educational curriculum, which is also well documented by Jackson and Chapman [26]. In their study, the perception of fresh graduates and the existing skill gap were asked to be rated by the employers. The result obtained clearly states that the graduates were proficient in non-technical skills which include soft skills, character skills, and generic skills and were deficient in essential hard skills (technical expertise and knowledge required for job). Though developing soft skills is essential for efficient work performance [27], growth and success at an international platform, effective delivery of systematic curriculum pattern, and the technical knowledge are more relevant in both global and local settings [28]. This encourages an educational module change and brings up an issue of who is responsible for designing it.

Pedagogy is the basis which is targeted in bridging the fresh graduate's aptitude and technical gap, and they are investigating to adopt and express the required competency in employability. The present study is framed on the basis of qualitative research design to explore the requisites of fresh graduates and the necessary steps for their "corporate-ready" state, enabling to identify and examine the precise nature of technical skill deficiency of fresh graduates graduating in India. The outcome of this study highlights the responsibility of the educational system those areas requiring curricular review and calls for the need for curricular reform and stakeholder's responsibility.

3 Objectives

- To frame a standardized questionnaire in order to measure the gap between "What is provided in terms of curriculum to what is produced to corporate world".
- To indentify the factors owing to un-employability.
- To open new avenues for further research in the domain of curriculum.

4 Research Design

4.1 Research Aim

This study follows a design of qualitative research to explore fresh engineering graduate's employability and skill sets requisite for their sustenance in the global market. Qualitative research necessarily begins with assumptions made using existing theoretical framework that informs the study of research problem caters to the need of the society. For such study, a qualitative approach begins with an inquiry and data analysis which concludes with an outcome that addresses the research problem. The final report includes the voices of participants, stakeholder's involvement in solving the research problem, the complexity in description and interpretation of the problem, and the required change in solving it. Figure 2 clearly depicts the employer's expectations in making an individual "corporate-ready." The first expectation here is the fresh graduate should satisfy the personal characteristics like age, personality trait, and gender to fit the requirement, and the second expectation is the preferred new talents who are self-motivated, creative, and innovative. The most important

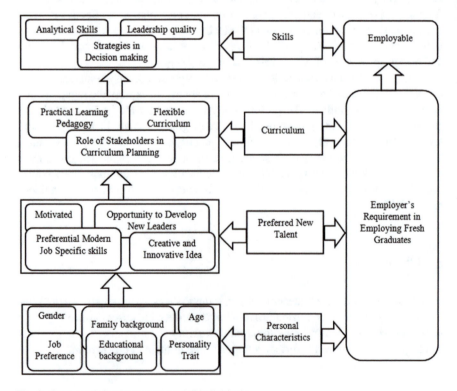

Fig. 2 Framework for "corporate-ready" individuals

expectation is the curriculum which should be flexible and should involve practical learning so that the fresh graduate can attain analytical skills, strategies in decision making and leadership qualities which makes them employable.

4.2 Research Methodology

A sample of respondents from alumni association of the academic institutions, stakeholder of corporate sector, academicians were chosen, and this study was carried out in four-phase format. A convenience statistical sampling technique was carried out using a visual analytical tool named Tableau to complete the study. A self-designed questionnaire rendering around employability, related modules in the curriculum, design of core curriculum, and incorporation of work-based learning within the compounds of curriculum was framed and analyzed. The data for this study was collected through circulation of questionnaire using social media, direct contacts and through telephonic survey. The questionnaire used is a semi-structured interview style with both open-ended and closed-ended format. The open-ended questions were framed in allowing further discussions and few answers from the responders led in asking more questions. This study solely consists of primary data based on which an inference is drawn.

4.3 Data Analysis

Qualitative dataset is obtained from the gathered relevant data and drawing conclusions out of it [29]. The data inventory was compiled based on the responses thus obtained and was analyzed in phase 4 as a framework to create evidence-based employability skill set. Skill sets for fresh graduates were generally categorized into employability skills, curricular changes required for employability, personal characteristics, and advantages of preferred new talents.

Phase 1: Questionnaire based on industries expectation of the skill sets for employability

A questionnaire on "Employability skill from the employer's perspective" was designed incorporating the set of possessed skill set by a graduating student and categorized into three sections pertaining to soft skills, hard skills, and others. Stakeholders of corporate sector were requested to select one option out of the three and there was also an option given to identify new skill set as required by them.

Phase 2: Questionnaire on existing curricular design

The existing curricular design for the engineering graduates was documented by formulating an inventory based on the survey of stakeholders of corporate sector from phase 1 and differentiated into various sections. There was also an option given for the alumni group of respondents to compile and make their own inventory so as to fill the gap in the existing curricular pattern.

Phase 3:Questionnaire on identifying deficiency on curricular design

A similar questionnaire to that of phase 2 was designed in making an inventory in assessing the deficiencies identified in the existing curricular pattern. The responders were given an option of suggesting their practical difficulties in securing and sustaining an employment in the corporate world.

Phase 4: Questionnaire revolving around probable solutions for the research problem

The questionnaire was prepared with a space for free response, and comments were appreciated on any matter relating toward preparing graduating students as "corporate-ready" individuals.

5 Result and Analysis

Phase 1: Questionnaire based on industries expectation of the skill sets for employability

The questionnaire was sent to 60 stakeholders of corporate sector, and the received responses were 51 which correspond to a response rate of 85%. Of the responses received, 26 were from EMC2, nine were from HP, and 16 were from Bosch. The stakeholders of corporate sector ranked the hard skill highly. The industries expectation of the skill sets for employability mean score for the importance of the various skills showed a strong positive correlation. The correlation coefficient was 0.9823 (Fig. 3).

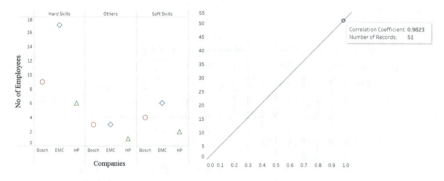

Fig. 3 Skill set that industry expects from fresh graduates

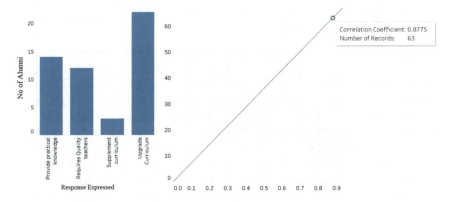

Fig. 4 Alumni inputs to make fresh graduates employable

Phase 2: Questionnaire on existing curricular design

The questionnaire was sent to 74 alumni, and the received responses were 63 which correspond to a response rate of 85%. Of the respondents, 21 were from MS Rama-iah Institute of Technology, and 42 were from Atria Institute of Technology. The respondents expressed their experience by stating that the curricula were designed to cater to the need of academically weaker section, a basic curricula which can be generalized for various strata of students not a technologically equipped one, no frequency in curriculum revision, and various other parameters. According to the statistical analysis, the questions pertaining to "curriculum revision is required" were highly ranked in comparison to all other above said parameters. The overall response from the alumni for the open question was to enrich the curriculum. The existing curricular design mean score for curriculum revision is required, showed a strong positive correlation. The correlation coefficient was 0.8775 (Fig. 4).

Phase 3: Questionnaire on identifying deficiency on curricular design

On the basis of the overall mean values of the assessment score from phase 1 and phase 2, the phase 3 was performed. The skill sets were placed in view with the deficient found in curricula. The questions were framed in view with (i) Awareness of the market demand by the members of the board of studies. (ii) Proficiency of the teachers in handling the revised curriculum. (iii) Technological deficiency. (iv) Insensitivity from the industry in providing inputs of the curriculum design. (v) various other factors. Statistically, an insignificant correlation was found among all the factor analyzed pointing that there is a dare necessity in supplementing the curriculum with an enrichment feature.

Phase 4: Questionnaire revolving around probable solutions for the research problem

The general response from the responders for the open question was to supplement curriculum (curriculum enrichment), provide practical knowledge/technical skills, and quality teachers.

6   Conclusion

It is evident from the findings carried out in four phases of the study that the stakeholders of corporate sector expect the fresh graduates to acquire hard skills but not ignoring the soft skills, and the alumni suggest for enriching the curriculum. The inputs from the stakeholders of corporate sector and the alumni highlights that curriculum enrichment is required based on the results obtained. Budding curriculum to meet the dynamic and challenging needs of industries should support the growth of new knowledge and new technologies in order to bridge the gap between the educational institutes and the industries to make fresh graduate individuals "Corporate-Ready." The results were very optimistic showing 85% of the study group members welcoming the prerequisite in comparison with only 15% for the control group members opting for other dimensions of skill sets.

Declaration Ethical approval: All procedures performed in studies involving human participants were in accordance with the ethical standards of the institutional and/or national research committee and with the 1964 Helsinki declaration and its later amendments or comparable ethical standards.

References

1. Gupta, B.L., Earnest, J.: Competency Based Curriculum. Mahamaya Publishing House, Jalandhar, Punjab (2008)
2. Balu S.A.: Curriculum implementation in the context of an organizational change: a working paper; Singapore (now Manila). Colombo Staff College for Technician Education (1982)
3. Pizika, N.: Embedding employability skills in computer and information science program curriculum. World Acad. Sci. Eng. Technol. Int. J. Soc. Behav. Educ. Econ. Bus. Ind. Eng. 8(2), 381–385 (2014)
4. Sharma, P.B.: Technical education in knowledge age. Univ. News J. AIU. New Delhi, July 2007
5. Kola, S.R., Chatarajupalli, S.: An effective framework for industry-academia collaboration. In: QScience Proceedings, vol. 38 (2015)
6. Montgomery, C., Porter, M.E. (eds.): Strategy: Seeking and Securing Competitive Advantage. Harvard Business School Publishing, Boston (1991)
7. Debabrata, G., Bhatnagar, D., Jancy, A., Saxena, N., Muneshwar, S.K.: Innovative mechanism to improve effectiveness of technical education—a case study of mission mode approach in India. Retrieved from www.indianjournal.com on 10 Oct 2009
8. Sanjay, M.: The task of shaping skills & employability, The Financial Express. 04 July 2009. Retrieved from www.finacialexpress.com/news/the-task-of-shaping-skills-&-employability/484760 on 09 Oct 2009
9. NACE Research: Job Outlook. National Association of Colleges and Employers. http://career.pages.tcnj.edu/files/2011/07/Job_Outlook_2011_Full_Report_PDF1.pdf (2011)
10. NACE Research: Job Outlook. National Association of Colleges and Employers. http://career.sdes.ucf.edu/docs/job_outlook_2013_spring.pdf
11. Great Expectations. What students want in an Employer and how federal agencies can deliver it? Partnership for Public Service & UNIVERSUM. http://www.ourpublicservce.org/OPS/publications/viewcontentdetails.php?id=131 (2009)

12. DuPre, C., Williams, K.: Undergraduates perceptions of employer expectations, J. Career Tech. Educ. **26**(1). http://scholar.lib.vt.edu/ejournals/JCTE/v26n1/pdf/dupre.pdf (2011). ISSN 1533–1 830
13. Udailal, P.: Educated youth and unemployment in Ethopia, Indian J. Commer. **62**(1) (2009)
14. Patil, M.R., Popker, T.M.: Business education: emerging challenges. Indian J. Commer. **51**(1) (1998)
15. Siememsma, F.: Hopes, tension & complexity: Indian students' reflections on the relationship of values to management education and future career options. J. Hum. Values. **4**(2). Sage Publication, New Delhi
16. Kagaari, J.R.K.: Evaluation of the effects of vocational choice and practical training on students' employability. J. Eur. Ind. Train. **31**(6), 449–471 (2007)
17. Mason, G., Williams, G., Cranmer, S.: Employability skills initiatives in higher education: what effects do they have on graduate labour market outcomes? Educ. Econ. **17**(1), 1–30 (2009)
18. Wellman, N.: Relating the curriculum to marketing competence: a conceptual framework. Mark. Rev. **10**(2), 119–134 (2010)
19. Pandiyan, A.: Employers' perspective of MBA curriculum in meeting the requirements of the industry. Manag. Labour Stud. **36**(2), 143–154 (2011)
20. McMahon, M., Patton, W. (eds.): Career Counseling: Constructivist Approaches. Routledge, London, UK (2006)
21. McMahon, M., Patton, W.: Using qualitative assessment in career counseling. Int. J. Educ. Vocat. Guid. **2**(1), 51–66 (2002)
22. McMahon, M., Patton, W., Watson, M.: Developing qualitative career assessment processes. Career Dev. Q. **51**(3), 194–202 (2003)
23. McMahon, M., Patton, W., Watson, M.: Creating career stories through reflection: an application of the systems theory framework of career development. Aust. J. Career Dev. **13**(3), 13–16 (2004)
24. McMahon, M., Watson, M., Patton, W.: Developing a qualitative career assessment process: the system of Career influences reflection activity. J. Career Assess. **13**, 476–490 (2005)
25. Kaur, D., Bhalla, G.S.: Perception of faculty towards college management: a case study. ICFAian J. Manag. Res. **8**(8) 2009
26. Jackson, D., Chapman, E.: Non-technical skills gaps in Australian business graduates, pp. 95–113. www.emeraldinsight.com/0040-0912.htm (2011)
27. Selvadurai, S., Choy, E., Maros, M.: Generic skills of prospective graduates from the employers' perspectives. Asian Soc. Sci. **8**(12), 295–303 (2012). ISSN 1911-2017, E-ISSN 1911-2025
28. Noronha, M.R.: Management education at crossroads in India. Asia Pac. J. Res. Bus. Manag. **2**(6), 87–101 (2011)
29. Caudle, S.L.: Qualitative data analysis. In: Wholey, J.S., Hatry, H.P., Newcomer, K.E. (eds.) Handbook of Practical Program Evaluation, pp. 417–438. Jossey-Bass, San Francisco (2004)

Cryptographic Algorithms to Mitigate the Risks of Database in the Management of a Smart City

Segundo Moisés Toapanta Toapanta⊙**, Félix Gustavo Mendoza Quimi**⊙**,**
Rubén Franklin Reina Salazar⊙ **and Luis Enrique Mafla Gallegos**⊙

Abstract It was analyzed the information of databases, encryption algorithms, and information security that are the main components of a Smart City. The problem is the lack of security of public or secret data that is stored in a repository, where there is access to the citizen, local government, and central government. The objective is to define a security prototype and adopt a cryptographic algorithm to mitigate the risks of the database in the management of a Smart City. It was used the deductive method and exploratory research to analyze the information of the reference articles. It turned out a general prototype of security; an algorithm to implement a database for Smart City; an algorithm of data protection expressed in flowcharts. It was concluded that information protected with an encryption algorithm, is a support to be more efficient, reduce costs, reduce the environmental footprint and improve the management of a Smart City.

Keywords Cryptographic algorithms · Database risks · Information security · Smart city · Security scheme

S. M. T. Toapanta (✉) · F. G. M. Quimi · R. F. R. Salazar
Department Computer Science, Universidad Politécnica Salesiana Sede Guayaquil,
Chamber 227 y 5 de junio, Guayaquil, Ecuador
e-mail: stoapanta@ups.edu.ec

F. G. M. Quimi
e-mail: fmendoza@ups.edu.ec

R. F. R. Salazar
e-mail: rreina@est.ups.edu.ec

L. E. M. Gallegos
Department Computer Science, Escuela Politécnica Nacional, Ladrón de Guevara E11-253,
Quito, Ecuador
e-mail: enrique.mafla@epn.edu.ec

© Springer Nature Singapore Pte Ltd. 2020   449
X.-S. Yang et al. (eds.), *Fourth International Congress on Information
and Communication Technology*, Advances in Intelligent Systems
and Computing 1027, https://doi.org/10.1007/978-981-32-9343-4_36

1 Introduction

In information and communication technologies (ICT), any information such as audio, text, and video is exposed to risks regardless of the means of storage or communication used; there are certain problems such as the storage capacity of devices, connections to networks, data traffic, information security, applications with security problems, non-accessible data, poor addresses, disconnected business processes; information security is more vulnerable; within ICT, each technology that solves problems, evolved independently; among these we name: cryptography, information security, databases (DB), and Smart City (SC); in practice, these converge and combine to increase security levels with access and authorizations for data manipulation. The security proposals are intended to eliminate, mitigate or prevent future threats, and these threats are given by people who steal or corrupt the data.

About cryptography, it is the knowledge of the protection of secret information, where there is secret-key cryptosystem and public-key cryptosystem [1]; in the encryption processes, the data takes more time to execute.

About DB, there are many DB proposals with different benefits, this accelerated offer results in a design of non-efficient DB; DB security and user privacy are in conflict when selecting information from the DB; DBs in the cloud have lower costs, but there are security doubts; also to be functional in security they must be in an extensible architecture and maintain a scalable design. In DB services in the cloud, there is no reliable audit scheme [2]; the science of data analysis includes pattern recognition, image analysis, data mining, information patterns, and scientific data; it must also allow greater security in data processing [3].

About information security, with the accelerated growth of data and communication networks, information security becomes more important; there is an increase in the use of cloud services. Which generates many doubts about the security level and privacy; encryption is considered as a way to increase security; other research analyzes cryptographic algorithms and mobile technologies documented from analog generation to digital generation, its evolution and improvement [4].

About Smart City, ICTs prepare the way for cities to undergo a process of urban transformation or evolution to reach SC; this change must be supported by preventive, detective or corrective functions to mitigate risks, vulnerabilities or threats that affect the information. Some of its components are: government, business, health, devices, citizens, buildings, houses, transportation, traffic control, physical security, networks, general services, and specific services; all these data generators are processed and converted into information that is stored in repositories. Research shows that more people live in cities than in rural areas, it is considered to increase to 66% by 2050; a Smart city is vulnerable to a series of security attacks, it is considered important to identify these threats and damages for the design of a solution [5].

Why is it necessary to define an information security mechanism to reduce the risks to which a database is exposed in the management of a Smart City?

To identify schemes, frameworks, algorithms, protocols or strategies that are used as a security mechanism, also define a mechanism to increase the level of information security of a DB and prevent factors that compromise the data.

The objective is to define a security prototype and adopt a cryptographic algorithm to mitigate the risks of the DB in the management of a Smart City.

The articles reviewed and related to DB, cryptographic algorithms, DB risks, Smart City, and information security are:

Implementing security technique on generic database [6], Design and Implementation of Database Encrypting Middleware [7], Database Encryption Using Asymmetric Keys: A Case Study [8], P-McDb: Privacy-preserving Search using Multi-cloud Encrypted Databases [9], Efficient Parallel Summation on Encrypted Database System [10], TrustedDB: A Trusted Hardware-Based Database with Privacy and Data Confidentiality [11], Efficient Format Preserving Encrypted Databases [12], ACAFP: Asymmetric Key-based Cryptographic Algorithm using Four Prime Numbers to Secure Message Communications. A Review on RSA Algorithm [13], Comparison of Cryptographic Algorithms in cloud and Local Environment using Quantum Cryptography [14], Parallel Implementation of Cryptographic Algorithm: AES Using OpenCL on GPUs [15], Improved DSA Cryptographic protocol and its comparative study with RSA protocol [16], Innovative infrastructure for Smart City/Smart Environment applications [17], High Speed Implementation of RSA Algorithm with Modified Keys Exchange [1], Securing smart cities using blockchain technology [5].

It is used deductive method and exploratory research to analyze the information of the reference articles.

The result is a general security prototype; an algorithm to implement a DB for Smart City; a data protection algorithm expressed in flowcharts.

It is concluded that information protected with an encryption algorithm, is a support to be more efficient, reduce costs, reduce the environmental footprint and improve the management of a Smart City.

2 Materials and Methods

2.1 Materials

In the first instance in materials, some works were reviewed in three areas: DB, encryption algorithms and SC; to have some relationship in terms of information security. In the second instance in methods, there were presented instruments that are used to propose a security prototype as: performance of SC services, components of information security, elements of SC, generation of SC data, and scope of the prototype.

About DB: The author presented an algorithm structure to solve the data entry and its storage; the architecture encrypts and converts the sensitive data, resolves the data storage in DB that manages the system; the architecture identifies the information that contains vulnerabilities and proceeds to protect the system [6]. To protect the confidentiality and usability of data in a DB; the authors implemented three schemes: encryption in relation to the client, encryption in relation to the server, and server encryption mix with client encryption; the process is not reversible when the attacker obtains an encryption index from which, based on analysis, the encryption pattern complies with the security; the authors concluded that it is not enough to protect the DB security, the adjustment and efficient allocation of the firewalls are necessary [7]. The authors showed an encryption with two types of keys for the intersection of own data; to make a DB with asymmetric keys have very private encryption for data and files; these types of algorithms were created with application interface, file and folder protection techniques [8]. The authors tried with the preservation of privacy through encrypted DBs of multiple clouds, which is useful when working with external providers; the scheme focuses on complete security: searches, insertions, deletions, without filtering information to the cloud; the proposed system is not vulnerable to access and other types of attacks; they use two clouds of servers that do not converge, where one server stores the data and the other is in charge of re-randomization [9]. The authors reviewed the CryptDB and MONOMI databases, which are systems that offer encryption; they made the comparison with their own system, where both DB perform different processes of decryption by blocks [10]. The authors developed DB-based hardware with privacy and reliable data; this prototype allows customers to perform SQL queries with restrictions, privacy, and without limitations in expressiveness; conducted falsification tests in a critical consultation and processing stages; the prototype removes any limitation in the type of supported queries; the attributes in the database are classified, the private attributes are encrypted and can only be deciphered by the client [11]. To guarantee the privacy of any sensitive field such as MAC addresses, email addresses, user names, among others; the authors reviewed cryptographic schemes such as AES and Blowfish, these do not preserve the same type of text when encrypting the data in the DB; the authors exposed the types of ciphers that the DB preserves [12].

About encryption algorithms; for end-to-end communication environments, where cryptographic keys are transmitted and received, from the server to the end user; the authors made modifications to the RSA, they basically used four dynamic primes; the work was divided into three parts: key generation, encryption, and decryption; they conclude in a very acceptable efficiency and greater security in flexible communications [13]. The authors described the quantum cryptography that occurs when combining quantum mechanics with cryptography, as well as a modern cryptography widely used in computer networks; compared the AES, DES, TDES, and Blowfish algorithms; they analyzed the results of several quantum cryptography techniques when the server is deployed in the cloud, they considered the performance of the metrics, encryption and decryption time and performance [14]. The authors tested the application of the GPU of the AES algorithm, where the execution time for GPU implementation decreases compared to the sequential execution; the AES algorithm is implemented through OpenCL and tested on GPU devices; they conclude in a performance improvement [15]. To give security to data that is shared over the internet for commercial or other reasons, the authors reviewed symmetric algorithms and asymmetric algorithms; in the results, the time of encryption and decryption by use the RSA algorithm technique shows that it is more effective [16]. The authors researched the security of networks and the Internet where cryptography was used for secret information, they used cryptosystem of secret key and public key; developed a new generation key method called RSAKey, to generate and store all key values in tables within the DB [1].

About Smart City: The authors proposed an integral and conceptual infrastructure, which considers four components: availability, high coverage area, a fast connection with the application and notifications; the components are flexible according to the field to be applied, but some principles have a topology of sensory information in addition to a DB distribution in their cloud [17]. The authors proposed a security framework for Smart City, it is formed by four layers; in the physical layer are the devices with sensors and are data collectors; in the communication layer standards are used for the exchange of information, here, are located the blockchain protocols that maintain data integrity; in the DB layer, it uses a larger, decentralized, and encrypted repository, with public and private access; the blockchain keeps the transaction complete; the interface layer has collaborative applications to make effective decisions [5].

2.2 Methods

From the Smart City annual ranking reports from 2014 to 2018, called Cities in Motion and prepared by the University of Navarra [18–22]; took the first 20 cities for each year to develop the Smart City frequency of the last five years; Figure 1 presents Tokyo, Seoul, Paris, New York, London, Chicago, and Amsterdam as cities with a high performance in the service indicators of the different quality of life dimensions of their habitants.

Figure 2 proposes the components named below, which converge and result in digital information security for the management of a Smart City.

Figure 3 shows the components of a Smart City, some are: government, business, health, devices, citizens, buildings, houses, transportation, traffic control, physical security, networks, general services, and specific services. All these elements generate data that will be stored in a DB. Part of this data will be Open Data; this data of free access should be segmented as follows: commerce, culture, population, health indicators, parking, public transport, traffic, tourism, and public budgets.

A report by CISCO, exposed that some elements the generate data such as: connected plane, connected factory, public safety, weather sensors, intelligent building, smart hospital, smart car, smart grid; they can generate up to 200 million gigabytes per day the year 2020 [23]. Figure 4 shows the amount of data generated in GB for each element in a Smart City; the public safety that generates more data is used or transmitted at a low percentage, the weather data that generates less data is used or transmitted in high percentage.

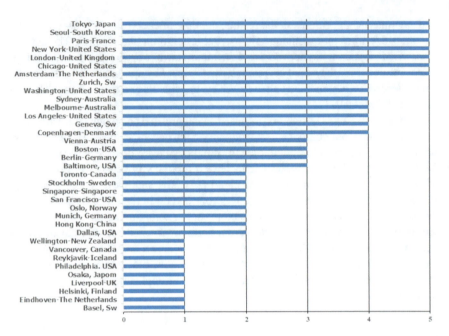

Fig. 1 Performance of smart cities services

Fig. 2 Information security components

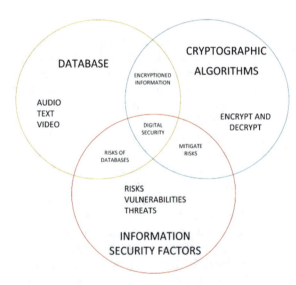

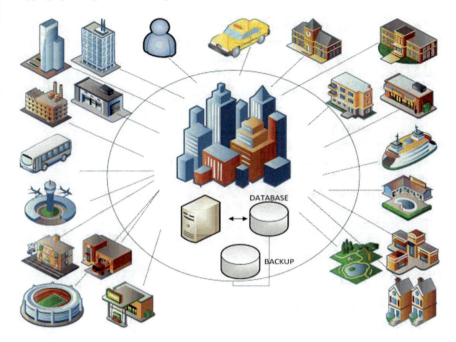

Fig. 3 Smart city, generation and storage of data

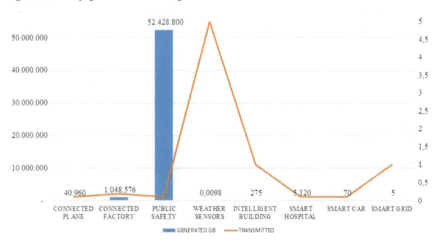

Fig. 4 Multiple applications create data

With these data: Smart Cities ranking, security components, and elements that generate data with their quantities; was determined the importance and motivation to propose a security measure of the information of a DB in the management of a Smart City.

A general prototype of information security was proposed with the following scope:

- Define a general algorithm for accessing the DB
- Security is applied to DB relational type
- Encryption is for plain text and at the level of columns in the tables of the DB
- Operations must be SQL ANSI Standards
- The DB to support DML operations: create, drop, alter, truncate
- The DB to support DDL operations: delete, insert, select, update, where
- The operations will be on the encrypted data
- Adopt encryption and decryption algorithms in the DB to apply to request, response, and SQL operations
- For the data encryption in the DB, the revised and tested RSA algorithm is adopted in [8]
- The result will be an encrypted DB.

  The prototype applies the three-layer model:

- Presentation layer: The Web application for presentation of management indicators is found, the mobile application for data request and response
- Business layer: Rules, SQL operations, encryption functions, and decryption functions are found.
- Data access layer: Encrypted data is found.

3   Results

In this phase, the following results were obtained:

- A general security prototype
- An algorithm to implement a DB for Smart City
- A data protection algorithm expressed in flowcharts.

3.1 A General Security Prototype

Figure 5 proposes the general prototype of information security of the DB, from the data generation, send to save, encrypted data for storage, data request, and decryption of data to send to the citizen. It was integrated the three-layer model for the control of interfaces, business rules, and data storage.

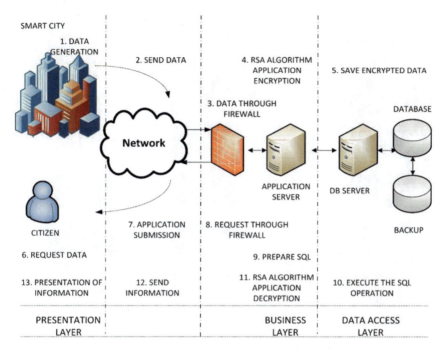

Fig. 5 Prototype of information security

3.2 An Algorithm to Implement a DB for Smart City

- Select the appropriate hardware, operating system, and DBMS
- Determine the uses and objectives of the information
- Determine accesses and authorizations
- Determine general and specific reports and access credentials
- Determine what information will be Open Data
- Review when releasing the DB.

3.3 A Data Protection Algorithm Expressed in Flowcharts

Data generation:

- All Smart City components generate data
- The data is sent over the Internet
- Packages pass through a firewall
- In the application server, the received data is encrypted with RSA algorithm
- The DB server stores the encrypted data.

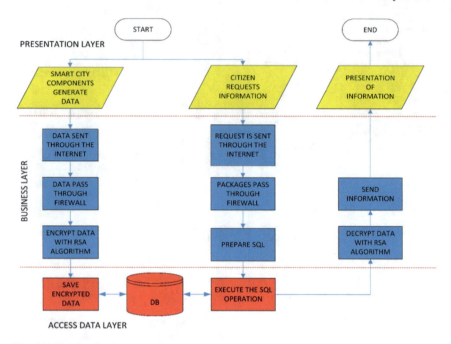

Fig. 6 Algorithm for data protection

Data request:

- The citizen requests information
- The request is sent through the Internet
- The packages pass through a firewall
- In the application server, the received request prepares the SQL statement
- In the DB server, to execute the SQL operation
- In the application server, the data is decrypted with RSA algorithm
- The data is sent to the citizen through the Internet
- The mobile or Web application interprets and presents data, indicators or reports.

Figure 6 shows the algorithm, expressed in flowcharts.

4 Discussion

- The RSA algorithm tested in another research was adopted [8], as it is more efficient.
- To maintain a secure communication platform, research [5]; which can be adopted.
- The information should be in the following categories: environment, society and welfare, public sector, economy, demography, culture, education, employment,

tourism, health, transport, science and technology, urban planning, rural areas, housing, commerce, legal, energy, physical security, industry and sport.

- Public or private policies are not considered, legal or social aspects, physical controls and user levels.
- The proposal does not consider security for audio and video.
- Was defined a prototype of information security for the management of a Smart City, the structure for the DB is not defined.

5   Future Work and Conclusion

As future work is foreseen, the generation of management indicators for a Smart City and its security for exposure in Web or mobile platforms.

1. It was concluded that information protected with an encryption algorithm is a support to be more efficient, reduce costs, reduce the environmental footprint, and improve the management of a Smart City.
2. Data with a higher level of security generates new services based on public data and increases citizen participation.
3. Sensitive data must be encrypted in private or public areas.

Acknowledgements The authors thank to Universidad Politécnica Salesiana del Ecuador, to the research group of the Guayaquil Headquarters "Computing, Security and Information Technology for a Globalized World" (CSITGW) created according to resolution 142-06-2017-07-19 and Secretaría de Educación Superior Ciencia, Tecnología e Innovación (Senescyt).

References

1. Nagar, S., Alshamma, S.: High speed implementation of RSA algorithm with modified keys exchange. In: 2012 6th International Conference on Sciences of Electronics, Technologies of Information and Telecommunications (SETIT), pp. 639–642 (2012). https://doi.org/10.1109/setit.2012.6481987
2. Hwang, G., Fu, S.: Proof of violation for trust and accountability of cloud database systems. In: 2016 16th IEEE/ACM International Symposium on Cluster, Cloud and Grid Computing (CCGrid), pp. 425–433 (2016). https://doi.org/10.1109/ccgrid.2016.27
3. Zhu, X., Li, J., Liu, Z., Yang, F.: A joint grid segmentation based affinity propagation clustering method for big data. In: 2016 IEEE 18th International Conference on High Performance Computing and Communications; IEEE 14th International Conference on Smart City; IEEE 2nd International Conference on Data Science and Systems (HPCC/SmartCity/DSS), pp. 144–151 (2016). https://doi.org/10.1109/hpcc-smartcity-dss.2016.0172
4. Jasim, K., Al-Shaikhli, I.: Mobile technology generations and cryptographic algorithms: analysis study. In: 2015 4th International Conference on Advanced Computer Science Applications and Technologies (ACSAT), pp. 50–55 (2015). https://doi.org/10.1109/acsat.2015.26

5. Biswas, K., Muthukkumarasamy, V.: Securing smart cities using blockchain technology. In: 2016 IEEE 18th International Conference on High Performance Computing and Communications; IEEE 14th International Conference on Smart City; IEEE 2nd International Conference on Data Science and Systems (HPCC/SmartCity/DSS), p. 1 (2016). Biswas, K., Muthukkumarasamy, V.: Securing smart cities using blockchain technology. In: Proceedings of the 18th IEEE International Conference on High Performance Computing and Communications; 14th IEEE International Conference on Smart City; 2nd IEEE International Conference on Data Science Systems; HPCC/SmartCity/DSS 2016, pp. 1392–1393 (2017). https://doi.org/10.1109/hpcc-smartcity-dss.2016.0198; https://doi.org/10.1109/hpcc-smartcity-dss.2016.0198

6. Dubey, G., Khurana, V., Sachdeva, S.: Implementing security technique on generic database. In: 2015 Eighth International Conference on Contemporary Computing (IC3), pp. 370–376 (2015). https://doi.org/10.1109/ic3.2015.7346709

7. Wenning, H., Hongjun, Z., Dengchao, H., Gang, C., Shuining, Z.: Design and implementation of database encrypting middleware. In: 2012 International Conference on Computer Science and Service System, pp. 1134–1137 (2012). https://doi.org/10.1109/csss.2012.287

8. Boicea, A., Radulescu, F., Truica, C., Costea, C.: Database encryption using asymmetric keys: a case study. In: 2017 21st International Conference on Control Systems and Computer Science (CSCS), pp. 317–323 (2017). https://doi.org/10.1109/cscs.2017.50

9. Cui, S., Asghar, M., Galbraith, S., Russello, G.: P-McDb: privacy-preserving search using multi-cloud encrypted databases. In: 2017 IEEE 10th International Conference on Cloud Computing (CLOUD), pp. 144–151 (2017). https://doi.org/10.1109/cloud.2017.50

10. Horio, K., Kawashima, H., Tatebe, O.: Efficient parallel summation on encrypted database system. In: 2017 IEEE International Conference on Big Data and Smart Compu-ting (BigComp), pp. 178–185 (2017). https://doi.org/10.1109/bigcomp.2017.7881735

11. Bajaj, S., Sion, R.: TrustedDB: a trusted hardware-based database with privacy and data confidentiality. IEEE Trans. Knowl. Data Eng. **26**(3), 752–765 (2014). https://doi.org/10.1109/tkde.2013.38

12. Chandrashekar, P., Dara, S., Muralidhara, V.: Efficient format preserving encrypted databases. In: 2015 IEEE International Conference on Electronics, Computing and Communication Technologies (CONECCT), pp. 1–4 (2015). https://doi.org/10.1109/conecct.2015.7383885

13. Chaudhury, P., Dhang, S., Roy, M., et al.: ACAFP: asymmetric key based cryptographic algorithm using four prime numbers to secure message communication. A review on RSA algorithm. In: 2017 8th Annual Industrial Automation and Electromechanical Engineering Conference (IEMECON), pp. 332–337 (2017). https://doi.org/10.1109/ie-mecon.2017.8079618

14. Murali, G., Sivaram Prasad, R.: Comparison of cryptographic algorithms in cloud and local environment using quantum cryptography. In: International Conference on Energy, Communication, Data Analytics and Soft Computing (ICECDS-2017) Comparison, pp. 55–69 (2017). https://doi.org/10.1109/icecds.2017.8390165

15. Inampudi, G., Shyamala, K., Ramachandram, S.: Parallel implementation of cryptographic algorithm: AES using OpenCL on GPUs. In: 2018 2nd International Conference on Inventive Systems and Control (ICISC), pp. 984–988 (2018). https://doi.org/10.1109/icisc.2018.8398949

16. Singh, D., Nand, P., Astya, R., Dixit, P.: Improved DSA cryptographic protocol and its comparative study with RSA protocol. In: International Conference on Computing, Communication & Automation, pp. 755–759 (2015). https://doi.org/10.1109/ccaa.2015.7148511

17. Aldoiu, S., Tapus, N.: Innovative infrastructure for SmartCity/SmartEnvironment applications. In: 2018 5th International Conference on Control, Decision and Information Technologies (CoDIT'18), pp. 767–772 (2018). https://doi.org/10.1109/codit.2018.8394822

18. IESE Cities in Motion Índice 2014. IESE Business School, Centro de Globalización y Estrategia, Barcelona, España (2014)

19. IESE Cities in Motion Index: Barcelona. Center for Globalization and Strategy, España (2015)

20. Engler, A., Tange, C., Frank-Bertoncelj, M., Gay, R., Gay, S., Ospelt, C.: Regulation and function of SIRT1 in rheumatoid arthritis synovial fibroblasts. J. Mol. Med. **94**(2), 1–74 (2016). https://doi.org/10.1007/s00109-015-1332-9

21. Berrone, P., Ricart, J., Carrasco, C., Duch, A.: Cátedra Schneider Electric de Sostenibilidad y Estrategia. IESE Cities Motion Index **2017**, 11–78 (2018). https://doi.org/10.15581/018.st-471
22. Berrone, P., Ricart, J., Carrasco, C., Duch, A.: Cátedra Schneider Electric de Sostenibilidad y Estrategia. IESE Cities Motion Index **2018** (2018). https://doi.org/10.15581/018.st-471
23. Barnett, T.: Cisco Global Cloud Index 2015–2020, p. 45. Cisco (2016)

Analysis of Adequate Bandwidths to Guarantee an Electoral Process in Ecuador

Segundo Moisés Toapanta Toapanta⑩, Johan Eduardo Aguilar Piguave⑩ and Luis Enrique Mafla Gallegos⑩

Abstract The analysis of the appropriate bandwidths was made based on the electoral processes in countries with availability of electronic voting and these have been adapted to the situation in the electoral processes of Ecuador. The problem is the low importance given and the recidivism in the scarce bandwidth used in the current electoral processes. The objective reflects the problems caused by a mediocre bandwidth for events of such magnitude and importance in any country and creates awareness of the appropriate conditions to guarantee all aspects of the electronic process. The method used is deductive to analyze the data that were used as parameters to calculate the bandwidth in the electoral process in Nigeria. The analysis shows that the voting processes up to the 2017 period have not been optimal, but they are sustainable and acceptable despite the fact that there were setbacks when issuing the results of the votes. It is concluded that there must be simulations to avoid failures at the time of the actual electoral process and improve resources for the implementation of electoral processes.

Keywords Bandwidth · Electoral process · Security information · Protocols security · Electoral processes

S. M. T. Toapanta (✉) · J. E. A. Piguave
Department of Computer Science, Universidad Politécnica Salesiana del Ecuador, Robles 107 Chambers, Guayaquil, Guayas, Ecuador
e-mail: stoapanta@ups.edu.ec

J. E. A. Piguave
e-mail: jaguilarp1@est.ups.edu.ec

L. E. M. Gallegos
School of Systems Engineering, Escuela Politécnica Nacional, Ladrón de Guevara E11-253, Quito, Pichincha, Ecuador
e-mail: enrique.mafla@epn.edu.ec

© Springer Nature Singapore Pte Ltd. 2020
X.-S. Yang et al. (eds.), *Fourth International Congress on Information and Communication Technology*, Advances in Intelligent Systems and Computing 1027, https://doi.org/10.1007/978-981-32-9343-4_37

1   Introduction

The National Electoral Council (CNE), in Ecuador, is in charge of formalizing the results of the electoral processes. But nevertheless, in the last electoral processes, these final results have been questioned [1] due to several inconveniences, one of them and the one that stands out most, is the so-called "Electro blackout."

Ecuador still does not take the step to electronic voting and prefers to keep the traditional by voting for ballots [2]. There are neighboring countries such as Colombia, which are conducting research regarding the security of electronic voting, in order to determine the right time to move to this type of process [3].

This problem arises from inadequate and insufficient bandwidth, which increases even more if they are countries with electronic voting already implemented in their electoral processes such as Brazil and Venezuela in Latin America, and there are even nations with sufficient resources to apply the electronic vote and these are not put at risk by different factors that have not yet been solved [4]. Something that is little talked about, and also has little documentation, is the existence of electoral processes that often suffer mishaps due to poor bandwidth or poor implementation of this.

Why should the analysis of bandwidth be carried out to guarantee an electoral process?

According to the analysis extended to the past problems in the electoral processes and considering factors such as the passage of a vote of vote to an electronic vote and the relative increase of the population of the country, we will have an adequate width of band for any electoral process.

This objective to reflect the problems caused by a mediocre bandwidth for events the popular election in the Ecuador; of such magnitude and importance in any country and raise awareness of the appropriate conditions to guarantee all aspects of the electronic process.

The articles analyzed to make this document are

Numerical Analysis of Ecuador's Electoral Register Integrity [1], From Piloting to Roll-out: Voting Experience and Trust in the First Full e-election in Argentina [2], Sistema de votación electrónico con características de seguridad SSL/TLS e IPsec en Colombia [3], Ensuring the Blind Signature for the Electoral System in a Distributed Environment [4], Bandwidth and Resource Allocation for Full Implementation of e-Election in Nigeria [5], Ancho de banda, crisis y crecimiento del PIB en países latinoamericanos en el periodo 2005–2015 [6], ICT for National Development in Nigeria: Creating an Enabling Environment [7], International Internet connectivity in Latin America and the Caribbean [8], Remote Internet Voting: Security and performance issues [9], Comparison of ID-based blind signatures from pairings for e-voting protocols [10], and Optimal Bandwidth Choice for the Regression Discontinuity Estimator [11].

The method used is the deductive one to analyze the data that was used as parameters to calculate the bandwidth in the electoral process in Nigeria [5].

It concludes by that an increase in bandwidth is needed, both for votes by ballots and for electronic votes, and it is also considered that if the implementation of a system by electronic voting is carried out, it should be contemplated even more bandwidth that is going to be destined to an electoral process in Ecuador.

2 Materials and Methods

In the first instance, in the Materials section, some determining works were reviewed such as: the different election systems, the average bandwidths in Latin American countries and the growth of this in Ecuador in different periods of time.

In the second instance, for the Methods section, parameters were proposed according to the Ecuadorian electoral process to generate a bandwidth calculation model.

2.1 Materials

Information provided by the International Telecommunications Union (ITU) was used because it has high relevance data on bandwidth in Latin America. In addition, the results of the simulations developed in Nigeria for the implementation of electronic voting were analyzed [5] and these were compared with the existing data in Ecuador to recreate the simulation with the amount of current population in the country.

The only countries that already implemented electronic voting in a real situation [4] have been placed in this table, to compare values, as bandwidth is concerned. The comparison with Colombia was included, considering that security in the implementation of the electronic process has been studied in this territory [3]. According to the description in Table 1.

The data taken as reference for the elaboration of this table correspond to the 2016 period; curiously we can notice that despite the crisis that the vast majority of Latin American countries have experienced and the penetration of bandwidth in them [6], there are countries that have already used electronic voting.

Table 1 International Internet bandwidth per user, kb/s

Country	Average Internet bandwidth	World ranking
Brazil	42.97	Position 60
Venezuela	14.40	Position 94
Colombia	34.99	Position 67
Ecuador	36.89	Position 65

Source International Telecommunications Union

Fig. 1 Average Internet bandwidth in Ecuador, kb/s. *Source* International Telecommunication Union

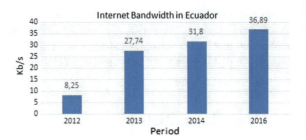

Venezuela is an example of this, and even being positioned near the bottom of the Latin American list with a bandwidth value well below Ecuador, it is one of the countries that has been able to carry out electoral processes through this system.

The ITU is an entity that is characterized by periodically providing information from all countries, by having these data we can be more specific in the analysis of the average Internet bandwidth in Ecuador and Nigeria over the past few years.

Figure 1 shows the values published by the ITU between the periods of 2012 and 2016 regarding the bandwidth in Ecuador.

The average value for Ecuador was 26.17 kilobits per second with a minimum of 8.25 kilobits per second in 2012 and a maximum of 36.89 kilobits per second in 2016.

The data granted by the ITU between the periods of 2012 and 2016 with respect to the bandwidth in Nigeria are (Fig. 2):

The average value for Nigeria during that period was 1.48 kilobits per second, with a minimum of 0.11 kilobits per second in 2012 and a maximum of 3.44 kilobits per second in 2016.

For this reason, research work in Nigeria was taken as the main reference, with a maximum average bandwidth per user of 3.15 kb/s in 2016 and a population index that exceeded 170 million inhabitants, leaving Ecuador behind with 17 million.

Fig. 2 Average Internet bandwidth in Nigeria, kb/s. *Source* International Telecommunication Union

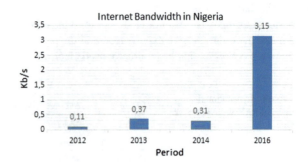

2.2 Methods

The scarcity of the bandwidth arises from the fact that many users are not considered connecting to it [7], so the method analyzed is the one studied in Nigeria for the implementation of electronic voting [5].

However, since ballot voting was used in Ecuador, it was considered as a method of less control over voting users. The approximate number of votes was determined taking into account the electoral laws in Ecuador, which indicate which are the people for the electoral process [1] and in this way those who fall within the voting range are identified.

Ecuadorians participating in this election process are all those between the ages of 16 and 65, including those who are abroad. The figures registered by the CNE of the number of voters in the last elections are 12,437,523 inhabitants within Ecuador and 378,292 inhabitants abroad giving a total of 12,815,815.

The research work in Nigeria was composed of something simple: a server which had the function of receiving the votes and ten laptops simulating voting machines [5], this launched as results: the average time between votes, lapses of delay of the whole process and also examined if there was any congestion in the ranks of voters during the duration of the drill.

We took the analysis that was developed, and we passed it to the voting system that is carried out in the electoral processes of our country, emphasizing that we focus on the bandwidth of the information recipient, that is, the National Electoral Council (CNE), in the Ecuador.

The parameters deduced are:

- CT: # of total candidates
- LV: Vote length
- VB: Value of votes in bytes
- VH: Votes per average hour
- TPA: The active voter population in Ecuador

These parameters were taken into account for the generation of calculations regarding the bandwidth in an Ecuadorian electoral process.

It is important to mention that the CT parameter is subject to special considerations with respect to the candidates to be elected, since in the same electoral process the ballot is cast on different ballots to postulants for various positions, not only to a specific category such as they are the president or vice president of the Republic. To be more explanatory, during the same election process the winning candidates are also selected to the position of ministers, assembly members, mayors, councilors, among others.

3 Results

Regarding the results obtained in the analysis it is stated that:

- The process is scalable so that while the number of candidates and the population in Ecuador are increasing, the servers that receive the votes should also be increased.
- By increasing the servers that host the information should increase the bandwidth depending on the increase that occurs.
- Ecuador exceeds Nigeria in terms of infrastructure and bandwidth, so the electronic method is a viable process, and if implemented, would greatly improve the current electoral process.
- As much as the electronic vote as a ballot vote must have the same level of privacy and security.
- Ecuador can take a leap to electronic voting from the perspective of reaching a national level and not just specific sectors.

1. Mathematical description of the bandwidth calculation

Among the various parameters mentioned, the following formulas are proposed for the calculation of the total candidates, the length of the vote and the value of votes in bytes.

- To determine the total candidates, we have the development of the following formula:

$$CT = numbers\ of\ candidates + 2 \tag{1}$$

Two are added to the value of the number of candidates since voting processes can also be voted on with whites or with nulls in the electoral processes.

- For the length of the votes the calculation is carried out using the formula below (Figs. 3 and 4):

```
START
    DATA:
        VARIABLES
                number_candidate    Numeric Entire
    ALGORITHM:
            Read number_candidate
            Total_candidates = number_candidate + 2
            Length_votes = (Total_candidates x 3) x 2
            Value_votes = (Length_votes / 245) x 256
    END
```

Fig. 3 Mathematical algorithm of the calculation of the vote value. *Source* Authors

```
START
    DATA:
        VARIABLES
            People_between_16_and_65        Numeric Entire
            People_foreign         Numeric Entire
    ALGORITHM:
        Read People_between_16_y_65
        Read People_foreign
        Total_active_population = People_between_16_and_65
                        + People_foreign
        Votes_per_average_hour = Total_active_population /
        12 Bandwidth = Votes_per_average_hour / Value_Votes
END
```

Fig. 4 Mathematical algorithm of bandwidth calculation. *Source* Authors

$$Length\ of\ vote = (CT \times 3) \times 2 \tag{2}$$

The multiplication by the value of 3 is fixed while the multiplication by 2 is given because it is stored as a data of short type and occupies 2 bytes.

- To obtain the vote value in bytes, it is calculated from the following formula:

$$Value\ of\ votes = \frac{Length\ of\ vote}{245} \times 256 \tag{3}$$

The division to 245 is given by the standard of encryption RSA (Rivest, Shamir and Adleman) that allows this maximum of encrypted bytes and the multiplication to 256 is by the total length of the block that stores the vote.

2. Prototype of the generic algorithms for the calculation of the bandwidth

Figure 5 describes the steps for calculating the vote value in bytes with the established parameters.

The steps of the first algorithm are detailed below:

- The number of total candidates included the blank and null vote.
- The length of the vote with the result of the previous calculation
- The vote value in bytes

Fig. 5 Flow chart of the
process of calculating the
vote value in bytes for an
electoral process

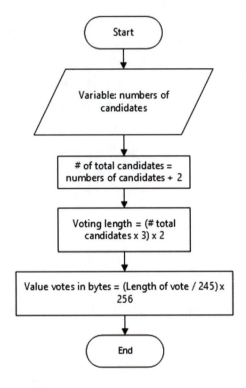

Figure 6 describes the steps for calculating the bandwidth with the result of the parameters of Fig. 5 and the other established parameters.

The steps of the second algorithm are detailed below:

- The total active population
- Average hourly votes
- The bandwidth of the electoral process

In this case, the description of the algorithms made using will be visualized first with programming language techniques and second with flow diagrams to present the prototype algorithms.

Now we proceed to show the flow diagrams of the resulting algorithm:

Also, as this article is based on electoral process in Ecuador; not only the data such as the number of eligible population for the vote should be taken into consideration but also add the state of the infrastructure in which the website that registers and publishes the votes is hosted. Therefore, safety parameters and contingency plans must be contemplated when carrying out this process.

Fig. 6 Flow chart of process of calculation of the bandwidth for an electoral process

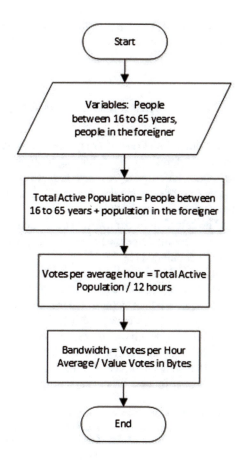

4 Discussion

From the analysis of the adequate bandwidth for an electoral process in Ecuador, the following points are proposed to be discussed:

- The results were obtained without having 100% guaranteed the integrity of their data, something that in the perception of the majority in Latin America should improve [8].
- The system must be available during the entire election process [9], that is, there must be no interruption or crash in the system or in the website where the information is hosted.
- The parameters considered are applicable to similar countries in Latin America that wish to put aside the transfer of vote count in paperwork and make an electronic vote [8].
- Equal or improve infrastructure in the country, since having reception servers does not mean improving bandwidth [6].

- The protocol used in the electoral process also affects bandwidth as a measure of greater security also consumes bandwidth resources [3, 10].

5 Work Futures and Conclusions

Carry out simulations of greater or lesser scale to avoid failures at the time of the actual electoral process, following as an example some of the works analyzed here [5], with the purpose of converting it into a clear process at all times and in any place, free of possible suspicions of fraud, in addition to that all information must be collected and published.

The selection of bandwidth should be based on the requirements of an electoral process, not the same measures if the approach of the voting changes is of greater or lesser magnitude [11].

It is concluded that there should be drills to avoid failures at the time of the actual electoral process, improve resources for the implementation of electoral processes.

It requires better resources to implement the electoral processes, even if they are not done through electronic voting, since we are on a par with other countries that have a similar bandwidth and these have carried out the electoral processes of this way, in addition to the website where the information is hosted has a good infrastructure.

Establish new proposals to standardize a better Internet quality for all of Ecuador, since it only focuses on the main cities of the country.

Acknowledgements The authors thank to Universidad Politécnica Salesiana del Ecuador, to the research group of the Guayaquil Headquarters "Computing, Security and Information Technology for a Globalized World" (CSITGW) created according to resolution 142-06-2017-07-19 and Secretaría de Educación Superior Ciencia, Tecnología e Innovación (Senescyt).

References

1. Mafla, E., Gallardo, C.: Numerical analysis of Ecuador's electoral register integrity. In: 2018 5th International Conference on eDemocracy and eGovernment, ICEDEG 2018, pp. 351–355 (2018). https://doi.org/10.1109/icedeg.2018.8372305
2. Pomares, J., Levin, I., Alvarez, R.M., Mirau, G.L., Ovejero, T.: From piloting to roll-out: voting experience and trust in the first full e-election in Argentina. In: Proceedings of the 6th International Conference on Electronic Voting (EVOTE 2014), pp. 33–42 (2014). https://doi.org/10.1109/evote.2014.7001136
3. Cardenas Urrea, S.E., Navarro Núñez, W., Sarmiento Osorio, H.E., Forero Paez, N.A., Bareño Gutierrez, R.: Sistema de votación electrónico con características de seguridad SSL/TLS e IPsec en Colombia. Rev. UIS Ing. **16**, 75–84 (2017). https://doi.org/10.18273/revuin.v16n1-2017008
4. Toapanta, S., Huilcapi, D., Cepeda, M., Mafla, L.: Ensuring the blind signature for the electoral system in a distributed environment. In: 2018 3rd International Conference on Computer

Science and Information Engineering (ICCSIE 2018) (2018). https://doi.org/10.23977/iccsie. 2018.1026

5. Akonjom, N.A., Ogbulezie, J.C.: Bandwidth and resource allocation for full implementation of e-election in Nigeria. Int. J. Eng. Trends Technol. **10**, 58–65 (2014). https://doi.org/10.14445/ 22315381/IJETT-V10P211

6. Pascual, G., Elvia, I., Reinaldo, A.: Broadband, crisis and GDP growth in Latin American countries in the 2005 2015 period I Ancho de banda, crisis y crecimiento del PIB en países latinoamericanos en el periodo 2005–2015. In: Iberian Conference on Information Systems and Technologies (2017). https://doi.org/10.23919/cisti.2017.7976041

7. Oghogho, I.: ICT for national development in Nigeria: creating an enabling environment. Int. J. Eng. Appl. Sci. **3** (2013)

8. Messano, O.: Study on International Internet Connectivity in Latin America and the Caribbean, pp. 1–56 (2012)

9. Ahmed, M.I., Abo-Rizka, M.: Remote Internet voting: security and performance issues. In: 2013 World Congress on Internet Security, WorldCIS 2013, pp. 56–64 (2013). https://doi.org/ 10.1109/worldcis.2013.6751017

10. Ribarski, P., Antovski, L.: Comparison of ID-based blind signatures from pairings for e-voting protocols. In: 2014 37th International Convention on Information and Communication Technology, Electronics and Microelectronics, MIPRO 2014—Proceedings, pp. 1394–1399 (2014). https://doi.org/10.1109/mipro.2014.6859785

11. Imbens, G., Kalyanaraman, K.: Optimal bandwidth choice for the regression discontinuity estimator. Rev. Econ. Stud. **79**, 933–959 (2012). https://doi.org/10.1093/restud/rdr043

Applied N F Interpolation Method for Recover Randomly Missing Values in Data Mining

Sanjay Gaur, Darshanaben D. Pandya and Manish Kumar Sharma

Abstract Data cleansing is a critical step for data preparation. The values lost in the database are a common problem faced by data analysts. Missing values in data mining are continual troubles that can grounds errors in data analysis. Randomly missing elements in the attribute/dataset make data analysis complicated and also confused to consolidated result. It affects the accuracy of the result and intermediate queries. By using statistical/numerical methods, one can recover the missing data and decrease the suspiciousness in the database. The present paper gives an applied approach of Newton forward interpolation (NFI) method to recover the missing values.

Keywords Data mining · Missing values · Interpolation · Newton forward · Random

1 Introduction

Generally, data in the database are kept in the tabular format. Dataset are basically attributes of the associated table, whereas the record set is rows of the table. Data in the dataset stay as basic element and are utilized for query and advance reports. While dataset are incomplete or have missing values, it straightforwardly has an effect on the concluding reports. In data mining, randomly missing values identification and recovery is till now important issue. Missing values forever reason of ambiguity and it influence error in final results. It degrades accuracy of query and reduces decision-making competence of authorities. It is compulsory to resolve such crisis earlier than

S. Gaur (✉)
CSE, Jaipur Engineering College and Research Center, Jaipur, India
e-mail: sanjay.since@gmail.com

D. D. Pandya
Department of Computer Science, Madhav University, Pindwara, Sirohi, Rajasthan, India
e-mail: pandya_darshana@rediffmail.com

M. K. Sharma
Maharaja College, University of Rajasthan, Jaipur, India

© Springer Nature Singapore Pte Ltd. 2020   475
X.-S. Yang et al. (eds.), *Fourth International Congress on Information and Communication Technology*, Advances in Intelligent Systems and Computing 1027, https://doi.org/10.1007/978-981-32-9343-4_38

moving for query and report preparation. To overcome such situation there is need of statistical or numerical techniques to recover the randomly missing values.

Newton forward interpolation is numerical method which can be used to generate artificial values in association with available data. The present paper is an effort to generate artificial value at the place of missing value to as recovery method. It works as closest fit approach through applied Newton forward interpolation to recover missing value. This is basically an application of the concept of Newton forward interpolation approach is used to recover the missing values.

2   Formulation of Problem

The projected numerical method is a simple approach for recovery of randomly missing value in dataset. It gives a way to work in direction of closest fit approach for recovery of missing data. In this, we first look at the complete attribute element for missing value cases. Subsequent to missing and observed values, attribute is separated in two parts as mentioned as observed and missing values. Although both are remaining in the same attribute, it is only logical demarcation.

Now looking for the missing values in attribute and search begin. At this point, we have two variable X and Y in proportion titled as year and dataset value. Variable X (year) is fixed for other attributes Y, which have missing values. Attributes for Y are changeable, whereas X is stable for present study and Y, has missing value. Here randomly missing values are available in the attribute Y. At this point the variable X is corresponding variable of Y, which does not have any missing values.

Construct loop, for $i = 1$ to $i \leq$ n.

$$X_1 = \text{value}(X_{i-1}) \tag{1.1}$$

X_1 previous value from X_i

$$X_2 = \text{value}(X_{i+1}) \tag{1.2}$$

X_2 first succeeding from X_i

$$X_3 = \text{value}(X_{i+2}) \tag{1.3}$$

X_3 second succeeding from X_i

$$X_4 = \text{value}(X_{i+3}) \tag{1.4}$$

X_4 third succeeding from X_i

$$X_5 = \text{value}(X_{i+4}) \tag{1.4}$$

X_5 fourth succeeding from X_i

$$Y_1 = \text{value}(Y_{i-1}) \tag{1.6}$$

Y_1 preceding from Y_i

$$Y_2 = \text{value}(Y_{i+1}) \tag{1.7}$$

Y_2 first succeeding from Y_i

$$Y_3 = \text{value}(Y_{i+2}) \tag{1.8}$$

Y_3 is the value of second succeeding of Y_i

$$Y_4 = \text{value}(Y_{i+3}) \tag{1.9}$$

Y_4, third succeeding from Y_i

$$Y_5 = \text{value}(Y_{i+4}) \tag{1.10}$$

Y_5, fourth succeeding from Y_i

$$X = \text{value}(X_i) \tag{1.11}$$

X is the consequent value from Y_i.
    whereas $X_1, X_2, X_3, X_4, X_5, Y_1, Y_2, Y_3, Y_4, Y_5, X \neq \text{NULL}$
    Now, initialize the variables

$$h = 0, \text{Diff} = 1, p, \text{Sum} = 0 \tag{1.12}$$

Here, variable h initialized with 0, to store the value of interval value.
Sum $= 0, p, \text{Diff} = 1$
Now, calculating interval value of X_i using below given formula

$$h = \text{value}(X_2) - \text{value}(X_1) \tag{1.13}$$

Then calculate difference of X_i divide by interval using the following equation

$$p = (\text{value}(X_2) - X)/h \tag{1.14}$$

Now, initialize first two dimensional arrays for difference assign to Sum, therefore

$$\text{Sum} = Y(1)(1)$$

Here, loop encountered for attribute. Thus, for $j = 2$ to n., the inner loop gets activated n ascending order.

For $i = 1$ to $(n - j) + 1$ then applied this approach for calculating difference table

$$\text{value}(Y_i)(Y_j) = \text{value}(Y_{i+1})(Y_{j-1}) - \text{value}(Y_i)(Y_{j-1}) \quad (1.15)$$

Then make increment in i counter, thus $i = i + 1$, then inner loop encountered till $i < (n - j) + 1$.

Here, inner loop is closed, after that increment j loop encountered, thus $j = j + 1$, loop is finished till $j \leq n$. here loop is completed.

Now calculate the value of h, Diff, p, Sum, and loop encounter in ascending order.

Thus, for $j = 2$ to n. Construct inner loop. Thus, for $i = 1$ to j, calculating h for division using below given formula.

$$h = h * i \quad (1.16)$$

Calculating difference of table using formula

$$\text{Diff} = \text{Diff} * (p - (j - 1)) \quad (1.17)$$

Now, calculation of Sum using formula, divided by interval h

$$\text{Sum} = \text{Sum} + \big(\text{value}(Y_1)(Y_j) * \text{Diff}\big) / h \quad (1.18)$$

Initialize variable h from 1. Then make increment in i counter, therefore $i = i + 1$. Now inner loop get finished, then j loop encountered, $j = j + 1$ this loop get finished till $j \leq n$. Here, loop is completed. After these process, estimated value is obtained Yest = Sum. Assigning estimated value to missing value place

$$\text{value}(Y_i) = Y_{\text{est}} \quad (1.19)$$

Assigning estimated value to missing value place. Then encounter loop i, $i = i + 1$. Here, main loop gets finished.

3 Algorithm

Attribute X = {X1 , ..., X_n}, Y = { Y_1 , ..., Y_n}

Where X = X_{obs} + X_{mis}

X_{obs} = { X_1 , ..., X_k} // Attribute values observed

X_{mis} = { X_{k+1} , ..., X_n} // Attribute values missing

Y = Y_{obs} + Y_{mis}

Y_{obs} = { Y_1 , ..., Y_k} // Attribute values observed

Y_{mis} = { Y_{k+1} , ..., Y_n} // Missing Attribute values.

array of (X) = { X_1 , ..., X_n } // Single dimensional array declaration

array of (Y) = { Y_1 , ..., Y_n }{ Y_1 , ..., Y_n }// Two dimensional array declaration

Read X = { X_1 , ..., X_n}, Y = { Y_1 , ..., Y_n } // missing data place detection

for i=1 to n, do // initialization of loop

If (value (Yi) = = NULL) then

X_1 = value(X_{i-1})   //preceding of Xi.

X_2 = value(X_{i+1}) // first succeeding from Xi.

X_3 = value(X_{i+2}) //second succeeding from Xi.

X_4 = value(X_{i+3}) //third succeeding from Xi.

X_5 = value(X_{i+4}) //fourth succeeding from Xi.

Y_1 = value(Y_{i-1})   // preceding from Yi.

Y_2 = value(Y_{i+1}) // value of first succeeding Yi.

Y_3 = value(Y_{i+2}) //second succeeding from Yi.

Y_4 = value(Y_{i+3}) //third succeeding from Yi.

Y_5 = value(Y_{i+4}) //fourth succeeding from Yi.

X = value(Xi)

where $X_1, X_2, X_3, X_4, X_5, Y_1, Y_2, Y_3, Y_4, Y_5, X \neq$ 'NULL'

h =0 , Diff = 1 , p , sum = 0   // Initialize the variables

h = value(X_2) - value(X_1) // interval value of Xi

p = (value(X_2) - X) / h // calculate difference of Xi, divide by interval

Sum = Y (1)(1) // initialize first two dimensional array for difference

for j=2 to n, do //

for i=1 to (n-j)+1 do

value(Y_i)(Y_j) = value(Y_{i+1})(Y_{j-1})- value(Y_i)(Y_{j-1}) // calculating difference table

        $i = i + 1$ // increase the i counter

        endfor // second inner loop closed.

          $j = j + 1$ // increase in j loop

     repeat-until $(j <= n)$, end for //loop closed.

for j=2 to n, do // construct loop

   for i =1 to j do // encounter i loop

      $h = h * i$ // calculating h for division

      $Diff = Diff * (p - (j-1))$ // calculating difference

      $Sum = Sum + (value(Y_1)(Y_j) * Diff) / h$ // calculation of sum of formula

      $h=1$ // Initialize the variable to 1

       $i = i + 1$ // Increment in the i counter

      endfor // second inner loop closed.

       $j=j+1$ // increase the j loop

    repeat-until $(j <= n)$, end for // inner loop finish

    $Y_{est} = Sum$ // predicted value

    value $(Yi) = Y_{est}$

    $i = i+1$

    repeat-until $(i <= n)$, endfor

    stop

4 Discussion of Results

Analysis [mean]: According to Table 1 the average value of carbon emissions from coal, oil, and natural gas are 2109, 2262 and 879, respectively. In the missing value condition, values are recorded as 2129 for coal, 2307 for oil, and 901 for natural gas. After filling of missing values from the calculated estimated values the results are 2111 for coal, 2261 for oil, and 879 for natural gas, respectively. Here, it is found that after estimation of missing value by proposed method, values are very close to original value. **Standard Deviation**: Here, it is originate that later than generation of missing value by proposed method, values are very close to original value and value of the standard deviation are almost equal to the standard deviation of original set values. **Coefficient of Variation**: It is found that after estimation of missing value by proposed method, values of the coefficient of variation are not very or we can say CV are similar to CV of original dataset.

Table 1 Applied N F interpolation method

Global carbon dioxide emission from fossil burning by fuel type (Carbon emission in million tones)

SN	Year	Standard dataset			Missing value dataset			Recovered dataset		
		Coal	Oil	Natural gas	Coal	Oil	Natural gas	Coal	Oil	Natural gas
1	1960	1410	849	235	1410	849	235	1410	849	235
2	1961	1349	904	254	1349		254	1349	**929**	254
3	1962	1351	980	277	1351	980	277	1351	980	277
4	1963	1396	1052	300	1396	1052	300	1396	1052	300
5	1964	1435	1137	328	1435	1137		1435	1137	**330**
6	1965	1460	1219	351	1460	1219	351	1460	1219	351
7	1966	1478	1323	380		1323	380	**1455**	1323	380
8	1967	1448	1423	410	1448		410	1448	**1469**	410
9	1968	1448	1551	446	1448	1551	446	1448	1551	446
10	1969	1486	1673	487	1486	1673	487	1486	1673	487
11	1970	1556	1839	516	1556	1839		1556	1839	**529**
12	1971	1559	1946	554	1559	1946	554	1559	1946	554
13	1972	1576	2055	583		2055	583	**1591**	2055	583
14	1973	1581	2240	608	1581		608	1581	**2263**	608
15	1974	1579	2244	618	1579	2244	618	1579	2244	618
16	1975	1673	2131	623	1673	2131	623	1673	2131	623
17	1976	1710	2313	650	1710	2313		1710	2313	**636**
18	1977	1766	2395	649	1766	2395	649	1766	2395	649
19	1978	1793	2392	677		2392	677	**1824**	2392	677

(continued)

Table 1 (continued)

Global carbon dioxide emission from fossil burning by fuel type (Carbon emission in million tones)

SN	Year	Standard dataset			Missing value dataset			Recovered dataset		
		Coal	Oil	Natural gas	Coal	Oil	Natural gas	Coal	Oil	Natural gas
20	1979	1887	2544	719	1887		719	1887	**2446**	719
21	1980	1947	2422	740	1947	2422	740	1947	2422	740
22	1981	1921	2289	756	1921	2289	756	1921	2289	756
23	1982	1992	2196	746	1992	2196		1992	2196	**746**
24	1983	1995	2177	745	1995	2177	745	1995	2177	745
25	1984	2094	2202	808		2202	808	**2157**	2202	808
26	1985	2237	2182	836	2237		836	2237	**2278**	836
27	1986	2300	2290	830	2300	2290	830	2300	2290	830
28	1987	2364	2302	893	2364	2302	893	2364	2302	893
29	1988	2414	2408	936	2414	2408		2414	2408	**937**
30	1989	2457	2455	972	2457	2455	972	2457	2455	972
31	1990	2409	2517	1026		2517	1026	**2369**	2517	1026
32	1991	2341	2627	1069	2341	2627	1069	2341	**2501**	1069
33	1992	2318	2506	1101	2318	2506	1101	2318	2506	1101
34	1993	2265	2537	1119	2265	2537	1119	2265	2537	1119
35	1994	2331	2562	1132	2331	2562		2331	2562	**1123**
36	1995	2414	2586	1153	2414	2586	1153	2414	2586	1153

(continued)

Table 1 (continued)

Global carbon dioxide emission from fossil burning by fuel type (Carbon emission in million tones)

SN	Year	Standard dataset			Missing value dataset			Recovered dataset		
		Coal	Oil	Natural gas	Coal	Oil	Natural gas	Coal	Oil	Natural gas
37	1996	2451	2624	1208		2624	1208	**2491**	2624	1208
38	1997	2480	2707	1211	2480	2707	1211	2480	2707	1211
39	1998	2376	2763	1245	2376	2763	1245	2376	2763	1245
40	1999	2329	2716	1272	2329	2716	1272	2329	2716	1272
41	2000	2342	2831	1291	2342	2831	1291	2342	2831	1291
42	2001	2460	2842	1314	2460	2842	1314	2460	2842	1314
43	2002	2487	2819	1349	2487	2819	1349	2487	2819	1349
44	2003	2638	2928	1399	2638	2928	1399	2638	2928	1399
45	2004	2850	3032	1436	2850	3032	1436	2850	3032	1436
46	2005	3032	3079	1479	3032	3079	1479	3032	3079	1479
47	2006	3193	3092	1527	3193	3092	1527	3193	3092	1527
48	2007	3295	3087	1551	3295	3087	1551	3295	3087	1551
49	2008	3401	3079	1589	3401	3079	1589	3401	3079	1589
50	2009	3393	3019	1552	3393	3019	1552	3393	3019	1552
	Mean	2109	2262	879	2129	2307	901	2111	2261	879
	S.D.	567.89	621.13	400.27	586.60	606.41	410.80	567.96	616.66	400.03
	C.V.	0.27	0.27	0.46	0.28	0.26	0.46	0.27	0.27	0.46

Source www.earth_policy.com

Analysis of Variance: We wish to test the hypothesis
H0: $\mu1 = \mu2 = \mu3$ against the alternative
H1: at least two μ different
For testing the hypothesis following arrangement have been done:
ANOVA test result for Coal

Source of variation	SS	df	MS	F	P-value	F crit
Between groups	10613.98	2	5306.991	0.016125	0.984006	3.060292
Within groups	46405090	141	329114.1			
Total	46415704	143				

Observed value at 5% level of significance = 0.0161, the F critical value is 3.06, so hypothesis/assumption is accepted.
ANOVA test result for Oil

Source of variation	SS	df	MS	F	P-value	F crit
Between groups	63223.13	2	31611.56	0.083563	0.919878	3.059831
Within groups	53717801	142	378294.4			
Total	53781024	144				

Observed value at 5% level of significance = 0.0835, the F critical value is 3.06, so hypothesis/assumption is accepted.
ANOVA test result for Natural Gas

Source of variation	SS	df	MS	F	P-value	F crit
Between groups	14822.4	2	7411.199	0.045537	0.955499	3.060292
Within groups	22948080	141	162752.3			
Total	22962902	143				

Observed value at 5% level of significance = 0.0455, the F critical value is 3.06, so hypothesis/assumption is accepted.

Decision and Conclusion: Given that F (observed/calculated) < 3.06 for coal, oil and natural gas ANOVA (One way) test. In case hypotheses are accepted in all cases, therefore, it is considerable that no significant difference found between groups regarding mean value **5**.

5 Conclusion

In general, it is unanimously acknowledged that there is no sent percent proficient technique to handle all types of missing values. The projected approach is significant for the integer values. This approach provides appropriate consequence for the consolidated report generated by the database. According to measurement of central tendency, SD, and CV result are significant. One way ANOVA test also gives significant result with acceptance of hypothesis. So it can be said that the results are statistically significant. Finally, it can be said that proposed techniques are significant for small database which consist of linear trends in the dataset.

References

1. Allison, P.D.: Estimation of linear models with incomplete data. In: Social Methodology, pp. 71–103. Jossey Bass, San Francisco (1987)
2. Allison, P.D.: Missing Data. Sage publication, Thousand Oaks, CA (2001)
3. Buck, S.F.: A method of estimation of missing values in multivariate data suitable for use with an electronic computer. J. R. Stat. Soc., Ser. B **2**, 302–306 (1960)
4. Chen, L., Drane, M.T., Valois, R.F., Drane, J.W.: Multiple imputation for missing ordinal data. J. Mod. Appl. Stat. Methods **4**(1), 288–299 (2005)
5. Gaur, S., Dulawat, M.S.: A perception of statistical inference in data mining. Int. J. Comput. Sci. Commun. **1**(2), 653–658 (2010)
6. Gaur, S., Dulawat, M.S.: Univariate analysis for data preparation in context of missing values. J. Comput. Math. Sci. **1**(5), 628–635 (2010)
7. Gaur, S., Dulawat, M.S.: A closest fit approach to missing attribute values in data mining. Int. J. Adv. Sci. Technol. **2**(4), 18–24 (2011)
8. Gaur, S.: Closest fit approach to handle odd size missing block values. Int. J. Math. Arch. **3**(7) (2012)
9. Grzymala-Busse, J.W.: Data with missing attribute values: Generalization of in-discernibility relation and rules induction. Trans. Rough Sets **1**, 8–95 (2004). (Lecture Notes in Computer Science Journal Subline, Springer-Verlag)
10. Kim, J.O., Curry, J.: The treatment of missing data in multivariate analysis. Soc. Methods Res. **6**, 215–240 (1977)
11. Rubin, D.B.: Inference and missing data. Biometrika **63**, 581–592 (1976)
12. Sharma, S., Gaur, S.: Contiguous agile approach to manage odd size missing block in data mining. Int. J. Adv. Res. Comput. Sci. **4**(11), 214–217 (2013)

ICT-Enabled Business Promotion Approach Through Search Engine Optimization

Sanjay Gaur, Hemant Sahu and Kulvinder Singh

Abstract In the present era, the world becomes global village, and all the business-related activities are now entered in the open market. Small businesses are also managing their product and services selling throughout the world with the help of online marketing. Such kind of product promotion activities with the help of internet and search engine is now subject of the digital marketing. At present, there are numerous methods and promoting schemes available with text, image, audio, and video promoting platform. These all activities are empowered by the information communication technology (ICT) and recent advance software development advancements. The present study provides a small insight into online product promoting scenario and its statistics. For that purpose, we are using Google Adwords search engine marketing journey analysis and its effect on Indian market. The whole process gives a view of search engine marketing and its growth.

Keywords Search engine · Google · E-commerce · Product · Market

1 Introduction

Nowadays, online searching is one of the most powerful tools to find out any product and services on a single click and is also most popular web service. With the development of information communication technology (ICT), the necessity supports are easily available. The internet users are now able to search anything by using search engine. Now, the search engines like Google, Yahoo, Bing, Ask, Yandex, Baidu, etc.,

S. Gaur (✉)
Jaipur Engineering College & Research Centre, Jaipur, India
e-mail: sanjay.since@gmail.com

H. Sahu
Geetanjali Institute of Technical Studies, Udaipur, India
e-mail: hemantsahu@gmail.com

K. Singh
Shri Khandelwal Vaish P G College, Jaipur, India
e-mail: kulvinder.hundal@gmail.com

© Springer Nature Singapore Pte Ltd. 2020
X.-S. Yang et al. (eds.), *Fourth International Congress on Information and Communication Technology*, Advances in Intelligent Systems and Computing 1027, https://doi.org/10.1007/978-981-32-9343-4_39

487

are very much popular web support. The outcome is normally accessible on search engine consequence pages ranked by relevancy and erstwhile factors resolute by the concern search algorithm.

In general and worldwide, Google is search market leader through a stable global market share, approximately 90 percent. Google is furthermore ranked top among unique visitors and core searches in the United States, and situation is same for India. The emerging trend and figures about online searching indicate just similar for European countries. The information search through web is big practice during the travel planning or product/service investigation. It also plays vital role for the expansion of m-commerce and shopping through smart phone.

The contemporary statistics illustrate that shoppers utilize their mobile devices for online search prior to and throughout the shopping, as mobile search assists them to find opening hours, costing assessments, and complete product information. These days, mobile search with smart gadgets has turn into a set part of brick-and-mortar retail.

2 Search Engine Perspective

Nowadays, shoppers amplify the visibility of their products/services, and it is necessary to make them appear at the top of pinnacle of search engine outcomes; therefore, organizations have to optimize their websites by search engine optimization (SEO). The search engine optimization is the oldest terminology which is applied in the ethical/organic web searching. With rapid growth of web development, huge amount of webs is available on every server as well on search engine. So, there is less possibility about ethical, organic searching, and search results. Therefore, if one wants to stand in top searching results, then have to go with search engine marketing (SEM). It is one of the commercial activities regarding searching product and services through search engines.

3 Google Search Engine

With multidimensional campaign setting, Google provides an efficient search engine marketing tool named as Google Adwords. If we are looking for worldwide use of search engine scenario, then we will definitely find that Google is world leader as web search engine (Fig. 1).

The given statistics shows that the revenue of Google's advertising from year the 2001 and the score of year 2018 was approximately 116.3 billion US dollars. On another side, the whole business from advertise was 136.2 billion US dollars. This growth is a continual growth, almost 71% of revenue from advertises are covered by Google, and this is the second highest income segment of Google with 16% share of company.

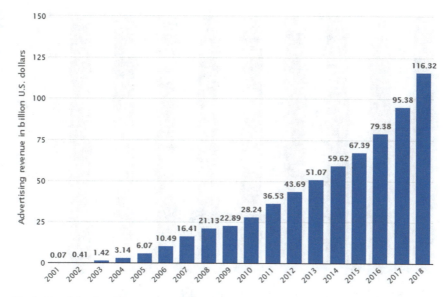

Fig. 1 Google's advertising revenue. *Source* Statistica

We can now see the statistics of leader of the global market in internet search engine. The leader "Google" has conquered the market of search engine by maintaining 86.02 percent share of July 2018. The mainstream of Google revenues is produced by advertising. The "Google" prolonged its area of service by mail, productivity tools, enterprise products, mobile devices, etc. In the year 2017, "Google" earned the highest revenue (110 billion US dollars approx.) as Tech Company also.

In October 2018, online search engine Bing accounted for 3.82% of the global search market. During the same month, Chinese search engine Baidu had a market share of 0.55% (Fig. 2).

This is one of the significant information as per the statistics that traffic on the Google's share of desktop search engine in India was 94.39% in the month June year 2018. This is one of big deals of Google via Indian users.

Other search engines are very ineffective in Indian market. So, it is very clear that this situation is also applicable for the Google's commercial product Adwords as well as Adwords Express (Fig. 3).

4 Pay-Per-Click Advertising

Google Adwords or Pay-per-click advertising is one of the finest marketing investments in India as well as worldwide. Google Ads are comparatively economical cost-per-click as compared to other. The Google Adwards is popular search engine because it reaches to the people, who want to reach them. The display ads shown on

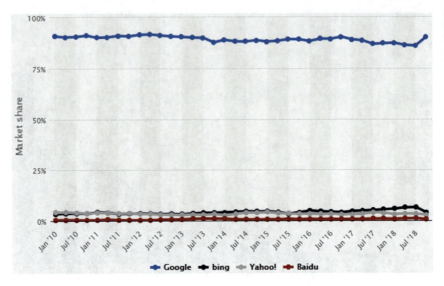

Fig. 2 Share of search engines worldwide in October 2018. *Source* Statistica

specific websites that are concern to one's business, also search ads come into view at the top of the search engine results (Fig. 4).

- Almost 41% clicks come on the aids which are shown on the top of the first page of the search engine.
- Almost 65% of clicks received by Google Adwords search results, whereas, the organic search acquires 35% of results.

5 Search Ad Statistics

The Google search ads target populace, when they are searching for product and service. They appear at the page of search engine results. This is intelligence of search engine's algorithms. The current statistics concern to Google searches says that the number of monthly search is approximately 160 + billion. It is found that in the year 2018, user who clicks on ad earlier to go on a store is 27% extra probable to buy something in store. Even the 51% of searches comprise more than four words. The current situation shows that almost 70% of entire online searches are through Google and 90% user believes on Google review and results. 66% of buyer-intent keywords are paid clicks, and 40% of store purchases start online.

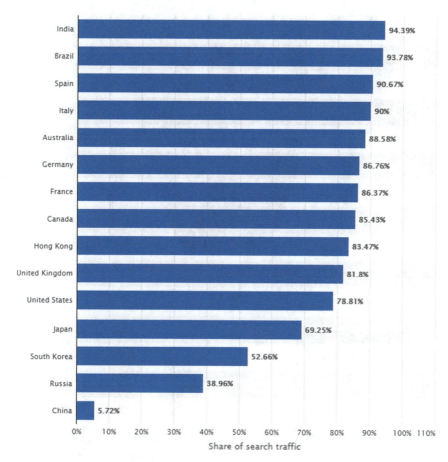

Fig. 3 Share of search traffic. *Source* Statistica

6 India as E-commerce Destination

Now, India becomes a hot e-commerce destination for world market. Each and every day, small business units and MNC's foraying into Indian e-commerce arena. Also, Google and RIL publicize their strategy to enter e-commerce through a hybrid model. This indicates that e-commerce has colossal prospective in India. As per Goldman Sachs, India's e-commerce arena will account for 2.5% of India's GDP by 2030 and is possibly to touch $300 billion.

According to "Walmart," entire trade will raise at a compounded-annual-growth-rate (CAGR) of about 9% during FY-2018–2023, whereas, Indian e-commerce market will cultivate about 36%. Sooner or later it will boost Indian e-commerce's circulation from 2.1% in 2017–18 to approximately 6.2% in 2022–23 (Fig. 5).

The factors that show India will be hot destination for e-commerce as compared to the rest of the world.

Fig. 4 block structure of PPC. *Source* Blue corona

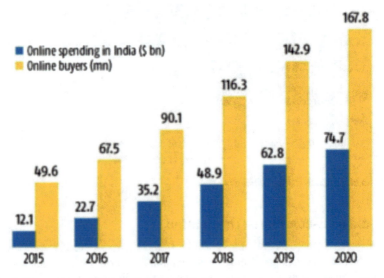

Fig. 5 Online retail spending in India 2015–2020. *Source* Forrester research online retail forecast, 2015–2020, Asia-pacific

6.1 Technologies Enable Purchasing

Now, shopping in India is digitally influenced activity. The idea of shopping "anything-anywhere-anytime" boosts e-commerce in India. Every one of this has been contributed by ICT technology. According to report of "Boston Consulting Group" by 2025, India is likely to have 850 million online users. Also, it is exposed that during the year2014 and 2016, online buyers have multiplied seven-fold.

6.2 Digital Payment

The digital currency and online money transaction are user friendly in India. Now, money in mobile wallets as PayTm, Ola Money, Mobiwik, BHIM, etc., are frequently in use, and online banking and electronic cards come in full swing. Also, digital India campaign and demonetization has spell speculate for the e-commerce industry.

6.3 Leveraging Small Shoppers

India has a considerable population in small towns and villages. According to "Red Seer," small-town Indians contribute about 41% of the entire online shoppers. The top online market players know the strength of rural India. In that direction "Amazon," with its project "Udaan," has aligned small stores in rural area of nation, in which shopkeepers' direct consumers to shop from Amazon. *The* "RedSeer" forecast that non-metro towns will account for 55% of the entire vigorous online shoppers in 2020 (Fig. 6).

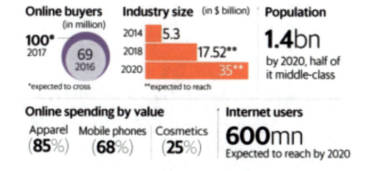

Fig. 6 E-commerce potential of India

6.4 Ample of Festive Seasons

Festivals times are boom time in India, and there are varieties of festivals running whole year. This is directly connected to the retail purchasing of verity of products. Now, e-commerce is popular for purchasing due to novelty and festivals discounts. As per "Red Seer" in 2017, the festivals time produces the highest monthly sales for e-commerce (India) at $3.2 billion.

6.5 Shoppers Smitten with Discounts

In India, online shopping flourishes cash backs and discounts on purchase. According to "Cuponation," India ranks top user of coupons leaving nations such as Australia, Brazil, Spain, and Singapore behind. With the rising access to a smart phone, 62% of Indians using coupons are on smart-phones. It is larger than Australia (28%), Brazil (30%), Spain (22%), and Singapore (37%).

7 Conclusion

The present study gives a clear picture of growth through information communication technology in the nation. Study provides an insight of online product promoting scenario and its statistics. We are using Google Ads search engine marketing journey analysis to show the whole process of search engine marketing and its growth. These all activities are empowered by ICT and recent advance software development. Overall, the Indian market now influences from online marketing and search engine. As per culture and tradition, trends predictors saying that in near future, India will be super power in digital marketing and e-commerce market.

References

1. Namrata, Anand, Vandana, Khetarpal: Growth of apparel industry in India: present and future aspects. Scholars World Int Refereed Multidisciplinary J. Contemp. Res. **2**(1), 64–70 (2014)
2. Vanmathi, A., Thangavel, N.: An investigation of media marketing on Online consumer behaviour. IJA BE R **14**(4), 2465–2475 (2016)
3. Gaur, S., Sharma, L., Pandya, D.D.: A perception of ICT for social media marketing in India. In: Proceedings of ICT4SD/IRSCNS. Springer, Goa, India (2018)
4. Gaur, S., Jain, L., Choudhary, N., Ojha, G.: An ICT insight of digitization of banking in India. In: Proceedings of ICT4SD/ IRSCNS. Springer Goa, India (2018)
5. Anukrati, Sharma: Study on E-commerce and online shopping: issues and influences. Int. J. Comput. Eng. Technol. **4**(1), 364–376 (2013)
6. https://digitalreport.wearesocial.com/

7. https://www.statista.com/
8. https://www.proschoolonline.com/
9. https://www.bluecorona.com/blog/pay-per-click-statistics

Infrared Versus Visible Image Matching for Multispectral Face Recognition

Wafa Waheeda Syed and Somaya Al-Maadeed

Abstract Multispectral face recognition has been an interesting area of research where images obtained from different bands are matched. There are many face image datasets available which contain infrared and visible images. In most face recognition applications, the IR image taken in different circumstances is matched against the visible image available in the application database. High computational cost is required for processing these images. In the literature, there is no guideline about the optimal number of features for dealing with multispectral face datasets. Thus, in this paper, we will perform image matching using infrared and visible images for face recognition and establish the threshold of the optimal number of features required for multispectral face recognition. The experiments conducted are on SCFace—surveillance cameras face database. The experimental setup for multispectral face recognition using LBP and PCA feature sets and the experimental results are discussed in the paper.

Keywords Multispectral face recognition · Image matching · LBP · PCA · Feature sets

1 Introduction

Image recognition applications in domains such as security, robotics, satellite imagery, medicine have the need to identify and process images taken at different wavelengths. The images taken at any other wavelength than visible spectrum have

This publication was made possible by NPRP grant # NPRP8-140-2-065 from the Qatar National Research Fund (a member of the Qatar Foundation). The statements made herein are solely the responsibility of the authors.

W. W. Syed (✉) · S. Al-Maadeed
Computer Science and Engineering Department, Qatar University, Doha, Qatar
e-mail: Wafawaheeda@qu.edu.qa

S. Al-Maadeed
e-mail: s_alali@qu.edu.qa

© Springer Nature Singapore Pte Ltd. 2020
X.-S. Yang et al. (eds.), *Fourth International Congress on Information and Communication Technology*, Advances in Intelligent Systems and Computing 1027, https://doi.org/10.1007/978-981-32-9343-4_40

absence of color. Infrared imagery is often used in applications where surveillance is required, and images are captured even when there is the absence of light. These images are called as multispectral images as they are captured in different bands, and often more information is obtained in form of data at specific wavelengths across the electromagnetic spectrum.

Applications dealing with multispectral images require high computational resources as they are always challenged with problem of scalability and increased computation time. This is due to factors such as the image size and large number of features involved while processing the visual content. Detection and recognition of certain regions between multispectral images and visible images are an existing research challenge. For instance, in most biometric applications, there are visible images taken prior and enrolled in databases known as enrollment database. And, the images captured in nearest infrared (NIR) band are matched for surveillance of a certain region in the images. This region of interest (ROI) detection and recognition from the enrollment database can be obtained once the NIR image is identified with its corresponding visible image stored in enrollment database—as both images share the same ROI. The identified information is used for surveillance and is found very useful for making decisions based on the domain and applications. But, the computation cost and scalability factors affect this process.

Dimensionality reduction, a well-known pre-processing strategy can be used to reduce the features and help with optimizing the consumption of computation resources. In the literature, approaches like image quantization [1] was also used to reduce the dimensionality of visual data where the color information representing the image was reduced to fewer colors. Dimensionality reduction techniques like PCA, LDA, and image quantization when applied on visible images facilitate reducing computing cost and scalability issues. But, since the NIR images do not have color information, image quantization cannot be applied. To the best of our knowledge, there is no optimal dimension reduction technique which when applied on both visible and multispectral images gives a threshold value to facilitate identification between those images. A series of experiments using feature extraction and dimensionality reduction methods need to be performed on a dataset to obtain a threshold.

Face recognition has been an interesting area of research which also deals with multispectral images. In multispectral face recognition, the face region of IR image is to be identified with the face region of visible image present in enrollment database. Various feature extraction and dimensionality reduction techniques have been applied and tested in the face recognition studies. However, no efforts have been made to optimize the feature sets aiding better recognition rate and guide other research studies. A threshold value for every face dataset can be found with series of experiments. In this paper, experiments are made for the SCFace database [2] as the face images are taken in naturalistic conditions. This face database has low recognition rate as the images reflect the real scenarios. In this paper, we conduct a series of experiments using approaches used by the studies using SCFace database in the literature. A related study using the same dataset will be used as a baseline system for comparing our results.

The rest of the paper is organized as follows: Sect. 2 will contain the literature review of dimensionality reduction efforts on multispectral images and face recognition using SCFace database. In Sect. 3, the methodology for face recognition using IR vs visible images is discussed. In Sect. 4, the experimental evaluation containing the experimental setup and results obtained is elaborated. The results obtained are also analyzed and discussed. Finally, the Sect. 5 contains conclusion for summarizing the paper and future work discussing the next steps.

2 Literature Review

2.1 Dimensionality Reduction

Multispectral images are used in variety of applications such as remote sensing involving aerial imagery [3–5]; weather [6], security [7], and biomedical [8, 9] use images taken at near infrared (NIR) and ultraviolet bands. There are applications focusing on vegetation, detection of injury in fruits [10], finding tenderness of meat [8, 9], and inspecting the ancient manuscripts [11] using multispectral images. The images taken in different bands contain more information and are used for surveillance by matching or classifying the ROI with the visible light images. A variety of classification techniques were used. Tarabalka and Benediktsson in [12] used pixelwise SVM classification and spectral-spatial classification to obtain 89.31% accuracy in classifying hyperspectral images. However, dimensionality reduction was not used to reduce the huge number of extracted features.

Various dimensionality reduction techniques such as PCA and LDA have been mostly used for surveillance purposes in such systems. The studies [13, 14] use SIFT for obtaining a feature set and apply PCA on images for dimensionality reduction on images. The studies also compare with existing dimension reduction techniques. But, the SIFT technique when applied on images proved to have a better matching accuracy. Clemmenson et al. in [15] study to measure the moisture content and differentiate between sand and concrete using dimension reduction methods on multispectral images compared and applied different dimension reduction methods. However, no clear explanation was given for the choice of dimension reduction methods and classification used by the study in the paper.

Linear [16] and nonlinear [7] approaches of dimensionality reduction were also used in the existing studies, but these methods showed almost the same accuracy in comparison with the approaches such as PCA for dimension reduction followed by SVM for classification. Neural network approaches such as in [5, 10, 17] were also used in classification and found to have slightly better accuracy. Efforts such as a global geometric framework [18] for nonlinear dimensionality reduction were also proposed. Many studies compared the dimension reduction techniques [4, 15, 19–22] for a dataset and proposed the accurate measure for that particular dataset. Thus, in the literature, we note a variety of approaches being used for dimensionality reduction and classification on various datasets from different domains. But, there is a clear lack of studies suggesting which feature extraction method for generating feature sets or dimensionality reduction and classification technique is best suited for different image datasets from different domains.

2.2 Face Recognition

Use of multispectral images in face recognition has attracted a lot of research in the recent years [23–26]. In biometric applications, infrared images were mostly used for surveillance. The SCFace database [2] is a surveillance face database consisting of 130 different user faces, which were obtained by capturing visible and IR images from five different surveillance cameras by M. Grgic et al. Many face recognition studies have been made using the SCFace database, due to it's high reflection to the real scenarios. In the literature, different techniques like the Gabor feature-based classification were applied [27, 28]. And, different dimension reduction techniques such as PCA, whitened principal component analysis (WPCA), and LDA were used for dimension reduction. We note that a variety of different principal component analysis (PCA) and local binary pattern (LBP) approaches are being used in the recent studies. Thus, we narrow down on PCA and LBP methods for conducting experiments on SCFace database.

3 Methodology

The IR vs visible image matching for SCFace multispectral face database is explained by elaborating on the following stages: image acquiring, segmentation, preprocessing, feature extraction and face recognition as shown in Fig. 1. The multispectral face recognition is carried out by first acquiring, segmenting, and then preprocessing using a set of preprocessing methods. The visible images are first preprocessed to form the enrollment database. Later, LBP and PCA methods are applied with experimentation under different parameters. Similar preprocessing and feature extraction methods are applied on the IR image which is to be matched to the visible images in enrollment database, using the same parameters and settings.

3.1 Image Acquiring

The visible and IR image sets one and three were chosen from SCFace database. The image set one contains a frontal mugshot of 130 subjects with one image per subject under visible light settings. And, the image set three is taken from camera eight in IR band, for 130 subjects with one image per subject.

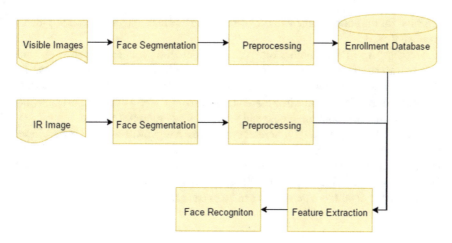

Fig. 1 IR versus visible image matching in multispectral face recognition

3.2 Segmentation

Segmentation is a process of separation or segmentation of the ROI from the images. This requires the ROIs to be first detected from the images for segmenting. In our experiments, the Voila–Jones face detection algorithm [29] is used to detect face region from the images. And, imcrop MATLAB function is used to segment the face region from the images.

3.3 Preprocessing

Histogram equalization and normalization were used as preprocessing methods for contrast stretching and eliminating noise from the face images. Histogram equalization is a commonly used technique for tweaking the image intensities by equalizing as it enhances the image contrast.

3.4 Feature Extraction

PCA and LBP were used as feature extraction methods. LBP is commonly used to get texture information from images. LBP refers to the ASCII value of binary pattern obtained for a pixel from its neighboring pixels. The binary pattern is obtained by marking 1 if the neighboring pixel is higher than the selected pixel and 0 if the neighboring pixel value is lower than the selected pixel. The window size of 3×3 is generally selected, and the middle pixel value is often selected. This process is repeated for the entire image without overlapping and selecting the next 3×3 matrix from an image.

Overlapping-LBP is another form of LBP where the parameters such as the window size and the overlapping window ratio are set. When the window is set to overlap 60% on a 3×3 window size for finding the LBP of an image, the window will be shifted by only one column while maintaining an overlap ratio of 60%. This generates more information about the image and thus gives rise to useful features.

PCA is a technique which generates eigenfaces or eigenvalues for an image, based on the co-variance of pixels in images. This is a commonly used technique, like LBP for face recognition. PCA is also used for dimensionality reduction. Here, the PCA is used to generate eigenfaces for an image, which are considered as features. And, the number of eigenfaces generated can be tweaked to improve the recognition rate.

3.5   Face Recognition

The features obtained from the IR image are recognized against the visible face images in enrollment database. Distance classification is often used for matching the IR image with visible image. Distances such as Euclidean, Jaccard, and Chebyshev are used for finding the similarity between the IR and visible images.

4   Experiments

The multispectral face recognition is performed by following the above methodology. In this section, we discuss in detail the experiments conducted to find the optimal number of features generated by the PCA, LBP, and overlapping-LBP methods. The experimental setup starting from image acquisition to face recognition is described, and the obtained results and evaluation are discussed below.

4.1   Experimental Setup

On image acquisition from SCFace database, the visual and IR images were processed separately as two modules. The objective of first module was to form an enrollment database of visible images. The visible images were first segmented and preprocessed using techniques described in Sect. 3 to form the enrollment database. In module two, the IR image which is to be matched for recognizing similar faces from the enrollment database is acquired, segmented, and preprocessed.

The IR image and the images in enrollment database are subject to similar feature extraction techniques. The corresponding feature sets obtained are then subject to distance classification for IR vs visible image matching. Interface is shown in Figure 2

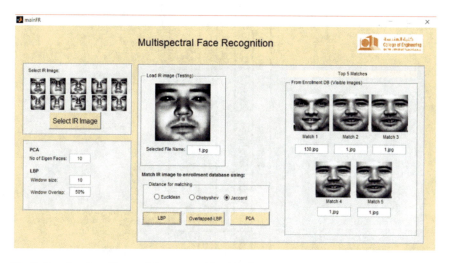

Fig. 2 Developed multispectral face recognition interface

was developed and used to conduct the experiments and visually verify the image matching process for face recognition. The features extracted using PCA, LBP, and overlapping-LBP were used for measuring similarity using distance classification methods. For PCA, top 130 eigenfaces were used for obtaining the feature set. In LBP, the uniform LBP was computed, and a histogram with ten bins was generated as a feature set. In overlapping-LBP, a window size of 10×10 matrix was used with a window overlap of 50%.

4.2 Experimental Results

The evaluation measures used to evaluate face recognition systems are known as the cumulative match score (CMS) [30]. The recognition rate is often measured as rank, in terms of the number of minimum distances considered for evaluation. On performing the experiments, the results obtained are evaluated using rank measure and summarized in Table 1. From Table 1, we note that the overlapping-LBP gives highest rank one recognition rate of 13.08%. This supports the hypothesis that on tweaking parameters and conducting experiments on datasets will help in obtaining better results. While 13.08% rank one recognition rate is still a low value, when taking into consideration the nature of SCFace database, and the highest rank one recognition rate in [2] was 10%, using PCA and cosine angle for measuring distance. This proves our obtained rank one recognition rate is comparatively good.

Table 1 Rank 20, rank ten, rank five, and rank one recognition rate in %

	Rank 20			Rank ten			Rank five			Rank one		
	Euclidean	Jaccard	Chebychev	Euclidean	Jaccard	Chebychev	Euclidean	Jaccard	Chebychev	Euclidean	Jaccard	Chebychev
PCA	13.8	15.3	13.1	6.9	7.0	3.8	2.3	3.8	2.3	0	0.8	0
LBP	18.4	14.6	21.5	10.7	6.9	10.7	7.7	3.1	8.46	1.5	0	1.54
Overlapping-LBP	18.4	36.9	16.9	8.4	26.9	5.4	4.6	23.1	3.85	1.5	13.1	1.54

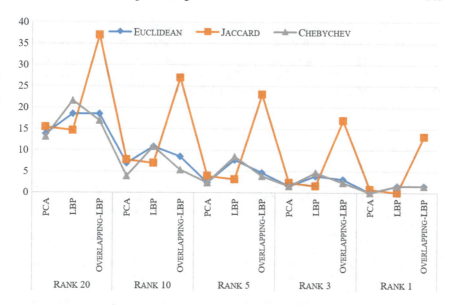

Fig. 3 Obtained Rank recognition rate in % for PCA, LBP, and overlapping-LBP

4.3 Analysis and Discussion

For the SCFace database, since one image per subject was involved in enrollment database used for training and one IR image was used for matching, the rank one results obtained might not seem satisfactory. From Fig. 3, we note the trend in ranks based on the feature set used and the distance classification applied. In terms of the feature set used, the LBP and overlapping-LBP always give better Rank recognition rate than PCA. The overlapping-LBP when used with Jaccard distance measurement gives highest recognition rates.

5   Conclusion and Future Work

The SCFace—surveillance cameras face database [2] was used for face recognition of IR vs visible images, performed using distance classification or matching. The multispectral face recognition was tackled using PCA dimensionality reduction method, and the tests were also done using the LBP feature extraction method. On experimenting with different conditions and parameters for these methods, one-to-many face recognition approach was applied to identify within the visible image enrollment database. Evaluation against the existing Rank one face recognition rate presents the SCFace dataset was done, and our experiments were found to give better results.

As a future work, we hope to continue the experiments on more multispectral face databases and use different feature sets and dimensionality reduction methods. By adopting the same methodology—we hope to identify the optimal features set and dimensionality reduction methods, which when applied to both visible and multispectral images will generate a reduced dimension set.

References

1. Mahmood, A., Uzair, M., Al-Maadeed, S.: Multi-order statistical descriptors for real-time face recognition and object classification. IEEE Access **6**, 12993–13004 (2018)
2. Grgic, M., Delac, K., Grgic, S.: SCface–surveillance cameras face database. Multimed. Tools Appl. (2011)
3. Griparis, A., Faur, D., Datcu, M.: Dimensionality reduction for visual data mining of earth observation archives. IEEE Geosci. Remote (2016)
4. Yang, H., Du, Q., Chen, G.: Particle swarm optimization-based hyperspectral dimensionality reduction for urban land cover classification. IEEE J. Sel. Top. (2012)
5. Pacifici, F., Chini, M., Emery, W.: A neural network approach using multi-scale textural metrics from very high-resolution panchromatic imagery for urban land-use classification. Remote Sens. Environ. (2009)
6. Han, H., Im, J., Kim, H.: Variations in ice velocities of Pine Island Glacier Ice Shelf evaluated using multispectral image matching of Landsat time series data. Remote Sens. Environ. (2016)
7. Uzair, M., Mahmood, A., Al-Maadeed, S.A.: Non-cooperative and occluded person identification using periocular region with visible, infra-red, and hyperspectral imaging. Biomet. Secur. Privacy 223–251 (2017)
8. Peyret, R., Bouridane, F Khelifi, A., Tahir, M., Al-Maadeed, S.: Automatic classification of colorectal and prostatic histologic tumor images using multiscale multispectral local binary pattern texture features and stacked generalization. Neurocomputing (2017)
9. Bhateja, V., Srivastava, A., Moin, A., Lay-Ekuakille, A.: Multispectral medical image fusion scheme based on hybrid contourlet and shearlet transform domains. Rev. Sci. Instrum. **89**(8) (2018)
10. ElMasry, G., Wang, N., Vigneault, C.: Detecting chilling injury in Red Delicious apple using hyperspectral imaging and neural networks. Postharvest Biol. Technol. (2009)
11. Hollaus, F., Diem, M., Fiel, S., Kleber, F.: Investigation of ancient manuscripts based on multispectral imaging. In: Proceedings of the 2015 ACM Symposium on Document Engineering (2015)
12. Tarabalka, Y., Benediktsson, J.: Spectral–spatial classification of hyperspectral imagery based on partitional clustering techniques. IEEE Trans. (2009)
13. Fotouhi, M., Kasaei, S., Mirsadeghi, S.: BSIFT: boosting SIFT using principal component analysis. In: 2014 22nd Iranian Conference on Electrical Engineering (ICEE) (2014)
14. Brown, M., Süsstrunk, S.: Multi-spectral SIFT for scene category recognition. Comput. Vis. Pattern (2011)
15. Clemmensen, L., Hansen, M., Ersbøll, B.: A comparison of dimension reduction methods with application to multi-spectral images of sand used in concrete. Mach. Vis. (2010)
16. Renard, N., Bourennane, S.: Denoising and dimensionality reduction using multilinear tools for hyperspectral images. IEEE Geosci. (2008)
17. Hinton, G., Salakhutdinov, R.: Reducing the dimensionality of data with neural networks. Science (80-) (2006)
18. Tenenbaum, J., De Silva, V., Langford, J.: A global geometric framework for nonlinear dimensionality reduction. Science (80-) (2000)

19. Hasanlou, M., Samadzadegan, F.: Comparative study of intrinsic dimensionality estimation and dimension reduction techniques on hyperspectral images using K-NN classifier. IEEE Geosci. Remote (2012)
20. Bostan, S., Ortak, M., Tuna, C., Akoguz, A.: Comparison of classification accuracy of co-located hyperspectral & multispectral images for agricultural purposes. Agro-Geoinformatics (2016)
21. Lei, T., Wan, S., Chou, T.: The comparison of PCA and discrete rough set for feature extraction of remote sensing image classification—a case study on rice classification, Taiwan. Comput. Geosci. (2008)
22. Bingham, E., Mannila, H.: Random projection in dimensionality reduction: applications to image and text data. In: Proceedings of seventh ACM SIGKDD (2001)
23. Li, S., Chu, R., Liao, S., Zhang, L.: Illumination invariant face recognition using near-infrared images. IEEE Trans. (2007)
24. Akhloufi, M., Bendada, A.: Probabilistic Bayesian framework for infrared face recognition. Mach. Vis. Image Process (2009)
25. Bendada, A., Akhloufi, M.: Multispectral face recognition in texture space. Comput. Robot Vis. (CRV) (2010)
26. Akhloufi, M., Bendada, A.: Multispectral face recognition using non linear dimensionality reduction. SPIE Defense (2009)
27. Choi, J., Ro, Y., Plataniotis, K.: Color local texture features for color face recognition. IEEE Trans. Image (2012)
28. Ahire, M., Dighe, D.: Web server based secure real time embedded system for ATM. ijaers.com
29. Viola, P., Jones, M.: Robust real-time face detection. Int. J. Comput. Vis. (2004)
30. Delac, K., Grgic, M., Grgic, S.: Statistics in face recognition: analyzing probability distributions of PCA, ICA and LDA performance results. In: Proceedings of the 4th International Symposium on Image and Signal Processing and Analysis (2005)

Using Machine Learning Advances to Unravel Patterns in Subject Areas and Performances of University Students with Special Educational Needs and Disabilities (MALSEND): A Conceptual Approach

Drishty Sobnath, Sakirulai Olufemi Isiaq, Ikram Ur Rehman and Moustafa M. Nasralla

Abstract Universities and colleges in the UK welcome about 30,000 students with special needs each year. Research shows that the dropout rate for disabled students is much higher at 31.5% when compared with about 12.3% for non-disabled students in the EU. Supporting young students with special educational needs while pursuing higher education is an ambitious and important role, which needs to be adopted by tertiary education providers worldwide. We propose, MALSEND, a conceptual platform based on human-machine intelligence (HMI), a collective intelligence of human and machine to understand patterns of learning of disabled students in higher education. This platform aims to accommodate and analyse data sets features of universities activities to discover trends in performances with regards to subject areas for autistic students, dyslexic students and students having attention deficit hyperactive disorder (ADHD), among others. Analysis of variables, such as students' performances in modules, courses and other engagement-indices will give new insights into research questions, career advice and institutional policymaking. This paper describes the developmental activities of the MALSEND concept in phases.

Keywords MALSEND · Machine learning · Special educational needs · Performance · Unsupervised learning

D. Sobnath (✉) · S. O. Isiaq
Solent University, Southampton, UK
e-mail: drishty.sobnath@solent.ac.uk

I. U. Rehman
School of Computing and Engineering, University of West London, London, UK

M. M. Nasralla
Prince Sultan University, Riyadh, Saudi Arabia

© Springer Nature Singapore Pte Ltd. 2020
X.-S. Yang et al. (eds.), *Fourth International Congress on Information and Communication Technology*, Advances in Intelligent Systems and Computing 1027, https://doi.org/10.1007/978-981-32-9343-4_41

1   Introduction

The term 'special educational needs and disabilities' (SEND) refers to students who have learning problems or disabilities that make it harder for them to learn than most of their peers. This may include physical, development disabilities, behavioural, emotional and communication disorders and learning deficiencies [1]. Universities and colleges in the UK welcome almost 30,000 disabled students each year [2]. As of today, only nine countries of the European Union, including France and the United Kingdom, have implemented policy plans to help SEND students in higher education [3]. Some of these plans include free transport to and from universities, special software to aid learning and teaching and other simple assistance to students with specific impairment. However, there is a lack of support and social inclusion for students having a learning disability [4]. There are a number of other concerns such as poor quality, wrong career advice or lack or guidance for students with a learning disability [5]. Research shows that the dropout from education in the EU for the disabled is at 31.5%, much higher, when compared to only 12.3% for non-disabled students [3]. Supporting young students who require special educational needs in pursuing higher education is an ambitious and necessary step that needs to be adopted by all tertiary education providers worldwide. Figure 1 shows the number of people in the UK with a learning disability.

MALSEND project aims at developing an intelligence platform using learning algorithms to identify learning patterns of disabled students and their corresponding chosen subject at university level. The platform intends to explore large data sets about students-universities activities to discover trends in subject areas and performance among autistic students, dyslexic students or students having attention deficit hyperactive disorder (ADHD) among others.

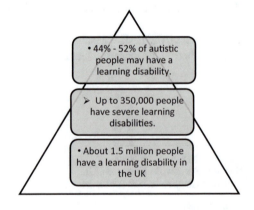

Fig. 1 Number of people with learning disabilities in the UK [6]

1.1 Statement of the Problem

Disability Rights UK, one of the leading charity for people with disability or health conditions, states that there are a number of issues they found out during their events and steering groups, among which were the issues of having poor quality or wrong-career advice for students with a learning disability. There is also a bad choice of subjects that do not match these students' aspirations and a lack of guidance for these students [5]. There is a lack of support and social inclusion for students with learning disabilities in going into higher education, participating in various domains and contributing to the economy of a country [4]. Subsequently, this results in problems such as having a regular income, good quality of life and access to the overall social and educational system. Pathways to tertiary education for SEND students depend on the type of learning disability, financial resources and self-motivation of students after leaving secondary schools. To reduce the rate of dropout and encourage students with learning disabilities, research must be carried out to identify effective and suitable options for these students in terms of subject areas, performance and their potential contribution to the society. However, to the authors' knowledge, there is no platform, framework or model available to detect any trends or patterns in the learning disability of a student, their corresponding performances and success in specific subject areas. Implementing such models can improve the possibility of making tertiary institution more conducive and hopefully reduces the dropout rate. Also, such models can help students with learning disabilities have a voice in HE and help other organisations in developing policies relating to SEND students.

1.2 Purpose of Inquiry and Inquiry Questions

This paper is a conceptual proposition for the discovery and identification of factors that affect the learning and performance of higher education students suffering from a learning disability. Using a collective intelligence approach, an extensive analysis of students' records will be utilised to determine relative trends. Therefore, the research questions for this study focus on the following: (1) Will a machine learning platform suitably identify patterns of learning disability, subject area and performance of students at tertiary education level? and (2) Can a machine learning platform accurately predict subject area performance for SEND students based on their learning disability? The next section of this paper discusses the related work.

2 Related Work

Machine learning (ML) is an aspect of artificial intelligence (AI) whereby programs use statistical models to give computer systems the ability to learn without being fully programmed [7]. Nowadays, the application of machine learning techniques is rampant throughout different areas including marketing improvement, decision-making in health care, manufacturing, education and also applied for financial analysis. For example, machine learning is being used to predict mortality rate associated with certain type of diseases, predict effectiveness of surgical procedures, to help physicians make better decision and to discover relationship among clinical and diagnosis data. Unsupervised algorithms attempt to overcome limitations of supervised learning algorithms by automatically identifying patterns and dependencies in the data. Therefore, this work can benefit from unsupervised learning allowing algorithms to look back for patterns that have not been previously considered. Consequently, new knowledge can be extracted from the observed data to build predictors. In statistical unsupervised learning pattern recognition, the data can be identified by finding clusters, for example, by using K-means algorithms or adaptive resonance theory (ART) algorithms [8, 9]. Dimensionality reduction or principle component analysis (PCA), is another method used to reduce the number of random variables under consideration by defining a set of principle variables [10]. It is important to understand the relation between identified clusters so that competitive learning algorithms can be applied to provide efficient solutions to problems.

2.1 Prediction of Student's Performance

Previously, work has been done in the area of predicting the performance of students as shown by a few studies [11–13]. Different machine learning techniques, such as matrix factorisation [14] or collaborative filtering [15], have been used to predict students' grades. The right research questions are important to understand the existing studies of predicting SEND students' area of expertise. Current studies make use of cumulative grade point average (CGPA), assignment mark, quizzes, lab work, class test and attendance to predict performance of students in general [16]. Other researchers considered gender, age, family background, disability, extra-curricular activities, social interaction and psychometric factors [16], to see how these affect the student's performance. However, the proposed work is looking at finding relationships between identified factors and SEND students' areas of expertise by analysing at least 15,000 student records since this was an identified gap in the literature.

3 MALSEND Platform

A composition of approaches with multifaceted techniques is adopted at different stages of the research work. At the initial stage, multiple anonymised data sets of student records from universities are to be examined to determine existing patterns. However, for the purpose of this conceptualisation, we are only considering students' data from 2 UK universities. Once the platform has been implemented and initial results have been obtained, the platform will be reinforced and trained with data sets from other universities. The following sections describe the activities of the development phases of this concept.

3.1 Ethical Approval

This work is being carried out under strict ethical standards, for example, in relation to students' privacy, confidentiality and university's consent. Ethical approval has therefore been obtained for this research project from the participating universities' ethics committee in November 2018. The data collected will be completely anonymised to prevent the identification of any student and to abide by the General Data Protection Regulations (GDPR) EU regulations. Another ethics application will be made in the second stage of the project when data from other universities will be required to reinforce and test the platform.

3.2 Data sets

At least 15,000 anonymised student records over the last eight years from two UK universities will be analysed for the first pilot study. Anonymised data for students who have been clinically diagnosed with one of the published learning disabilities (dyslexia, dyspraxia, ADHD, Asperger's syndrome, other autistic spectrum disorder), as shown in Table 1, is being collected in a spreadsheet. In the UK, higher education

Table 1 Type of learning disability recorded by HE institutions in UK [17]

Code	Label
0	No known disability
8	Two or more impairments and/or disabling medical conditions
51	A specific learning difficulty such as dyslexia, dyspraxia or ADHD
53	A social/communication impairment such as Asperger's syndrome/other autistic spectrum disorder
54	A long-standing illness or health condition such as cancer, HIV, diabetes, chronic heart disease or epilepsy
55	A mental health condition, such as depression, schizophrenia or anxiety disorder
56	A physical impairment or mobility issues, such as difficulty using arms or using a wheelchair or crutches
57	Deaf or a serious hearing impairment
58	Blind or a serious visual impairment uncorrected by glasses
96	A disability, impairment or medical condition that is not listed above

institutions use the following standard codes to classify disabilities, a coding frame introduced by the HESA and the Disability Rights Commission (DRC) [17].

Data collected consist of age range (e.g. 19–21 years old, 22–25 years old), sex, status (full time, part time, distance learners) and type of learning disability (autistic, ADHD, dyslexic, dysgraphia). Also, included are entry type (foundation/A-level/diploma), A-level of students (UCAS points and subjects), module grades, number of sittings, no of credits, module type, course code and description, module code and academic level (undergraduate or postgraduate). In addition, grades of 1st, 2nd, 3rd year or postgraduate results, alumni information (career path, job position after graduation) and other related parameters are considered to be analysed using suitable algorithms as explained in the next section.

3.3 Analysis

Dimensionality reduction

Following the scikit-learn (software machine library for Python) [18], dimensionality reduction algorithm will be utilised to reduce the number of meaningful variables to simplify the data without losing much information. In addition, other groups of algorithms can be adopted to remove unneeded data, outliers and other non-useful data. Dimensionality reduction will be performed to free storage space on our server and improve the performance of our machine learning system. It will also help the researchers to visualise the data [19]. An anomaly detection algorithm can also automatically remove outliers from the data sets.

K-Means Clustering

Clustering (K-means) and visualisations algorithms can then be applied to the data set to identify clusters and unsuspected patterns [8]. Finally, another method of unsupervised learning, known as association rule learning algorithm will be used to discover interesting relations among other attributes.

3.4 Components of MALSEND Platform

The findings will be evaluated with the second data set in the next stage of the project. The prototype can also be further developed with new data to predict subject areas of SEND students in future. Figure 2 shows the components of the proposed MALSEND platform.

Key algorithms help in model creation to determine patterns, correlations and clusters from the data. The objective for unsupervised learning is to model the fundamental organisation or scattering in the data in order to learn more about it.

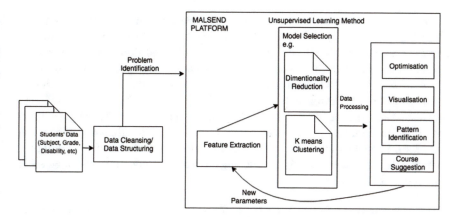

Fig. 2 Components of the proposed MALSEND platform

4 Limitations

There are a few identified limitations for this pilot study which can affect the results. For example, since the sample data is taken from only two universities in the first phase, the results do not represent the demographics of the population. Moreover, the fact that some universities provide more specific courses than other universities, the results may show certain patterns and high correlations among a specific learning disability and a course. Hence, there is a need to carry out the second part of the project to test any hypothesis and conclude any findings. More importantly, there are other factors such as social, economic factors or family background that can also affect the results of this study.

5 Conclusion

We expect the machine learning platform to generate new knowledge by identifying patterns that could help solve some of the social challenges such as high dropout rates from education (31.5% compared to 12.3% for non-disabled students), low employability (only 6% of adults with a learning disability in England are in paid work) or depression among SEND students in future. The findings will create new research questions and help bring other universities together through collective intelligence to find similar patterns regarding other health conditions (visual or hearing impairment, epilepsy). The results in wider applicability could also be used to support career advisers in schools, colleges and communities by providing course suggestions tools. The findings of this study might open new opportunities and act as a guide to those having a learning disability and who are planning to pursue higher education studies.

As to the best of our knowledge, no such system has yet been implemented to help SEND students in the UK to find out which are the subject areas they will most likely to be successful in, based on past students who had similar learning disabilities. Future AI and machine learning prediction models can provide an extra set of eyes and ears for SEND students, therapists, teachers as well as for parents. Further analysis of this data can lead to pinpointing social success factors and assessing a student's strengths and weaknesses. Finally, the findings could assist the government and other institutions for the development of policies, curriculum and educational practices. Educators would be better able to understand the students' learning and emotional development, slowly introducing them to more complex and varied social environments over time.

Acknowledgements Dr. Drishty Sobnath would like to acknowledge the Research, Innovation and Enterprise department of Solent University for supporting this work. Dr. Ikram Ur Rehman would like to thank the CU Coventry for its support. Dr. Moustafa Nasralla would like to acknowledge the management of Prince Sultan University (PSU) for the valued support, fund and research environmental provision which have led to complete this work.

References

1. Kryszewska, H.: Teaching students with special needs in inclusive classrooms special educational needs. ELT J. **71**(4), 525–528 (2017)
2. UCAS: Disabled Students. Advice and Financial Support. UCAS https://www.ucas.com/undergraduate/applying-university/individual-needs/disabled-students (2018). Accessed 19 Sep 2018
3. Limbach-Reich, A., Powell, J.: Young adults with special educational needs (SEN) (2016)
4. Kim, K.M., Shin, Y.R., Yu, D.C., Kim, D.K.: The Meaning of Social Inclusion for People with Disabilities in South Korea, vol. 64, no. pp. 19–32 (2016)
5. Disability Rights UK: Careers Guidance and Advice for Disabled Young People (2017)
6. NHS: Learning disabilities—NHS. https://www.nhs.uk/conditions/learning-disabilities/ (2018). Accessed 23 Aug 2018
7. Jordan, M.I., Mitchell, T.M.: Machine learning: trends, perspectives, and prospects. Science (80-) **349**(6245), 255 LP-260 (2015)
8. Syakur, M.A., Khotimah, B.K., Rochman, E.M.S., Satoto, B.D.: Integration K-means clustering method and elbow method for identification of the best customer profile cluster. In: IOP Conference Series: Materials Science and Engineering (2018)
9. Carpenter, G.A., Grossberg, S.: Adaptive resonance theory. CAS/CNS Technical Report Series (2010)
10. Wagstaff, K., Cardie, C., Rogers, S., Schroedl, S.: Constrained K-means clustering with background knowledge. International Conference on Machine Learning (2008)
11. Cortez, P., Silva, A.: Using data mining to predict secondary school student performance. In: Proceedings of 5th Annual Future Bus Based Technology Conference (2008)
12. Thiede, K.W., et al.: Can teachers accurately predict student performance? Teach. Teach. Educ. (2015)
13. Chamorro-Premuzic, T., Furnham, A.: Personality predicts academic performance: evidence from two longitudinal university samples. J. Res. Pers. (2003)
14. Thai-Nghe, N., Horváth, T., Schmidt-Thieme, L.: Factorization models for forecasting student performance. In: Proceedings of the 4th International Conference on Educational Data Mining (2011)

15. Toscher, A., Jahrer, M.: EDM-59: collaborative filtering applied to educational data mining. Austria—KDD Cup (2010)
16. Mohamed Shahiri, A., Husain, W., Abdul Rashid, A.: ScienceDirect the third information systems international conference a review on predicting student's performance using data mining techniques. Procedia Comput. Sci. **72**, 414–422 (2015)
17. HESA: Fields required from institutions in all fields disability. HESA. https://www.hesa.ac.uk/collection/c16051/a/disable (2016). Accessed 10 Jan 2019
18. Géron, A.: Géron—2017—Hands-on machine learning with scikit-learn and Tensorflow.pdf. In: Hands-on Machine Learning with Scikit-Learn and TensorFlow (2017)
19. Hurwitz J., Kirsch, D., Machine Learning For Dummies, IBM Limited Edition Published. Wiley (2018)

Image-Based Ciphering of Video Streams and Object Recognition for Urban and Vehicular Surveillance Services

Karim Hammoudi, Mohammed Abu Taha, Halim Benhabiles, Mahmoud Melkemi, Feryal Windal, Safwan El Assad and Audrey Queudet

Abstract Nowadays, urban and vehicular surveillance systems are collecting large amounts of image data for feeding recognition systems, for example, toward proposing localization or navigation services. In many cases, these image data cannot directly be processed in situ by the acquisition systems in reason of their low computational capabilities. The acquired images are transferred to remote computing servers through various computer networks, and then analyzed in details toward object recognition. The objective of this paper is twofold (i) presenting image-based ciphering methods that can efficiently be applied for securing the image transfer against consequences of image interceptions (e.g., man-in-the-middle attacks) (ii) presenting generic image-based analysis techniques that can be exploited for ob-

K. Hammoudi (✉) · M. Melkemi
Université de Haute-Alsace, IRIMAS EA 7499, 68100 Mulhouse, France
e-mail: karim.hammoudi@uha.fr

M. Melkemi
e-mail: mahmoud.melkemi@uha.fr

K. Hammoudi · M. Melkemi
Université de Strasbourg, Strasbourg, France

M. Abu Taha
Palestine Polytechnic University, Hebron, Palestine
e-mail: m_abutaha@ppu.edu

H. Benhabiles · F. Windal
ISEN-Lille, Yncréa Hauts-de-France, Lille, France
e-mail: halim.benhabiles@yncrea.fr

F. Windal
e-mail: feryal.windal@yncrea.fr

S. El Assad
Institut d'Electronique et de Télécommunications de Rennes (IETR),
Université de Nantes, Nantes, France
e-mail: safwan.el-assad@univ-nantes.fr

A. Queudet
Laboratoire des Sciences du Numérique de Nantes (LS2N),
Université de Nantes, Nantes, France
e-mail: audrey.queudet@univ-nantes.fr

© Springer Nature Singapore Pte Ltd. 2020
X.-S. Yang et al. (eds.), *Fourth International Congress on Information and Communication Technology*, Advances in Intelligent Systems and Computing 1027, https://doi.org/10.1007/978-981-32-9343-4_42

ject recognition. Experimental results show end-to-end image-based solutions for fostering developments of surveillance systems and services in urban and vehicular environments.

Keywords Real-time video services · Surveillance systems · Image ciphering · Image analysis · Object recognition · Machine learning · PRNG

1 Introduction and Motivation

Throughout the world, urban and vehicular surveillance systems are more and more necessary for ensuring the security of citizens as well as for supporting surrounding video-based services such as those contributing to the daily comfort of citizens (e.g., video-based navigation systems). Surveillance systems can be either static such as those commonly installed in public space or mobile such as mobile mapping systems, unmanned aerial vehicles (UAVs), and new generations of general public vehicles (e.g., cars or buses equipped of cameras). Current trends consist of analyzing the acquired video streams in real time [6]. However, the surveillance systems have to be highly secured during the data transmission for avoiding unauthorized actors or entities to access to the acquired data toward intrusive or malicious uses.

In traditional remote video surveillance systems, the transfer of images from the acquisition node to the analyzing node (e.g., monitoring or recognition node) can be particularly unsecured in the sense that the image data transmission can be done through a succession of different computer networks and network layers; for example, through wireless and wire networks. Hence, the pathing of the data flow can completely be unknown when managed by third-party routing systems making then a gap for the control and the security of the data. Such routing systems can make the surveillance systems particularly vulnerable to man-in-the-middle attacks [5, 10] as one can observe on Fig. 1.

In this paper, we present an image-based solution that can provide an additional level of security to the security protocols of third-party computer network systems. It consists of directly securing the image data by making it unreadable during the image transfer that is operated between the acquisition node and the analyzing node. To this end, an efficient image-based ciphering/deciphering algorithm is highlighted.

Additionally, the paper presents a generic object recognition workflow that can be exploited on an analysis node of a surveillance system. In the field of image recognition, image descriptors permit to characterize specific properties of an image. Hence, image descriptors can be employed to automatically recognize objects thanks to machine-learning processes. In this sense, the paper highlights a technique for computing texture descriptors as well as a straight machine-learning principle that are known for their efficiency.

Video surveillance systems can be used in urban and vehicular environment for developing a large spectrum of road-monitoring services. In our case, we are particularly focused on emerging road services relying on the analysis of traffic flows

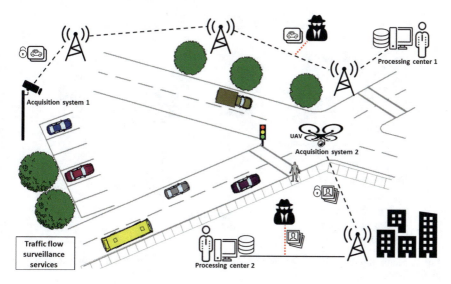

Fig. 1 Architecture of traffic flow surveillance services and risks related to image interceptions by man-in-the-middle attacks

(pedestrian or vehicle displacements [7]). In this context, static surveillance systems fitted with real-time object recognition capabilities can permit to facilitate the navigation of the drivers through the vision-based detection of available parking spaces (e.g., along street corridors) or through the vision-based regulation of traffic lights (e.g., in crossroads). Related sets of image data will then be employed for illustrating the efficiency of the presented works.

The remainder of this paper is organized as follows: an image-based ciphering method is presented in Sect. 2 for improving the security of video streams during the data transmission. In Sect. 3, we present a flexible image recognition workflow. Experimental results and performance evaluations are discussed in Sect. 4. We present our conclusions in Sect. 5.

2 An Image Ciphering for Secured Transmissions

Figure 2 illustrates a conventional principle for the image encryption/decryption of a data stream through a ciphering loop. Considering the acquisition of grayscale image by the surveillance system, each image pixel is converted into a data stream. Then, a keystream is randomly generated and exploited with an encryption function (e.g., an XOR function) for producing a ciphered image. Such stream ciphering method consists of adding a chaotic signal to the transferred data in order to transform it into unusable data. This processing is supposed to be operated by acquisition systems themselves such as those illustrated in Fig. 1 (stage applied before the transfer of

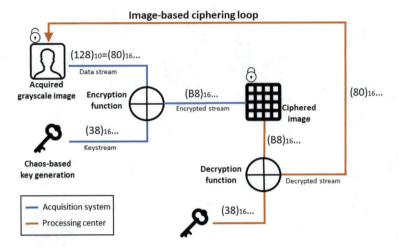

Fig. 2 Conventional principle for the image encryption/decryption of a data stream through a ciphering loop

the ciphered images). Once the transmission done, the processing center reuses a keystream with a decryption function in order to recover the original image.

An efficient image-based method that relies on such a ciphering loop has been proposed in [1–3]. This considered ciphering method belongs to the family of chaos-based encrypting/decrypting. More precisely, the encryption system relies on a pseudo random number generator (PRNG), which consists of an internal state and output function. The internal state, which contains the main cryptographic complexity of the system, is formed by two recursive filters. The first recursive cell contains a discrete skew tent map and the second one contains a discrete piecewise linear chaotic map. The output function depends on a chaotic multiplexer to produce the final keystream. The produced keystream value is XORed with the plaintext to cipher the data. For each call of the encryption system, a new keystream value is generated until encryption of the whole data.

3   An Image Recognition Workflow for Video Services

Machine-learning processes are intensively exploited in the development of image recognition systems. A general machine-learning pipeline is illustrated in Fig. 3. Such a pipeline can be used for identifying the nature or the category of an image from a query image. In this pipeline, two object categories are considered (e.g., vehicles and other objects) and represented by training image sets.

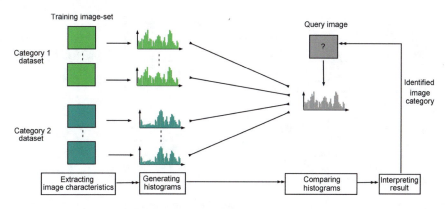

Fig. 3 A global machine-learning pipeline for image detection [9]

For each image set, a set of corresponding image descriptors is then generated (e.g., texture descriptors). Image descriptors are compact vectors of values embedding the most representative characteristics of the targeted image category. In this context, local binary pattern (LBP) is a method that permits to generate image descriptors through a pixel-by-pixel analysis of the concerned image [11]. For each pixel, its 8 neighbors are considered for generating the identifier of a pattern (8-bits sequence). Each pattern is then converted into a decimal value (value belonging to [0; 255]) which votes for the generation of a histogram of pattern frequencies. Finally, considering a query image (e.g., an image of the video stream that is acquired through a surveillance or recognition system), its analysis is operated by generating its image descriptor and by comparing this descriptor (i.e., histogram of targeted pattern frequencies) with all the descriptors that have been computed for the two image categories. This can be done by using a classifier (e.g., 1-nearest neighbor) with a specific metric[1] for identifying the image that is the most similar to the query image. By this way, a category is assigned to the query image.

Compact-mean LBP is a compact variant of the LBP descriptor. Its objective is twofold: (i) producing an image descriptor (vector of characteristic values) having a reduced size (e.g., 16 instead of 256) for fostering the speediness in the descriptor computation toward real-time image analysis (ii) enhancing the robustness of the descriptor to the low illumination changes that can occur under various conditions of acquisition. This principle of computation that is associated with this descriptor is illustrated in detail in Fig. 4. For more details, one can refer to [8].

---

[1]https://docs.opencv.org/2.4/doc/tutorials/imgproc/histograms/histogram_comparison/histogram_comparison.html.

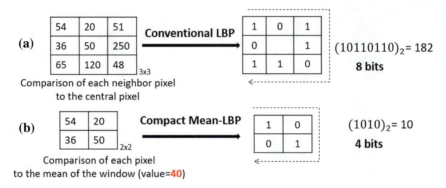

Fig. 4 A principle for fast generation of a texture-based image descriptor [8]

4 Experimental Results

Figure 5 illustrates a street image containing recognizable pedestrians as well as its corresponding ciphered image in row 1. Moreover, the associated histograms are presented in row 2. As can be seen, the histogram of the ciphered image is quasi uniform. Similarly, we observed that quasi-uniform histograms are obtained when this ciphering method is applied to other images; for example, with free images of varied natures coming from the unsplash website.[2] This means that this chaos-based ciphering method prevents from statistical attacks.

Table 1 highlights results that have been obtained by using 3000 grayscale images of parking slots. The images have been extracted from the PKLot dataset [4]. The training set was composed of 500 images of empty slots and 500 images of occupied slots (including a parked vehicle). The test set was composed of 2000 unknown images. The processing has been done by using a HP Elitebook 840 workstation (i5 2.3 GHz, 8 GB of RAM). It can be observed that the recognition rate is relatively close and high for both descriptors (approx. 86%) although the training set is only composed of 1000 images. Moreover, compact-mean is fastest than LBP. This latter point has also been observed through other recognition tests using image sets of varied sizes [8].

Hence, experimental results show that the presented ciphering method, as well as the presented image recognition workflow, can be combined for supporting the developments of secured and real-time-oriented video surveillance systems.

---

[2]https://unsplash.com/search/photos/urban.

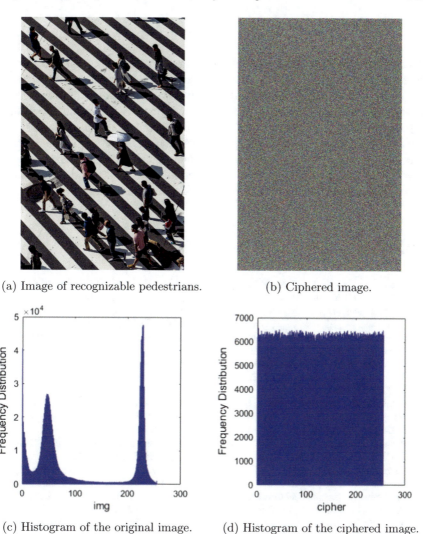

(a) Image of recognizable pedestrians.

(b) Ciphered image.

(c) Histogram of the original image.

(d) Histogram of the ciphered image.

Fig. 5 Row 1: A street image and its associated ciphered image. Row 2: Corresponding histograms

Table 1 Comparative table presenting performances of conventional LBP and Compact-mean LBP (result sample of [8])

Descriptor	Bin (nb)	Accuracy (%)	Comput. time for 1 img 93 × 154 (s)
LBP	256	86.95	0.022
CMLBP	16	86.40	0.014

5 Conclusion

This paper regroups an image-based ciphering method as well an image recognition workflow that fits well for fostering developments of secured video surveillance systems and services. In particular, the highlighted image-based ciphering method brings a double level of security. Indeed, it secures the image data in the applicative layer in addition to the security protocols that are provided and managed during the transfer by more low-level computer network layers. Moreover, a straight image recognition workflow is highlighted for facilitating video stream analysis. Experiments show image-based end-to-end controlled solutions for supporting the creation of secured and real-time urban and vehicular surveillance systems.

References

1. Abu Taha, M., El Assad, S., Jallouli, O., Queudet, A., Deforges, O.: Design of a pseudo-chaotic number generator as a random number generator. In: International Conference on Communications (COMM) (2016). https://doi.org/10.1109/ICComm.2016.7528291
2. Abu Taha, M., El Assad, S., Queudet, A., Deforges, O.: Design and efficient implementation of a chaos-based stream cipher. Int. J. Internet Technol. Secured Trans. **7**(2) (2017). https://doi.org/10.1504/IJITST.2017.087131
3. Abu Taha, M.: Real-Time and Portable Chaos-based Crypto-Compression Systems for Efficient Embedded Architectures, Ph.D. Thesis, Université de Nantes (2017). https://www.theses.fr/2017NANT4010
4. Almeida, P., Oliveira, L.S., Silva, Jr., E., Britto, Jr., A., Koerich, A.: PKLot A robust dataset for parking lot classification. Expert Syst. Appl. **42**(11), 4937–4949 (2015). https://doi.org/10.1016/j.eswa.2015.02.009
5. Conti, M., Dragoni, N., Lesyk, V.: A survey of man in the middle attacks. IEEE Commun. Surv. Tutorials **18**(3) (2016). https://doi.org/10.1109/COMST.2016.2548426
6. Hammoudi, K., Benhabiles, H., Melkemi, M., Dornaika, F.: Detection systems for improving the citizen security and comfort from urban and vehicular surveillance technologies: an overview. In: International Conference on ICT Infrastructures and Services for Smart Cities. Lecture Notes of the Institute for Computer Sciences, Social Informatics and Telecommunications Engineering, vol. 189. Springer, Berlin (2018). https://doi.org/10.1007/978-3-319-67636-4_5
7. Hammoudi, K., Cabani, A., Melkemi, M., Benhabiles, H., Windal, F.: Towards a model of car parking assistance system using camera networks: slot analysis and communication management. In: IEEE International Conference on Smart City 2018
8. Hammoudi, K., Melkemi, M., Dornaika, F., Benhabiles, H., Windal, F., Taoufik, O.: A comparative study of 2 resolution-level LBP descriptors and compact versions for visual analysis. In: FutureTech. Advanced Multimedia and Ubiquitous Engineering, LNEE. Springer, Berlin (2019). https://doi.org/10.1007/978-981-13-1328-8_28
9. Hammoudi, K., Melkemi, M., Dornaika, F., Phan, T.D.A., Taoufik O.: Computing multi-purpose image-based descriptors for object detection: powerfulness of LBP and its variants. In: International Congress on Information and Communication Technology. Advances in Intelligent Systems and Computing, vol. 797. Springer, Berlin (2018). https://doi.org/10.1007/978-981-13-1165-9_90

10. Huang, Y., Jin, L., Li, N., Zhong, Z., Xu, X.: Secret key generation based on private pilot under man-in-the-middle attack. In: Science China Information Sciences, Springer (2017). https://doi.org/10.1007/s11432-017-9195-3
11. Liu, L., Fieguth, P., Guo, Y., Wang, X., Pietikainen, M.: Local binary features for texture classification: taxonomy and experimental study. In: Pattern Recognition, vol. 62 (2017)

Augmented Reality—A Tool for Mediated Communication: A Case Study of Teen Pregnancy in Contexts like India

Suparna Dutta and Niket Mehta

Abstract Mediated communication has shrunk the world into "customized cottages". At the click of a button, we laugh, work, and even think together. Such digital interaction allows a fantastic mix of media usage. The best appears to be the combination of folk and mass media, where we have the intimacy and familiarity of the former and the outreach of the latter. Yet, challenges like illiteracy and problems of denial and deprivation of marginalized existence rampant in developing contexts necessitate the adoption of the futuristic "immersive technology" to communicate critical social messages.

Keywords Multimedia communication · Communication for social cause · Immersive technology · Mix media · Marginalized existence

1 Introduction: Denial, Deprivation, and Sustainable Behavior Change Communication

Development is a planned initiative for the uplift of the life conditions of the target population. Poignancy in the process is attained when the application of these initiatives targets those living in conditions of denial and deprivation. Denial is a more "human-constructed" situation, whereas deprivation appears to be a more natural existential condition. For instance, living in the remote high-altitude areas of Leigh and Ladakh in the North of India will naturally entail some deprivation due to the very geography of the area which the communities living in mainland India is not likely to endure. Similarly, some tribal communities of the North East India are naturally deprived of many developmental advantages due to their topography.

However, the fundamental majoritarianism of the democratic fabric of India has also unfortunately created and ushered in instances of denial that some have borne for

S. Dutta (✉) · N. Mehta
Birla Institute of Technology, Mesra, Noida Campus, Ranchi, India
e-mail: s.dutta@bitmesra.ac.in

N. Mehta
e-mail: niket@bitmesra.ac.in

© Springer Nature Singapore Pte Ltd. 2020
X.-S. Yang et al. (eds.), *Fourth International Congress on Information and Communication Technology*, Advances in Intelligent Systems and Computing 1027, https://doi.org/10.1007/978-981-32-9343-4_43

a very long period. Electoral politics, the one man, one vote principle of parliamentary elections, multi-party politics of India, embedded in a highly diverse and plural society have often also paradoxically led to the exploitation of many which entails some form of denial to rightful progress and often in a subtle manner. Interestingly, the subtler the denial, the more distressing is the effect on the suffering community. Further, the hidden denials also make redressal more difficult. As a consequence of such apathy, endurance frequently becomes a way of life for the victims and this leads often to a condition of resigned acceptance. This passive acceptance, unfortunately to the unsuspecting, may appear as a natural and spontaneous way of life! We often mistake submission as culture and tradition. In worse cases, it may just become so. Hence, when we talk of behavior change for sustainable development; it often has to contend with such paradoxes, contradictions, inconsistencies which compound the already-existing challenges of poverty and illiteracy in contexts like India.

In such conditions of almost double jeopardy, tools and principles of communication need to be applied both strategically and judiciously. Forgone is the conclusion that "communication" is critical in the process of development right from the stage of inception of an initiative to the point of commissioning of the same. It is also an inescapable reality that the principles of communication remain universal, but innovation should predominantly happen in the "tools" of communication with the understanding that new wars cannot be fought with old weapons. Innovation would often mean the adaption of the existing as well as the adoption of new and emerging technology.

1.1 Mediated Communication

Mediated communication is a norm today and the emerging trend in communication is to maximize the use and application of information and communication technology (ICT) options. Today, due to the mindless progress of the telecommunication network and service in India, aggressive competition among the service providers and the ever-increasing reach of the Internet, the never-ending cost-cutting bonanza and aggressive pricing offered by the manufacturers of smartphones [1, 2] the propensity of using smartphones is phenomenally increasing in every segment of the Indian society cutting across the nation's geography and segments of her plural society. As a result, technology is also changing at a reckless pace and so is the manner of communication. This is creating the new need to experiment, mix, and match media practices to arrive at the best and optimized options as mediated communication offer a plethora of choice, especially in the design of communication. This holds good, especially when efforts are made to respond directly to a well-understood problem plaguing the society. In accordance to this paradigm, an Action Research was designed with the help of students enrolled in the Management and Animation & Multimedia program of our university to study the hyper-exposed malady of "teen pregnancy". The target audiences were the marginalized slum dwellers of Maharajpur (Ghaziabad) in the near vicinity of New Delhi.

1.2 Demographic Dividend and the Urban Slums in India

In India, by the age of 15 approximately as many as 26% of females are married and by the age of 18, this figure rises to 54% [3, 4]. Since most "childbirth" in India is expected to occur within marriage, so the lower age at the time of marriage automatically links to an early onset of sexual activity, and thereby fertility mostly due to custom and social practice in this marginalized and bottom of the pyramid Indian population and most often not by choice as is the case with the middle and the affluent class of the Indian society today which is in tandem with the developing countries.

According to World Health Organization, during the year 1998–99 in India, the birth rate per 1000 females in the age group of 15–19 years was 107 with considerable difference between rural and urban regions: in rural areas, the adolescent birth rate was 121 out of 1000, while in Delhi it was 36 out of 1000 [5]. This also in the same line of argument indicates that most of the young mothers from rural India were in wedlock approved and supported by their society! And, it is a noted reality that most marriages in rural India are still arranged. So, in this specific context, the significant causality is "sustainable development" in terms of health and hygiene of Indian women, especially the young teenage women of India.

From the researcher's point of view, poverty, illiteracy, and lack of development still mark life in the rural area of India and a reflection of this is seen in the urban slums which hosts a steady migration of this bottom of the pyramid rural population. Perhaps the condition of the slums is even worse as in the slums, the population falls greater prey to the apathy of the city that makes them become more marginalized by being denied the support and sympathy of the folks and kin from their villages and the psychological comforts of a home they were used to living in for generations in their native place. In addition, the slum dwellers often have to also meet the demands of a language or a host of languages they are not familiar with and a lifestyle that is alien to them in the city. The aggression of the urban police and authorities, the snigger of the urban elite, and the hostile demands of a city life compounds the already impoverished simple people with meager belongings, little protection, inadequate sense of belonging in the city but carrying great expectations of making some meaning out of their highly challenged life. Here, given the smallest of space to barely survive in the slums of our progressive cities, these migrant population by and large are found to continue with the traditional lifestyle that they have been used to and simultaneously stretch themselves in every possible manner to survive in the city and the slums. This creates havoc in their existence. Further, the slums often provide less amenities than what they were used to in their villages, especially in terms of security and culture and tradition that mark like in India. So, any little similarity and sympathy for their lifestyle from their fellow slum dwellers makes them create a pattern of existence that emulates their life back in the villages. Otherwise, it is frequently seen that the migrants try to stay with fellow migrants from the same village or area carrying on similar traditions—at least the first generation does this.

1.3   Women, Health, and Future

As educated, privileged urban Indians, we are highly excited to both witness and relish a rapid and obvious economic progress around us. The smart urban infrastructure and the rising per capita income that we are experiencing make us believe that all is well in India. We feel that, as a nation, India is on a fast-track progress from being a developing economy a generation ago to emerging as an economic superpower today. Most of us yet fail to notice the chinks in this "all is well paradigm" and realize that this development that is enamouring us is highly skewed. Added to this, the patriarchal nature of our society further makes us wear blinkers that often make us blind to the needs of our own neighbors across the streets and many living in the back alleys. Here, we are also turning a blind eye to the needs of the poor and the marginalized women who are doubly jeopardized. Frankly, all this is not esoteric knowledge.

Therefore, we designed the following study where women experience unique healthcare challenges and are more likely to be both denied as well as get deprived of certain healthcare initiatives as victims of poverty, social apathy, and crunching traditions that emerging India is ashamed of. Many women-oriented diseases and conditions such as teen pregnancy, female health, and sanitation are the leading causes of death for women, especially in urban slums despite the fact that there is a perceptible change for the better with greater attention from the government, non-governmental organizations (NGOs), and the civic society. The same is the story in most of the villages in India today. The action research was designed to understand the attitude of the slum dwellers, create an awareness among them about progressive ideas and available services, and ironically sensitize the urban elites (students involved in the study) in the existing contradictions of our society and nation.

2   Action Research Conducted at Maharjpur (Sahibabad Industrial Area, Ghaziabad, Uttar Pradesh, India)

The Maharajpur site is at Sahibabad Industrial Area which is a group of industrial, residential, and commercial areas within the jurisdiction of District Ghaziabad, Uttar Pradesh, India, and falls in the national capital region as it is in the close vicinity of capital of India—New Delhi. The area of Maharajpur is divided into two units under the supervision of Urban Public Health Centers (UPHC). Hence, Maharajpur has all the amenities in its vicinity which an urban-developed area should generally possess like malls, multiplexes, a metro station at "Vaishali" and a five-star hotel—The Raddison Blu. In this neighborhood, there lies two small villages: Bhowapur and Maharajpur where this study was conducted.

The site includes migrants from various Indian states of Bihar, Jharkhand and Uttar Pradesh. It also houses foreign migrants from neighboring countries like Nepal and Bangladesh who hide their identity and frequently change their location to remain untraceable. The livelihood for most in this slum also locally referred as *jhuggies* is Garbage picking. A common belief among most here is that the more number of children they have will bring more number of garbage pickers in the family, and hence more members of earning for the family. Maharajpur Unit II area comprises 80% of Muslim population and refugees from Nepal, Bangladesh, and other neighboring countries. The livelihood of people residing in this area is also garbage picking, micro-food stalls selling local pani-puri, Chinese food, and momos (Dimsums). The womenfolk work as domestic-aids in the residential complexes of the neighboring residential colonies.

The official demographic census (2011) [6] shows that total population of Maharajpur is 122,975. Whereas, Bhowalpur and Maharajpur Unit II has total population of 1405, which comprises of 768 of male and 637 women (sources: UPHC) [6]. After the survey was conducted, the team could trace 21 cases of teen pregnancy in the area. This was much to the discomfort of the local authorities and was a surprise realization for the researchers that all the instance of underage pregnancies may not have been officially documented since the legal age of marriage for girls is 18. Hence, these teens were shown to have attained the legal age of 18 years, in the government documents by their family members.

After generating data using quantitative techniques like the gatekeeper interview, opinion leader interview, focal group discussions, successive informal interface with the target audience, the first level of coding arrived at the broad themes that led to the major thematic analysis and conclusions thereof.

The survey and research indicated and reaffirmed the researcher's prior assumption that marriage at an early age provided social recognition and approval of a sexual relation and consequently of pregnancy as well to this class of people. Here, since the girls that were married off by their family were mostly teenagers, the resultant pregnancy automatically was teen or underage pregnancy. Unfortunately, awareness that an early marriage and pregnancy at a tender and vulnerable and sensitive age, especially among girls shortly after their menarche can be greatly disadvantageous for their health and fitness was disregarded and appeared non-existent due to an arrogant attitude of the male population in the target audience. As these migrant population were primarily more focussed and concerned about earning their daily wages, it most often escaped their attention that such premature and early marriages and the consequent pregnancies affect the demographic dividend of a nation [7–10]. However, the population in question was found to be equally apathetic to both the education and health issues of the mothers of these young girls as well who were found to be young women in their early 30s at best but were worn and hardly in good health.

The major reason for teen pregnancy in India is lack of sex education and limited exposure to information.

3 Need to Explore Multimedia Communication Options

In India, the penetration of and access to mobile phones is very striking and still growing. Today, at least one member in almost every family has a (smart) phone in the urban areas. Rural India is not lagging far behind. Further, Android technology has made access to countless application or apps available for even the new and unaccustomed users. This is also propagating knowledge and disseminating information on perhaps every aspect of life including their health issues. For instance, Hannah Nichols (2018) has compiled a list of apps which helps to keep track of the (menstrual) period cycles. Such helpful and relevant applications may easily be developed for similar issues, challenges, and problems like teen pregnancy in contexts like India. Big problems represent even bigger opportunities… if customized apps can be developed with the help of stakeholders like the local health officials, the district administrations, and the local academic institutions using the basic competence and available skill sets, lingering problems such as this may begin to get addressed on a sustainable basis with tangible results.

4 Role of Augmented Reality and Virtual Reality (ARVR)

Immersive technology is under the global scanner and speculation. It has around it a global hyper-expectancy and this exciting opportunity for research and patents should not be lost to us now. Huge work is going on in the field of education, medicine, defense, and many more areas using AR and VR techniques. And where teen pregnancy is concerned, both the developed and developing countries are facing major challenges and problems. This is the opportunity to respond in a multidisciplinary approach. For instance, we can create some short films or capsule or simple games through AR and VR techniques in which women can get involved. This will not only give them a theoretical knowledge, but they will experience the facts so closely that it will leave some impression on their mind. Government organization such as *Anganwadis* may be given such multimedia installations so that women can make best use of it.

In such teaching sessions, we can also use the "Collaborative Augmented Reality" method. The best reason to use augmented reality in women education is its ability to remove the barriers of language, culture, and geographic distance. Through the potential of AR, students and educators can understand each other better than ever before. It opens doors to communication and learning that were, until now, firmly shut.

There is a strong need to break the stereotypical formats of communication. Taking the clues from the outcomes of the earlier research by authors where they communicated the message about ill effects of female foeticide to target audience through digital games [8, 10], authors recommend that now is the time to go one step further and use immersive technology augmented reality (multimedia-rich communication) to communicate health awareness issues. Using these technologies, one can understand the awareness messages better than conventional mediums. It is easier to see, hear, and experience something than have it explained to people, and frequently women just need to be taken out of a regular classroom environment and dropped into an immersive world; where they can see the virtual character around them who can teach different lessons related to early pregnancy.

5 Conclusion

Behavior change communication has emerged as a major tool to usher sustainable development. It works on the participatory principles of communication and frequently the intervention principles are qualitative in nature as quantification of behavioral manifestations is difficult and often misleading. However, progressively it is also being realized that computer-mediated and multimedia-enabled communication is most suitable for this communication paradigm. This is because multimedia communication is most capable of replacing "text" with attractive and even customized "images" that will carry the utmost "infotainment" value and potency. Further, multimedia communication is very compatible with the smartphones and the Android technology that has become very common across the global population. Literacy is not a challenge for either and this is what behavioral change communication needs to be leveraged to gain the maximum benefits in contexts like India.

References

1. Niket, M., Suparna, D.: Emotions and 21st century communication. Int. J. Eng. Technol. ISSN No.2227-524X (2018). www.sciencepubco.com/index.php/IJET
2. Mehta, N., Suparna, D.: Multimedia for effective communication. In: Proceedings of International Conference on Signal Processing, Communication, Power and Embedded System (SCOPES)—2016, Centurion University of Technology and Management (CUTM), Paralakhemundi Campus, Dist: Gajapati, Odisa, India, IEEE Xplore Digital Library (2016). http://scopes.co.in/index.php/proceedings
3. National Family Health Survey, Accessed from www.rchiips.org/nfhs/
4. Roy, D., Debnath, A.: Issues on Health And Healthcare in India, 2018. Springer, On the Determinants of Child Health in India: Does Teen Pregnancy Matter? https://link.springer.com/chapter/10.1007/978-981-10-6104-2_4
5. WHO (World Health Organization). Adolescent pregnancy. Accessed from www.who.int/mediacentre/factsheets/fs364/en/
6. Census Data from UPHC, Ghaziabad, Uttar Pradesh, India
7. Mehta, N., Suparna, D., Asit, B.: Social awareness through new media. Int. J. Emerg. Technol. Comput. Appl. Sci. 3(11), 258–261, December 2014–February 2015. http://iasir.net/journals.html
8. Suparna, D., Niket, M., Rachana, P.: Humane digital route to customize communication for sustainable development. In: 7th International Technology, Education and Development Conference [March 4th–6th, 2013, Valencia (Spain)] (2013). ISBN: 978-84-616-2661-8, ISSN: 2340-1079, Digital Library, IATED http://library.iated.org/view/dutta2013hum

9. Niket, M., Suparna, D., Asit, B.: Social awareness through new media. Int. J. Emerg. Technol. Computat. Appl. Sci. **3**(11), 258–261, December 2014–February 2015. http://iasir.net/journals. html
10. Niket, M., Suparna, D.: Overcome the challenges of social message communication in digital age through games. In: 16th International Conference on Remote Engineering and Virtual Instrumentation REV2019 (2019)

Author Index

© Springer Nature Singapore Pte Ltd. 2020
X.-S. Yang et al. (eds.), *Fourth International Congress on Information and Communication Technology*, Advances in Intelligent Systems and Computing 1027, https://doi.org/10.1007/978-981-32-9343-4

Printed in the United States
By Bookmasters